Applications of Biochar for Environmental Safety

Edited by Ahmed A. Abdelhafez and Mohammed H. H. Abbas

Published in London, United Kingdom

IntechOpen

Supporting open minds since 2005

Applications of Biochar for Environmental Safety
http://dx.doi.org/10.5772/intechopen.87828
Edited by Ahmed A. Abdelhafez and Mohammed H. H. Abbas

Contributors
Stephen Yeboah, Patricia Oteng-Darko, Joseph Adomako, Abdul-Rauf M Alhassan, Hanuman Singh Jatav, Surendra Singh Jatav, Manoj Parihar, Vishnu Rajput, Satish Singh, Sonu Kumar Mahawer, Rajesh Singhal, Sukirtee Chejara, Suzana Yusup, Nor Adilla Rashidi, Joao Antonangelo, Hailin Zhang, Khalid Alaboudi, Donald Rockwood, Martin Ellis, Ruliang Liu, Fengliang Zhao, Kyle Fabbro, Zhenli He, David Derbowski, Bishnu Acharya, Abhishek Pokharel, Aitazaz Farooque, Dongdong Feng, Yu Zhang, Yijun Zhao, Shaozeng Sun, Ana Carolina Feitosa De Vasconcelos, Ahmed Abdelhafez, Aicha Slassi Sennou, Ghasem Shahbazi, Shuangning Xiu, Godfrey Omulo, Senthilkumar Ramalingham, Reddy Prasad D.M., Tomáš Weidlich, Barbora Kamenická, Pavel Matějíček, Michael Pohořelý, Oladapo Okareh, Alaba Gbadebo, Xu Zhang, Li Zhou, Guoyan Zou, Naxin Cui, Mohammed H.H. Abbas, Mahdy H. Hamed

First published in London, United Kingdom, 2020 by IntechOpen
IntechOpen is the global imprint of INTECHOPEN LIMITED, registered in England and Wales, registration number: 11086078, 7th floor, 10 Lower Thames Street, London, EC3R 6AF, United Kingdom
Printed in Croatia

British Library Cataloguing-in-Publication Data
A catalogue record for this book is available from the British Library

Additional hard and PDF copies can be obtained from orders@intechopen.com

Applications of Biochar for Environmental Safety
Edited by Ahmed A. Abdelhafez and Mohammed H. H. Abbas
p. cm.
Print ISBN 978-1-78985-895-2
Online ISBN 978-1-78985-896-9
eBook (PDF) ISBN 978-1-83880-241-7

For EU product safety concerns:
IN TECH d.o.o., Prolaz Marije Krucifikse Kozulić 3, 51000 Rijeka, Croatia, info@intechopen.com or visit our website at intechopen.com.

Meet the editors

Ahmed A. Abdelhafez, PhD, is an associate professor of the Department of Soils and Water Science, Faculty of Agriculture, New Valley University. He is one of the leading scientists in the field of biochar in the Arab region. He worked as a researcher at the Department of Environmental Researches, Agricultural Research Center (ARC), Egypt, for more than 10 years. Dr. Abdelhafez focuses mainly on agricultural production, environmental contamination control, risk assessment and biochar technology. He is a member of the National Committee of Soil Sciences and the Academy of Scientific Research & Technology, Egypt. He has published several research papers related to environmental contamination, risk assessment and potential remediation technologies.

Mohamed H.H. Abbas, PhD, is Professor of Soil Chemistry at the Faculty of Agriculture, Benha University, Egypt. He began his academic career as a demonstrator at Zagarig University (Benha Branch) in 2000. He received his MA and PhD from Benha University. Currently, he is a board member for the *Egyptian Journal of Soil Science*. His research focuses on sorption/desorption of potentially toxic elements in soil, their environmental pathways, related risk assessment and environmental contamination control. He also conducts research on biochar, land reclamation, and the chemistry of organic matter in soil and plant nutrition. He is a reviewer of several national and international journals.

Contents

Preface

Biochar is a carbon-rich material produced from the pyrolysis of organic materials from agricultural and forestry biomass at a relatively low temperature in the absence of oxygen. Such a process turns agricultural wastes into more useful products. Several studies highlighted the beneficial roles of biochar in solving many agricultural and environmental problems. This product is characterized by the presence of relatively stable carbon forms and therefore it can persist in soils for years. Its application effectively raises soil fertility, especially light-textured soil, with low negative impacts on the global warming threat. Moreover, it can be used successfully to ameliorate contaminated soils and wastewaters. Thus, biochar has become a main focus of research activities over the last few years. Many of these researches bring promises and hopes about the potentiality of using biochar to achieve environmental sustainability goals. In this book, we summarize the latest developments in biochar research activities.

This book includes five sections: "Introduction," "Production and Legislation of Biochar," "Applications of Biochar for Soil Fertility Improvement," "Role of Biochar for Soil Remediation and Ameliorating Salinity Effects" and "Applications of Biochar for Water Treatment." Chapters in the mentioned sections discuss the pros and cons of handling this organic product to sustain the environment and the possible negative effects that come from its extensive use; production process of biochar and its development from basic techniques to highly advanced producing plants; the characteristics of biochar generated from different pyrolysis methods and its role for the remediation of potentially toxic-element contaminated soils and to alleviate adverse salinity effects in salt-affected soils; as well as the potentiality of biochar applications for treating water contaminated with organic and inorganic pollutants.

The editors would like to thank all authors for their contributions. We hope this book will be an invaluable resource for all researchers interested in biochar and environmental studies.

Ahmed A. Abdelhafez (Editor)
Eco-environmental Protection Research Institute,
Shanghai Academy of Agricultural Science (SAAS),
China

Faculty of Agriculture,
Soils and Water Department,
New Valley University,
Egypt

Mohammed H. H. Abbas (Co-Editor)
Faculty of Agriculture,
Soils and Water Department,
Benha University,
Egypt

Section 1

Introduction

Chapter 1

Introductory Chapter: Is Biochar Safe?

Ahmed A. Abdelhafez, Xu Zhang, Li Zhou, Guoyan Zou, Naxin Cui, Mohammed H.H. Abbas and Mahdy H. Hamed

1. Introduction

Biochar is a carbon-rich product resulted from the pyrolysis of organic biomass in the absence of oxygen or at relatively low-oxygen conditions [1]. Such a process transforms the easily oxidized carbon fractions presented in the organic residues into more stable forms [2] that can persist in soils for years, probably 7 [3] to 10 [4] years. This amendment reduces, therefore, the emissions of greenhouses gases [5] and can be considered as a climate change mitigation [6]. On the other hand, required amounts of this conditioner to improve soil productivity might be less comparable with compost or other organic amendments on the long run. Accordingly, biochar is presented as a promising soil amendment of high economic and environmental value. It is also named as "the black diamond" [7]. However, many environmental aspects should be considered while using this amendment. The first one considers its manufacturing process. During the pyrolysis process of biochar, significant emissions of CO_2 are produced, and this probably raises the levels of greenhouse gases (GHGs) in air [8]. The second important issue is related to the biochar degradation in soil. Under warm climate conditions, its degradation was reported to be relatively high [9, 10], and therefore, further emissions of greenhouse gases might take place from biochar-amended soils. The third topic concerns ethylene, which is a by-product of the pyrolysis process of biochar [11]. This gas is increased considerably in biochar-amended soils to suppress several soil microbial processes [12]. Many researches considered this point a positive one that increases the stability of biochar in soil while reduces the emissions of greenhouse gases produced upon its degradation in soil [13], yet biochar affects negatively soil biota [14]. This is because this product contains a small part of bioavailable C [15] as the labile C is already degraded [16]. Thus, the sustainability of crop production in soil referred by soil health (or soil quality) which "reflects the capacity of a soil to provide ecosystem services" [17] may also be affected. Soil biota not only affects the physical and chemical properties of soil but also improves plant health [18]. Further pros and cons of amending soils with biochar will be discussed briefly in the following section.

Several studies demonstrated the positive impacts of amending soils with biochar on increasing crop productivity. For example, amending soils with biochar improves significantly macro- and micronutrients availability [9], in spite of the fact that many biochar additives have an alkaline nature [19], and consequently raises soil pH [7–20]. Nevertheless, soil nutrients strongly are adsorbed on biochar which serves as a slow release fertilizer [21]. It is then thought that the better utility of biochar can be detected on acid soils, rather than alkaline or calcareous soils. Moreover, this amendment reduces NO_3^- loss through leaching as well as the gaseous loss through release of nitrous oxide [22]; hence; this amendment can positively enhance plant growth [23]. Also, this product, which is characterized by its porous structure and high surface area [24], recorded indirect impacts on soil physical characteristics; for example, this amendment increases significantly water retention [10–25], hydraulic conductivity [26], and the total porosity of sandy soils while decreasing soil bulk density [27]. However, the impacts of amending soils with biochar are not always the same and depend mainly on the characteristics of the used biochar such as its grain size and pyrolysis temperature. According to [28], fine biochar decreases soil hydraulic conductivity, while the coarse biochar (particles were coarser than sand) did not affect the hydraulic conductivity of soils. Also, the pyrolysis temperature seems to have a significant impact on ash content in biochar, its pH, EC, and basic functional groups as well as carbon stability which increases in biochar with increasing pyrolysis temperature [29]. Generally, the effect of biochar on soil physical properties was comparable with the effect of compost [30]. On the other hand, other reports indicate that this amendment recorded unfavorable changes in chemical, physical, and biological properties of soil and consequently reduced crop yield [31]. Also, its application to soil hinders root penetration into soil depth [32]. Moreover, its negative impacts were also considerable on earthworm populations even on the short range [33]. It seems that the environmental and health risks due to biochar applications in agricultural soils are not well explored.

Another positive point for using biochar as a soil conditioner is related to its success to mitigate salinization of arable lands [34]. Additionally, biochar plays positive significant impacts on controlling the contaminants presented in water and soils [35, 36]. However, many contaminants may also originate from biochar [37]. Moreover, herbicides, e.g., atrazine and acetochlor, are sorbed on biochar [8], and this may reduce its efficacy [38] .

Furthermore, biochar has a remarkable effect on minimizing the emissions of greenhouse gases, especially CO_2 [39, 40] vs. the traditional organic amendments [13]. Although biochar played important positive roles on sustaining the environment, there is a lake of knowledge concerning the recommended application rates of biochar to soils to avoid its negative potential impacts on the environment.

In this book, we will investigate the major techniques followed in the production and characterization of biochars. Their roles in sustaining agricultural productivity and environmental cleanup will be also a matter of concern. Finally, we will try to draw a legalization mode of biochar applications to the environment in order to ensure its safe applications.

Acknowledgements

This study was supported by Shanghai Science and Technology Committee (Grant No. 29218230742600)-China.

Author details

Ahmed A. Abdelhafez[1,2]*, Xu Zhang[1], Li Zhou[1], Guoyan Zou[1]*, Naxin Cui[1], Mohammed H.H. Abbas[3] and Mahdy H. Hamed[2]

1 Eco-environmental Protection Research Institute, Shanghai Academy of Agricultural Science (SAAS), China

2 New Valley University, Faculty of Agriculture, Soils and Water Department, Egypt

3 Benha University, Faculty of Agriculture, Soils and Water Department, Egypt

*Address all correspondence to: ahmed.aziz@aun.edu.eg and zouguoyan@263.net

References

[1] Pariyar P et al. Evaluation of change in biochar properties derived from different feedstock and pyrolysis temperature for environmental and agricultural application. Science of the Total Environment. 2020;**713**:136433

[2] Song X et al. Effect of cotton straw-derived materials on native soil organic carbon. Science of the Total Environment. 2019;**663**:38-44

[3] Giagnoni L et al. Long-term soil biological fertility, volatile organic compounds and chemical properties in a vineyard soil after biochar amendment. Geoderma. 2019;**344**:127-136

[4] Kätterer T et al. Biochar addition persistently increased soil fertility and yields in maize-soybean rotations over 10 years in sub-humid regions of Kenya. Field Crops Research. 2019;**235**:18-26

[5] Awad YM et al. Chapter one - Biochar effects on rice paddy: Meta-analysis. In: Sparks DL, editor. Advances in Agronomy. Academic Press Inc. 2018. pp. 1-32

[6] Leng L et al. Biochar stability assessment methods: A review. Science of the Total Environment. 2019;**647**:210-222

[7] Abdelhafez AA, Abbas MHH, Li J. Biochar: The black diamond for soil sustainability, contamination control and agricultural production. In: Engineering Applications of Biochar. Rijeka: InTechOpen; 2017

[8] Spokas KA et al. Impacts of woodchip biochar additions on greenhouse gas production and sorption/degradation of two herbicides in a Minnesota soil. Chemosphere. 2009;**77**(4):574-581

[9] Elshony M et al. Ameliorating a Sandy soil using biochar and compost amendments and their implications as slow release fertilizers on plant growth. Egyptian Journal of Soil Science. 2019;**59**(4):305-322

[10] Bassouny M, Abbas M. Role of biochar in managing the irrigation water requirements of maize plants: The pyramid model signifying the soil hydro-physical and environmental markers. Egyptian Journal of Soil Science. 2019;**59**(2):99-115

[11] Fulton W et al. A simple technique to eliminate ethylene emissions from biochar amendment in agriculture. Agronomy for Sustainable Development. 2013;**33**(3):469-474

[12] Spokas KA, Baker JM, Reicosky DC. Ethylene: Potential key for biochar amendment impacts. Plant and Soil. 2010;**333**:443-452

[13] Rahman GKMM et al. Biochar and organic amendments for sustainable soil carbon and soil health. In: Datta R et al., editors. Carbon and Nitrogen Cycling in Soil. Singapore: Springer; 2020. pp. 45-85

[14] Zheng H et al. Biochar-induced negative carbon mineralization priming effects in a coastal wetland soil: Roles of soil aggregation and microbial modulation. Science of the Total Environment. 2018;**610-611**:951-960

[15] Wang J, Xiong Z, Kuzyakov Y. Biochar stability in soil: Meta-analysis of decomposition and priming effects. GCB Bioenergy. 2016;**8**(3):512-523

[16] Kuzyakov Y, Bogomolova I, Glaser B. Biochar stability in soil: Decomposition during eight years and transformation as assessed by compound-specific 14C analysis. Soil Biology and Biochemistry. 2014;**70**:229-236

[17] Williams H, Colombi T, Keller T. The influence of soil management on soil health: An on-farm study in southern Sweden. Geoderma. 2020;**360**:114010

[18] Nakajima T. Soil health and carbon sequestration in urban farmland. In: Tojo S, editor. Recycle Based Organic Agriculture in a City. Singapore: Singapore: Springer; 2020. pp. 147-158

[19] Sarfaraz Q et al. Characterization and carbon mineralization of biochars produced from different animal manures and plant residues. Scientific Reports. 2020;**10**(1):955

[20] Abdelhafez AA, Li J, Abbas MHH. Feasibility of biochar manufactured from organic wastes on the stabilization of heavy metals in a metal smelter contaminated soil. Chemosphere. 2014;**117**:66-71

[21] Dai Y et al. Utilization of biochar for the removal of nitrogen and phosphorus. Journal of Cleaner Production. 2020;**257**:120573

[22] Prendergast-Miller MT, Duvall M, Sohi SP. Localisation of nitrate in the rhizosphere of biochar-amended soils. Soil Biology and Biochemistry. 2011;**43**(11):2243-2246

[23] Manolikaki I, Diamadopoulos E. Agronomic potential of biochar prepared from brewery byproducts. Journal of Environmental Management. 2020;**255**:109856

[24] Brassard P, Godbout S, Raghavan V. Soil biochar amendment as a climate change mitigation tool: Key parameters and mechanisms involved. Journal of Environmental Management. 2016;**181**:484-497

[25] Werdin J et al. Biochar made from low density wood has greater plant available water than biochar made from high density wood. Science of the Total Environment. 2020;**705**:135856

[26] Wiersma W et al. No effect of pyrolysis temperature and feedstock type on hydraulic properties of biochar and amended sandy soil. Geoderma. 2020;**364**:114209

[27] Abel S et al. Impact of biochar and hydrochar addition on water retention and water repellency of sandy soil. Geoderma. 2013;**202-203**:183-191

[28] Liu Z et al. Impacts of biochar concentration and particle size on hydraulic conductivity and DOC leaching of biochar–sand mixtures. Journal of Hydrology. 2016;**533**:461-472

[29] Al-Wabel MI et al. Pyrolysis temperature induced changes in characteristics and chemical composition of biochar produced from conocarpus wastes. Bioresource Technology. 2013;**131**:374-379

[30] Somerville PD et al. Biochar and compost equally improve urban soil physical and biological properties and tree growth, with no added benefit in combination. Science of the Total Environment. 2020;**706**:135736

[31] Mukherjee A, Lal R. The biochar dilemma. Soil Research. 2014;**52**:217-230

[32] Abdelhafez AA, Abbas MHH, Hamed MH. Biochar: A solution for soil Pb pollution. In: The 8th International Conference for Development and the Environment in the Arab World. Egypt: Assiut University; 2016. pp. 89-103

[33] Weyers SL, Spokas KA. Impact of biochar on earthworm populations: A review. Applied and Environmental Soil Science. 2011;**2011**:541592

[34] Thomas SC et al. Biochar mitigates negative effects of salt additions on two herbaceous plant species. Journal of Environmental Management. 2013;**129**:62-68

[35] García-Jaramillo M et al. An examination of the role of biochar and biochar water-extractable substances on the sorption of ionizable herbicides in rice paddy soils. Science of the Total Environment. 2010;**706**:135682

[36] Simiele M et al. Assisted phytoremediation of a former mine soil using biochar and iron sulphate: Effects on As soil immobilization and accumulation in three Salicaceae species. Science of the Total Environment. 2020;**710**:136203

[37] Zhang C et al. Biochar for environmental management: Mitigating greenhouse gas emissions, contaminant treatment, and potential negative impacts. Chemical Engineering Journal. 2019;**373**:902-922

[38] Graber ER et al. High surface area biochar negatively impacts herbicide efficacy. Plant and Soil. 2012;**353**:95-106

[39] Puga AP et al. Biochar-based nitrogen fertilizers: Greenhouse gas emissions, use efficiency, and maize yield in tropical soils. Science of the Total Environment. 2020;**704**:135375

[40] Wang H et al. Biochar mitigates greenhouse gas emissions from an acidic tea soil. Polish Journal of Environmental Studies. 2020;**29**:323-330

Section 2

Production and Legislation of Biochar

Chapter 2

A Mini Review of Biochar Synthesis, Characterization, and Related Standardization and Legislation

Nor Adilla Rashidi and Suzana Yusup

Abstract

The abundance of biomass in Malaysia creates an avenue for growth of bio-economic sector through the research and development (R&D) activities on the biochar production. Biochar that is described as a carbonaceous material derived from the thermochemical process at temperature of usually lower than 700°C is promising due to its applicability in wider range of applications, such as in soil amendment (fertilizer) and as a low-cost adsorbent for the pollution remediation, apart from minimizing the solid waste disposal problems. Therefore, this chapter discusses the current trends on various production techniques of biochar from both the lignocellulosic (plantation based waste materials) and non-lignocellulosic sources, as well as the physiochemical characteristics of the resulting biochar. In addition, overview of the biochar industry in Malaysia is presented in this chapter. Lastly, recap of standardization and legislation particularly related to the biochar utilization as a soil amendment agent is included to grasp readers' attention prior to the large scale applications.

Keywords: biochar, biomass, environmental standardization and legislation, pyrolysis, soil amendment, waste management

1. Introduction

Biochar, which is a subset of carbon-rich and black powder, is generally defined as a porous solid that is produced from biomass via pyrolysis process and in the absence of oxygen (O_2) [1]. Nevertheless, based on literatures, there are various definitions of the biochar [2–4]; accordingly, Sohi et al. [5] reported that the term of biochar remains ill-defined. Thus, the International Biochar Initiative (IBI) standardized the biochar "as a solid material obtained from the thermochemical conversion of biomass in O_2-limited environments." While the production route of biochar and charcoal is similar where both materials are derived from the carbonaceous feedstock through the pyrolysis process [6], but the distinct features that can distinguish these two materials lies in their starting material and end application. Biochar that contains high porosity, high nutrient content, and water-storage-capability is applied for soil amelioration or an adsorbent, whereas charcoal that is usually derived from the petroleum-based feedstock is used for heat generation (energy/fuel) purposes [3, 7]. In a nutshell, Mesa et al. [2] reported that the term biochar is not applicable for the charred materials used as a solid fuel, and to exclude the black carbon produced from non-renewable resources such as coal and petroleum. Besides, Abdelhafez et al. [8] reported that biochar contains lower ash compounds as compared to charcoal, due to an incomplete carbonization process. Further, due to wider application of biochar in both agronomic sector as well as in environmental management, Verheijen et al. [9] reported that the global market of biochar is rapidly growing, with the global market price is estimated around $80–13,480/oven dried metric ton (ODMT). In addition, Hersh et al. [10] reported that the global biochar market is projected to increase up to $3.14 billion by 2025, and expand at an average rate of 13.1% annually [11]. Due to the growing interest of the biochar production and application, number of scientific publications related to the biochar is gradually increasing (as presented in **Figure 1**), where most of these publications (since 2016) are from Republic of China, USA, Australia, South Korea, and India. Herein, this chapter aims to highlight the recent advancement of the biochar production from various processing techniques, as well as an overview on the biochar standardization (quality standard) and legislation, particularly for its application as soil amendment agent.

So far, research work on the biochar-related field in Malaysia is extensive in local universities and research institutes, where Universiti Putra Malaysia (UPM) is the leading organization in the biochar research. Being a pioneer in biochar research, UPM researchers in collaboration with Nasmech Technology has successfully built the first large scale biochar production plant within the region (as shown in **Figure 2**) in January 2010 [12, 13], where the carbonator is capable to accommodate up to 20 tons of different types of waste materials daily for the biochar production. Hypothetically, opportunities of biochar industry in Malaysia can be attributed to lower labor cost, low or no cost incurred of biomass, large agricultural industry, as well as fast-growing biomass. In fact, Ozturk et al. [14] reported that Malaysia produces about 168 million tons of biomass annually. Nevertheless, Kong et al. [15]

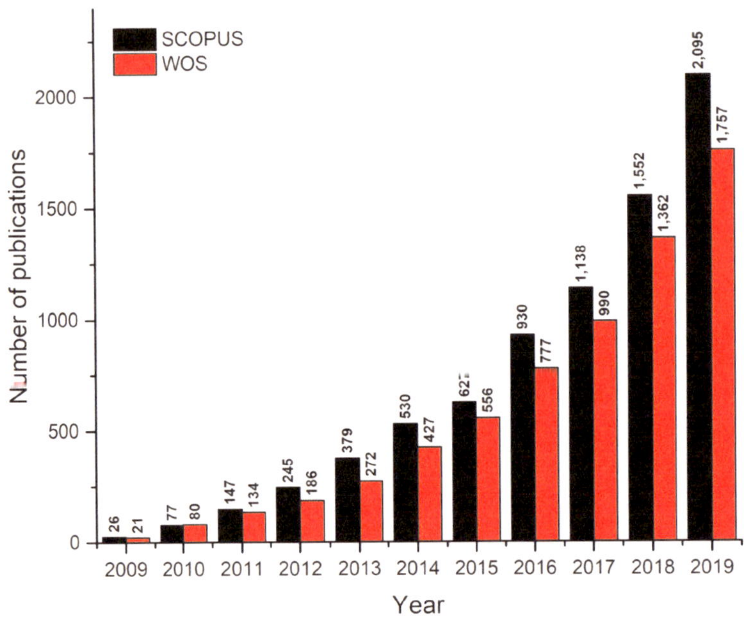

Figure 1.
Number of biochar-related publications from 2009 to 2019 from Web of Science (WOS) and SCOPUS database.

Figure 2.
Biochar plant in Dengkil, Selangor, Malaysia [18].

reported that the main challenge in biochar production in Malaysia is due to the physiochemical nature of biomass (particularly oil palm biomass) itself; where wet biomass will result in transportation problems from the source to production sites, thus need an additional drying process apart from normal pre-treatments such as chopping, shredding, and grinding stages. Consequently, this will increase both production cost and equipment's capital investment. Besides, difficulty in gaining a long-term contract basis between the biomass suppliers, producers, investors, and potential end users, is one of the major barriers in the biochar production in Malaysia [15, 16]. Due to these problems and the lack of key players along the value chain, biochar's production is rather costly, accordingly, Tang et al. [17] reported that commercialization of biochar in Malaysia is relatively new and still at an early stage. Based on literatures, biochar providers in Malaysia include the following: Global Green Synergy Sdn. Bhd., Pakar Go Green Sdn. Bhd., Usaha Strategik Sdn. Bhd., and CH Biotech Sdn. Bhd. In addition, realizing the prominence of the biochar industry toward the socio-environmental economy, Biochar Association Malaysia (BMA) has been established in 2014 with the missions are to promote the biochar production and application in both agricultural and industrial sector, to stimulate publics' awareness on the role of biochar as a carbon sequester, and as a platform for idea and information exchange in promoting the biochar industry in Malaysia. In addition, to further promote the advancement of biochar industry in Malaysia, key players including researchers, authorities, and business analysts should work closely together.

2. Production of biochar

Biochar can be produced from various types of biomass which include the lignocellulosic (i.e., bioenergy crop, agricultural waste, forestry residues) and non-lignocellulosic groups (i.e., manure, sewage sludge, microalgae) [19, 20]. To date, agricultural waste is the primary feedstock used for the biochar production, as confirmed in **Table 1**. Regardless of the different types of feedstock, the biochar's

Biomass	Process conditions		Findings		Ref.
	Temp (°C)	*time (min)*	*Yield (%)*	*Capacity*	
Heavy metals removal (i.e., cadmium, copper, lead, zinc, etc.)					
Cocoa pod	500	120	n/a	69.9 mg/g	[24]
EFB	615	128	25.49	15.18 mg/g	[25]
EFB	300	180	n/a	85 mg/g	[26]
Sludge	n/a	60	n/a	19 mg/g	[27]
Sludge	400	90	64.2	48.8 mg/g	[28]
Color/dyes removal (i.e., methylene blue, malachite green)					
Cassava stem	500	120	11.94	40.5 mg/g	[29]
Coconut frond	800	240	n/a	126.58 mg/g	[30]
Palm shell	700 W	25	33	48 mg/g	[31]
Seaweed	800	90	n/a	512.67 mg/g	[32]
Sugarcane bagasse	600	120	n/a	99.47%	[33]
Phenolic compounds removal					
EFB	500	80.27	n/a	7.38%	[34]
Gas/vapor adsorption (i.e., CO_2, mercury, sulfur dioxide)					
Coconut pith	900	60	27.76	6067.49 μg/g	[35]
Coconut pith	700	60	31.42	10 mmol/g	[36]
Sludge	405	88	54.25	9.75 mg/g	[37]
Wood sawdust	650	60	n/a	18 mg/g	[38]
Soil-based application (herbicides/pesticides removal, fertilizer)					
EFB	300	60	n/a	4.497	[39]
Rice husk	300	180	n/a	4.742	[39]
Palm shell	700 W	25	33	450 g	[31]

EFB, empty fruit bunches.

Table 1.
Summary of recent biochar production in Malaysia from local biomass and the corresponding optimum conditions.

skeleton is primarily comprised of carbon and ash, where the overall compositions and characteristics of each biochar varied, depending on the types of feedstock and the process conditions. Filiberto et al. [20] reported that the significant difference between the nutrient-rich feedstocks such as animal manure and sewage sludge, compared to the lignin-rich biomass feedstock is that the former materials contain considerably high nutrient and mineral compositions (i.e., nitrogen, phosphorus, potassium, etc.). In context of heavy metal removal application, Zhao et al. [21] reported that the sewage sludge biochar that has higher mineral contents (161 g/kg) compared to corn biochar (28.6 g/kg) and poplar wood biochar (19.5 g/kg) contributes to higher heavy metal removals from wastewater (sewage sludge > corn > poplar wood), thus implies the importance of the mineral compositions in heavy metal adsorption process. Likewise, for the soil amendment application (in terms of element supplementation and liming effect), Zhang et al. [22] also agreed that the biochar should contain a sufficient mineral composition. Meanwhile in context of the process technologies, biochar can be produced from four thermochemical routes that include pyrolysis, torrefaction, hydrothermal carbonization, as well as gasification, [23]; which is thoroughly described in the following subsections.

2.1 Pyrolysis

By definition, pyrolysis is the thermal conversion process conducted in absence of O_2; producing biochar, condensable liquid (i.e., bio-oil), and non-condensable gas (i.e., syngas). The yield distribution depends on the type of pyrolysis process—slow, fast, and flash pyrolysis; where it differs in terms of reaction temperature, heating rate, and holding time (as summarized in **Table 2**).

Referring to **Table 2**, the ideal route for the biochar production is through slow pyrolysis, also known as conventional carbonization, as compared to fast or flash pyrolysis that targets bio-oil production. Recently, Yuan et al. [42] confirmed that walnut shell biochar obtained through slow pyrolysis process has greater biochar yield as compared to the fast pyrolysis, irrespective of reaction temperature, thus it confirms the effectiveness of the slow pyrolysis mechanism toward the biochar production. Furthermore, slow pyrolysis for the biochar production is promising due to lower capital investment as compared to fast pyrolysis scheme ($132 vs. $200 million) [43]. Basically, Daful et al. [44] reported that biochar from slow pyrolysis route refers to primary and secondary char, where the mechanism of the process is simplified in Eqs. (1)–(3) [45]. The pre-pyrolysis reaction [Eq. (1)] involves the water elimination and evaporation from the biomass structure. During the primary reaction, devolatilization process including the dehydration, decarboxylation, and dehydrogenation occurs. Then upon the completion of primary decomposition, the secondary reaction (at high temperature) that refers to cracking of heavy organic compounds as well as repolymerization ensues, producing a stable and carbon-dense solid product (i.e., biochar) and non-condensable syngas such as methylene (CH_2), methane (CH_4), carbon monoxide (CO), and carbon dioxide (CO_2) [45–48].

$$\text{Biomass} \rightarrow \text{water} + \text{unreacted residue} \left(\text{Pre-pyrolysis}\right) \quad (1)$$

$$\text{Unreacted residue} \rightarrow (\text{volatiles} + \text{gases}) + \text{char} \left(\text{Primary reaction}\right) \quad (2)$$

$$\text{Char} \rightarrow (\text{volatiles} + \text{gases}) + \text{char} \left(\text{Secondary reaction}\right) \quad (3)$$

2.2 Torrefaction

Torrefaction or known as a mild pyrolysis refers to the thermochemical process at temperature of 200–300°C at atmospheric pressure and inert atmosphere, heating rate of ≤50°C/min, with residence time of 30 min to 2 h [44, 49].

Conditions		Slow pyrolysis	Fast pyrolysis	Flash pyrolysis
Temperature (°C)		300–700	550–1000	800–1100
Heating rate (°C/sec)		0.1–1	10–200	>1000
Vapor residence time (sec)		450–550	0.5–10	<0.5
Particle size (mm)		5–50	<1	<0.2
Yield (wt. %)	Biochar	**35**	20	12
	Bio-oil	30	**50**	**75**
	Syngas	35	30	13

Bold value refers to the highest product yield of slow pyrolysis, fast pyrolysis and flash pyrolysis. In summary, for the slow pyrolysis, the bold value is for biochar, while for fast and flash pyrolysis, the bold value is for the bio-oil. In other words, the slow pyrolysis favors the biochar production, and both fast and flash pyrolysis targets the bio-oil.

Table 2.
Process conditions for slow (conventional), fast, and flash pyrolysis and product distribution [40, 41].

Nevertheless, it has been reported that the torrefaction process is not a promising technique for the biochar production, regardless of higher product yield (70–80 wt. %), since the torrefied biomass still contains a significant fraction of volatile components from the raw biomass, and the physiochemical properties are in between raw biomass and biochar [44, 50]. For example, oxygen to carbon (O/C) ratio of the torrefied biomass which is >0.4 contradicts with the European Biochar Certification (EBC) of biochar [44]. Therefore, this torrefaction process is often being applied as a pre-treatment process for moisture removal, biomass densification, and to improve the biomass properties. Besides, while the torrefaction process alone cannot be used for biochar production, combination of torrefaction pretreatment and pyrolysis is feasible for the exceptional biochar production (in terms of yield) in addition to the physiochemical characteristics (i.e., surface area) [51–54].

2.3 Hydrothermal carbonization

Opposite to the slow pyrolysis and torrefaction process that is normally carried out under dry atmosphere, hydrothermal carbonization can also be referred as wet pyrolysis or wet torrefaction; since this process is performed in a biomass-water solution at temperature of 180–250°C at high pressure (subcritical condition) for several hours [50, 55–57]. Similar to pyrolysis, this hydrothermal carbonization produces 50–80 wt. % solid char (termed as hydrochar), bio-oil and water mixture (5–20 wt. %), and synthetic gas that is mainly CO_2 (2–5 wt. %) [58]. The great interest in this hydrothermal technology for the biochar production is that it can avoid the preliminary energy-intensive drying process that is usually required for the conventional pyrolysis, and thus it will minimize the operational costs. Besides, Oktaviananda et al. [59] agreed that such process is convenient for the biomass having >50 wt. % moisture content. On top of that, it has been reported that the energy requirement for hydrothermal carbonization and pyrolysis process for 1 kg of feedstock of 80% moisture content is 2.5 and 3.20 MJ, respectively [60]. Moreover, this hydrothermal technology offers the lowest reaction temperature as compared to other thermochemical conversion techniques. During the process, water (H_2O) acts as a solvent, reactant, catalyst, and as a medium for both mass and energy transfer [61], where it will facilitate the hydrolysis, dehydration, decarboxylation and depolymerization process [62]. Besides, at temperature of 200–280°C, H_2O that possesses similar behavior to mild acid and mild base at the same time results in an acceleration of biomass decomposition [61, 63]. Specifically, Libra et al. [64] reported that during the hydrothermal carbonization, hemicellulose decomposes at temperature of 180–200°C, lignin decomposition takes place at 180–220°C, whereas cellulose decomposition occurs at 220°C. However, most often, the hydrochar cannot be described as biochar since the reaction temperature is too low, low carbon contents, as well as an intolerable O/C and hydrogen to carbon (H/C) ratio [65, 66]. Yet, recent work shows that integration of this hydrothermal carbonization with pyrolysis process positively contributes toward the high-quality biochar production and can stabilizes the heavy metal in solid products [67].
For example, by referring to the experimental findings by Olszewski et al. [68], the preliminary hydrothermal treatment of brewery spent grains (that contains 70–90 wt. % moisture) prior to the pyrolysis process produces biochar with greater product yield and carbon contents as well as reduced ash compositions; where the corresponding value is varied, subjected to the intensity of the hydrothermal carbonization process. Likewise, Garlapalli et al. [69] confirmed that the carbon compositions of biochar from the combined hydrothermal and pyrolysis process (at 260 and 800°C, respectively) increases to 82 wt. % compared to standalone hydrothermal process, where the carbon contents is merely 70 wt. %. Moreover,

such combined processes also show an improvement of the surface area (63.48 m^2/g vs. 2.93 m^2/g). In overall, the upgrading of hydrochar is crucial since the hydrochar that possesses low surface area (<30 m^2/g), low porosity, and presence of noxious chemicals (i.e., furan, furfural, and phenolic compounds) limits its application in soil amelioration [69].

2.4 Gasification

The gasification process takes place at the temperature range of 600-1200°C, heating rate of 50–100°C/min, with vapor residence time of 10–20 s. Unlike the pyrolysis, gasification process is carried out in the presence of O_2 (including O_2, air, steam, CO_2, or mixture of the gases) and primarily used for the syngas production (i.e., CO, CO_2, CH_4, hydrogen [H_2]) instead of the biochar production. Due to this, the biochar yield is minimal (<10 wt. %) [44, 56]. With regards to this limitation, there are limited research works on the feasibility of biochar from the gasification process especially for soil amendment purpose [70]. In addition, Wang and Wang [71] reported that the charred product from the gasification process do not satisfy the biochar's definition; in addition to presence of hazardous polycyclic aromatic hydrocarbons (PAHs) as well as alkaline and alkaline heavy metals within the structure [55, 56].

3. Biochar's characterization, standardization, and legislations

The detailed characterization of biochar prior to any applications is significant in order to determine the relationship between nature and operating conditions with the physiochemical properties of biochar, to evaluate the suitability of biochar in desired target application, and to examine the presence of contaminants and eco-toxicology properties [72]. The overall characterization techniques that have been applied for biochar are summarized in **Table 3**.

Given that the biochar's characteristics is mainly influenced by various parameters such as feedstocks' type, technology (i.e., process type, reactor configuration), and process condition (i.e., temperature, heating rate, residence

Characterization	Detailed analysis
Physical property	• Surface area, pore volume and size (*N_2 gas sorption*) • Particle size distribution (*Laser sizing*) • Density (*Mercury porosity, Pycnometer*)
Chemical property	• pH (*pH meter*) • Electrical conductivity (*Conductivity meter*) • Cation exchange capacity (*Ion chromatography*) • Biochar compositions (*CHNS, EDS, XPS*) • Metallic/ash contents (*XRD, ICP, XRF*) • Proximate analysis (*Muffle furnace, TGA*) • Surface functionality (*FTIR, Raman*) • Surface acidity/alkalinity (*Boehm titration*) • Surface aromaticity (*^{13}C NMR, Raman spectroscopy*)
Surface structure & morphology	• SEM/FESEM • TEM • Crystallinity (*XRD, Raman*)
Stability behavior	• TGA-DSC

Table 3.
Summary of biochar's detailed characterization [19, 46, 64, 71–74].

Property	IBI-BS	EBC		BQM	
		Basic	*Premium*	*Standard*	*High gr.*
Organic C (wt. %)	≥10	≥50		≥10	
H:C molar ratio	≤0.7	≤0.7		≤0.7	
O:C molar ratio	—	≤0.4		—	
Moisture	—	≥30		≥20	
Total ash (wt. %)	✓	✓		✓	
Conductivity	✓	✓		Optional	
Liming equiv.	✓	—		—	
pH	✓	✓		✓	
Particle size distr.	✓	—		✓	
Surface area	—	✓		Optional	
Water holding capacity	—	Optional		✓	
Volatile matter (%)	Optional	✓		—	
Germination test	Pass/fail	Optional		—	
Macro-nutrients (wt. %)					
Total N	✓	✓		✓	
Total P, K, Mg, Ca	Optional	✓		✓ (Total P & K)	
Organic pollutants (mg/kg)					
PAH (US EPA 16)	6–300	<12	<4	<20	<20
B(a) P toxic equi.	≤3	—	—	—	—
PCB	0.2–0.5	<0.2		<0.5	
PCDDs/Fs	<17	<20		<20	
Heavy metals (mg/kg)—maximum limit					
Arsenic	12–100	—	—	100	10
Cadmium	1.4–39	1.5	1	39	3
Chromium	64–1200	90	80	100	15
Cobalt	40–150	—	—	—	—
Copper	63–1500	100	100	1500	40
Lead	70–500	150	120	500	60
Mercury	1–17	1	1	17	1
Manganese	—	—	—	n/a	3500
Molybdenum	5–20	—	—	75	10
Nickel	47–600	50	30	600	10
Selenium	2–36	—	—	100	5
Zinc	200–7000	400	400	2800	150
Boron	✓	—		—	
Chlorine	✓	—		—	
Sodium	✓	—		—	

Note: ✓ symbol refers to the required analysis for biochar (declaration).

Table 4.
Summary of biochar certification based on IBI-BS (Ver. 2.0), EBC (Ver. 4.8), and BQM Ver. 1.0.

time, pressure, carrier gas); the corresponding properties of biochar are widely varied. Therefore, the standardization of biochar prior to applications is significant as their performance can be generalized and predicted [64, 75]. To date, the biochar standards have been established by the International Biochar Initiative (IBI-BS), European Biochar Foundation (European Biochar Certificate, EBC); as well as the British Biochar Foundation (Biochar Quality Mandate, BQM) [76–78]. Referring to Verheijen et al. [9], the common objectives of these certifications are to provide the quality and safety indicator for biochar utilization as a soil amendment agent, to promote the biochar's industrial growth and commercialization, as well as for future legislative or regulations. Besides, development of such certifications assists in improving the confidence level of consumers and regulators of the biochar's safe application [79]. Thereby, the parameters and their corresponding threshold values in each biochar certificate are tabulated in **Table 4**.

However, Gelardi et al. [80] reported that variation between these certifications will led to inconsistencies in both scientific and legislative framework, accordingly, there is an urgent need to come out with a unified regulations that can benefit the communication in academics field and in the biochar market. In addition, it should be noted that these certifications are only applicable for the biochar categorization and their suitability as soil amendment agent, and to exclude the hydrochar [65]. Hence, more data and research work toward the hydrochar characterization and appropriate certificates that enable commercial hydrochar utilization is strongly recommended. In addition, since these certifications are only valid for the biochar usage in soil application, it is recommended to produce a detailed assessment and guideline for the biochar utilization in other environmental applications too [74].

4. Conclusions and future outlook

Biomass valorization to biochar materials has gained a significant attention due to its exceptional characteristics—high surface area, high pore volume, long-term stability, and presence of various surface functionalities, as well as wider potential application including energy and biomaterial development, agronomy sector (i.e., soil amelioration, fertilization), and environment pollution control; among others. Given the slow pyrolysis process is the most promising technique for the biochar production, more research studies on the various types of biomass need to be considered as the biochar field is rather a non-exhaustive subject, in addition to the continuous advancement toward cleaner, simpler, and inexpensive biochar production. In addition, a comprehensive analysis on different types of biomass (including agricultural, aquaculture, forestry, human and animal waste, as well as industrial waste) will result in a complete database; mainly focus on the influence of operating parameters toward the process performance, in terms of reaction rate and underlying mechanism, yield, selectivity, biochar's characteristics, as well as energy and mass balance; which are useful for practitioners and future researchers. In addition, from the databases, it is practical for ranking the biomass suitability for the biochar production for specific applications, accordingly facilitates a proper planning on biomass utilization in biochar industry. Besides, the recent work on both the biochar production and utilization is limited to the laboratory scale, thus upscaling the research work to a larger scale is necessary in order to determine the practicality. Finally, techno-economic analysis as well as life cycle assessment of the biochar production through various technologies is recommended. Overall, viability of the biochar industrial sector needs to incorporate the social, technical, economic, and environmental aspects to ensure its sustainability.

Acknowledgements

The authors would like to acknowledge the Ministry of Education Malaysia (MOE) for awarding the National Higher Institution Centre of Excellence (HICoE) award to Centre for Biofuel and Biochemical Research, Universiti Teknologi PETRONAS (UTP).

Conflict of interest

The authors declare no conflict of interest.

Author details

Nor Adilla Rashidi and Suzana Yusup*
Chemical Engineering Department, Higher Institution of Center of Excellence–Centre for Biofuel and Biochemical Research (CBBR), Institute of Self-Sustainable Building, Universiti Teknologi PETRONAS, Seri Iskandar, Perak, Malaysia

*Address all correspondence to: drsuzana_yusuf@utp.edu.my

References

[1] Zhang H, Tu Y-J, Duan Y-P, Liu J, Zhi W, Tang Y, et al. Production of biochar from waste sludge/leaf for fast and efficient removal of diclofenac. Journal of Molecular Liquids. 2020;**299**: 112193

[2] Mesa AC, Spokas KA. Impacts of biochar (black carbon) additions on the sorption and efficacy of herbicides. Herbicides and Environment. 2011;**13**: 315-340

[3] Ahmad M, Rajapaksha AU, Lim JE, Zhang M, Bolan N, Mohan D, et al. Biochar as a sorbent for contaminant management in soil and water: A review. Chemosphere. 2014;**99**:19-33

[4] Rajapaksha AU, Mohan D, Igalavithana AD, Lee SS, Ok Y. Definitions and fundamentals of biochar. In: Ok YS, Uchimiya SM, Chang SX, Bolan N, editors. Biochar: Production, Characterization, and Applications. New York: CRC Press; 2016. pp. 4-17

[5] Sohi SP, Krull E, Lopez-Capel E, Bol R. A review of biochar and its use and function in soil. In: Sparks DL, editor. Advances in Agronomy. United States: Academic Press; 2010. pp. 47-82

[6] Aller MF. Biochar properties: Transport, fate, and impact. Critical Reviews in Environmental Science and Technology. 2016;**46**:1183-1296

[7] Akdeniz N. A systematic review of biochar use in animal waste composting. Waste Management. 2019;**88**:291-300

[8] Abdelhafez AA, Abbas MH, Li J. Biochar: The black diamond for soil sustainability, contamination control and agricultural production. In: Huang WJ, editor. Engineering Applications of Biochar. Croatia: IntechOpen; 2017. pp. 7-27

[9] Verheijen FG, Bastos AC, Schmidt H-P, Jeffery S. Biochar and certification 1. In: Vogt M, editor. Sustainability Certification Schemes in the Agricultural and Natural Resource Sectors: Outcomes for Society and the Environment. London and New York: Routledge; 2019. pp. 113-136

[10] Hersh B, Mirkouei A, Sessions J, Rezaie B, You Y. A review and future directions on enhancing sustainability benefits across food-energy-water systems: The potential role of biochar-derived products. AIMS Environmental Science. 2019;**6**:379-416

[11] Guo M, Xiao P, Li H. Valorization of agricultural byproducts through conversion to biochar and bio-oil. In: Simpson BK, Aryee AN, Toldrá F, editors. Byproducts from Agriculture and Fisheries: Adding Value for Food, Feed, Pharma, and Fuels. USA: John Wiley & Sons; 2020. pp. 501-522

[12] Abdullah N, Sulaiman F. The oil palm wastes in Malaysia. In: Matovic MD, editor. Biomass Now: Sustainable Growth and Use. Croatia: IntechOpen; 2013. pp. 75-93

[13] Harsono SS, Grundman P, Lau LH, Hansen A, Salleh MAM, Meyer-Aurich A, et al. Energy balances, greenhouse gas emissions and economics of biochar production from palm oil empty fruit bunches. Resources, Conservation and Recycling. 2013;**77**:108-115

[14] Ozturk M, Saba N, Altay V, Iqbal R, Hakeem KR, Jawaid M, et al. Biomass and bioenergy: An overview of the development potential in Turkey and Malaysia. Renewable and Sustainable Energy Reviews. 2017;**79**:1285-1302

[15] Kong S-H, Loh S-K, Bachmann RT, Rahim SA, Salimon J. Biochar from oil palm biomass: A review of its potential and challenges. Renewable and

Sustainable Energy Reviews. 2014;**39**: 729-739

[16] Norli I, Fazilah A, Pazli IM. Agricultural biomass utilisation as a key driver for Malaysian bioeconomy. In: Dabbert S, Lewandowski I, Weiss J, Pyka A, editors. Knowledge-Driven Developments in the Bioeconomy: Technological and Economic Perspectives. Switzerland: Springer; 2017. p. 141–159.

[17] Tang KM, Ibrahim WA, Kadir WR. Towards environmental and economic sustainability via the biomass industry: The Malaysian case study. In: Bruckman VJ, Varol EA, Uzun BB, Liu J, editors. Biochar: A Regional Supply Chain Approach in View of Climate Change Mitigation. United Kingdom: Cambridge University Press; 2016. pp. 162-183

[18] UPM-Nasmech Carbonator Pilot Plant, Dengkil Selangor [Internet]. 2010. Available from: http://biocharmalaysia.blogspot.com/2010/01/carbonator-pilot-plant.html [Accessed: 23 March 2020]

[19] Nartey OD, Zhao B. Biochar preparation, characterization, and adsorptive capacity and its effect on bioavailability of contaminants: An overview. Advances in Materials Science and Engineering. 2014;**2014**:1-12

[20] Filiberto DM, Gaunt JL. Practicality of biochar additions to enhance soil and crop productivity. Agriculture. 2013: 715-725

[21] Zhao J, Shen X-J, Domene X, Alcañiz J-M, Liao X, Palet C. Comparison of biochars derived from different types of feedstock and their potential for heavy metal removal in multiple-metal solutions. Scientific Reports. 2019;**9**:1-12

[22] Zhang H, Chen C, Gray EM, Boyd SE. Effect of feedstock and pyrolysis temperature on properties of biochar governing end use efficacy. Biomass and Bioenergy. 2017;**105**: 136-146

[23] Bartoli M, Giorcelli M, Jagdale P, Rovere M, Tagliaferro A. A review of non-soil biochar applications. Materials. 2020;**13**:1-35

[24] Yong SK, Leyom J, Tay CC, Talib SA. Sorption of lead from aqueous system using cocoa pod husk biochar: Kinetic and isotherm studies. International Journal of Engineering & Technology. 2018;**7**:241-244

[25] Zamani SA, Yunus R, Samsuri A, Salleh M, Asady B. Removal of zinc from aqueous solution by optimized oil palm empty fruit bunches biochar as low cost adsorbent. Bioinorganic Chemistry and Applications. 2017;**2017**: 1-9

[26] Sadegh-Zadeh F, Samsuri AW, Seh-Bardan BJ, Emadi M. The effects of acidic functional groups and particle size of biochar on Cd adsorption from aqueous solutions. Desalination and Water Treatment. 2017;**66**:309-319

[27] Lee XJ, Lee L, Hiew B, Gan S, Thangalazhy-Gopakumar S. Evaluation of the effectiveness of low cost adsorbents from oil palm wastes for wastewater treatment. Chemical Engineering Transactions. 2017;**56**: 937-942

[28] Goh CL, Sethupathi S, Bashir MJ, Ahmed W. Adsorptive behaviour of palm oil mill sludge biochar pyrolyzed at low temperature for copper and cadmium removal. Journal of Environmental Management. 2019;**237**: 281-288

[29] Zaid M, Jamion N, Omar Q, Yong S. Sorption of malachite green (MG) by cassava stem biochar (CSB) kinetic and isotherm studies. Journal of

Fundamental and Applied Sciences. 2017;**9**:273-287

[30] Mohammad R, Rajoo AT, Mohamad M. Coconut fronds as adsorbent in the removal of malachite green dye. ARPN Journal of Engineering and Applied Sciences. 2017;**12**:996-1001

[31] Liew RK, Nam WL, Chong MY, Phang XY, Su MH, Yek PNY, et al. Oil palm waste: An abundant and promising feedstock for microwave pyrolysis conversion into good quality biochar with potential multi-applications. Process Safety and Environment Protection. 2018;**115**:57-69

[32] Ahmed M, Okoye P, Hummadi E, Hameed B. High-performance porous biochar from the pyrolysis of natural and renewable seaweed (*Gelidiella acerosa*) and its application for the adsorption of methylene blue. Bioresource Technology. 2019;**278**: 159-164

[33] Mohamad M, Mohammad R, May TS, Wei LJ. Removal of malachite green by sugarcane bagasse biochar using response surface methodology. AIP Conference Proceedings. 2019; **2068**:1-6

[34] Arshad SHM, Ngadi N, Wong S, Amin NS, Razmi FA, Mohamed NB, et al. Optimization of phenol adsorption onto biochar from oil palm empty fruit bunch (EFB). Malaysian Journal of Fundamental and Applied Sciences. 2019;**15**:1-5

[35] Johari K, Saman N, Song ST, Cheu SC, Kong H, Mat H. Development of coconut pith chars towards high elemental mercury adsorption performance—Effect of pyrolysis temperatures. Chemosphere. 2016;**156**:56-68

[36] Rahim ARA, Kuanaseaan K, Shehzad N, Rabat NE, Johari K, Mat H. Synthesis and characterization of coconut pith char adsorbents for carbon dioxide capture. Malaysian Journal of Fundamental and Applied Sciences. 2019;**15**:803-805

[37] Iberahim N, Sethupathi S, Bashir MJ. Optimization of palm oil mill sludge biochar preparation for sulfur dioxide removal. Environmental Science and Pollution Research. 2018;**25**(26): 25702-25714

[38] Ghani W, Azlina W, Da Silva G. Sawdust-derived biochar: Characterization and CO_2 adsorption/ desorption study. Journal of Applied Sciences. 2014;**14**:1450-1454

[39] Yavari S, Malakahmad A, Sapari NB, Yavari S. Synthesis optimization of oil palm empty fruit bunch and rice husk biochars for removal of imazapic and imazapyr herbicides. Journal of Environmental Management. 2017;**193**:201-210

[40] Uddin M, Techato K, Taweekun J, Rahman MM, Rasul M, Mahlia T, et al. An overview of recent developments in biomass pyrolysis technologies. Energies. 2018;**11**:3115

[41] Luque R, Menendez JA, Arenillas A, Cot J. Microwave-assisted pyrolysis of biomass feedstocks: The way forward? Energy & Environmental Science. 2012; **5**:5481-5488

[42] Yuan T, He W, Yin G, Xu S. Comparison of bio-chars formation derived from fast and slow pyrolysis of walnut shell. Fuel. 2020;**261**:116450

[43] Brown TR, Wright MM, Brown RC. Estimating profitability of two biochar production scenarios: Slow pyrolysis vs fast pyrolysis. Biofuels, Bioproducts and Biorefining. 2011;**5**(1):54-68

[44] Daful AG, Chandraratne MR. Biochar production from biomass waste-derived material. In: Choudhury I, Hashmi S, editors. Encyclopedia of Renewable and

Sustainable Materials. 1st ed. New York: Elsevier; 2018. pp. 370-378

[45] Oochit D, Selvarajoo A, Arumugasamy SK. Pyrolysis of biomass. In: Singh L, Kalia VC, editors. Waste Biomass Management—A Holistic Approach. New York: Springer; 2017. pp. 215-229

[46] Kan T, Strezov V, Evans TJ. Lignocellulosic biomass pyrolysis: A review of product properties and effects of pyrolysis parameters. Renewable and Sustainable Energy Reviews. 2016;**57**: 1126-1140

[47] Foong SY, Liew RK, Yang Y, Cheng YW, Yek PNY, Mahari WAW, et al. Valorization of biomass waste to engineered activated biochar by microwave pyrolysis: Progress, challenges, and future directions. Chemical Engineering Journal. 2020; **389**:124401

[48] Tripathi M, Sahu JN, Ganesan P. Effect of process parameters on production of biochar from biomass waste through pyrolysis: A review. Renewable and Sustainable Energy Reviews. 2016;**55**:467-481

[49] Chen Z, Wang M, Ren Y, Jiang E, Jiang Y, Li W. Biomass torrefaction: A promising pretreatment technology for biomass utilization. IOP Conference Series: Earth and Environmental Science. 2018;**113**:012201

[50] Bamdad H, Hawboldt K, MacQuarrie S. A review on common adsorbents for acid gases removal: Focus on biochar. Renewable and Sustainable Energy Reviews. 2018;**81**:1705-1720

[51] Zhu X, Luo Z, Diao R, Zhu X. Combining torrefaction pretreatment and co-pyrolysis to upgrade biochar derived from bio-oil distillation residue and walnut shell. Energy Conversion and Management. 2019;**199**:111970

[52] Chen D, Mei J, Li H, Li Y, Lu M, Ma T, et al. Combined pretreatment with torrefaction and washing using torrefaction liquid products to yield upgraded biomass and pyrolysis products. Bioresource Technology. 2017; **228**:62-68

[53] Zeng K, He X, Yang H, Wang X, Chen H. The effect of combined pretreatments on the pyrolysis of corn stalk. Bioresource Technology. 2019; **281**:309-317

[54] Aliyu AS, Abdullahi N, Sulaiman F. Pyrolysis of torrefied oil palm wastes for better biochar. Malaysian Journal of Fundamental and Applied Sciences. 2017;**13**:124-128

[55] Zhang Z, Zhu Z, Shen B, Liu L. Insights into biochar and hydrochar production and applications: A review. Energy. 2019;**171**:581-598

[56] Lee J, Sarmah AK, Kwon EE. Production and formation of biochar. In: Ok YS, Tsang DC, Bolan N, Novak JM, editors. Biochar from Biomass and Waste: Fundamentals and Applications. Amsterdam: Elsevier; 2019. pp. 3-18

[57] Gan YY, Ong HC, Show PL, Ling TC, Chen W-H, Yu KL, et al. Torrefaction of microalgal biochar as potential coal fuel and application as bio-adsorbent. Energy Conversion and Management. 2018;**165**:152-162

[58] Saqib NU, Sharma HB, Baroutian S, Dubey B, Sarmah AK. Valorisation of food waste via hydrothermal carbonisation and techno-economic feasibility assessment. The Science of the Total Environment. 2019;**690**: 261-276

[59] Oktaviananda C, Rahmawati RF, Prasetya A, Purnomo CW, Yuliansyah AT, Cahyono RB. Effect of temperature and biomass-water ratio to yield and product characteristics of hydrothermal treatment of biomass. AIP Conference Proceedings. 1823;**2017**:1-7

[60] Polprasert C, Koottatep T. Bioenergy production. In: Polprasert C, Koottatep T, editors. Organic Waste Recycling. Technology, Management and Sustainability. London: IWA Publishing; 2017. pp. 157-251

[61] Krylova AY, Zaitchenko V. Hydrothermal carbonization of biomass: A review. Solid Fuel Chemistry. 2018;**52**: 91-103

[62] Guiotoku M, Maia CM, Rambo CR, Hotza D. Synthesis of carbon-based materials by microwave hydrothermal processing. In: Chandra U, editor. Microwave Heating. New York: IntechOpen; 2011. pp. 291-308

[63] Xu Y, Xia M, Jiang Y, Li F, Xue B. Opal promotes hydrothermal carbonization of hydroxypropyl methyl cellulose and formation of carbon nanospheres. RSC Advances. 2018;**8**: 20095-20107

[64] Libra JA, Ro KS, Kammann C, Funke A, Berge ND, Neubauer Y, et al. Hydrothermal carbonization of biomass residuals: A comparative review of the chemistry, processes and applications of wet and dry pyrolysis. Biofuels. 2011;**2**: 71-106

[65] Wiedner K, Rumpel C, Steiner C, Pozzi A, Maas R, Glaser B. Chemical evaluation of chars produced by thermochemical conversion (gasification, pyrolysis and hydrothermal carbonization) of agro-industrial biomass on a commercial scale. Biomass and Bioenergy. 2013;**59**: 264-278

[66] Kwapinski W. Char production technology. In: Jeguirim M, Limousy L, editors. Char and Carbon Materials Derived from Biomass: Production, Characterization and Applications. Amsterdam: Elsevier; 2019. pp. 39-68

[67] Wang X, Chi Q, Liu X, Wang Y. Influence of pyrolysis temperature on characteristics and environmental risk of heavy metals in pyrolyzed biochar made from hydrothermally treated sewage sludge. Chemosphere. 2019;**216**: 698-706

[68] Olszewski M, Arauzo P, Wądrzyk M, Kruse A. Py-GC-MS of hydrochars produced from brewer's spent grains. Journal of Analytical and Applied Pyrolysis. 2019;**140**:255-263

[69] Garlapalli RK, Wirth B, Reza MT. Pyrolysis of hydrochar from digestate: Effect of hydrothermal carbonization and pyrolysis temperatures on pyrochar formation. Bioresource Technology. 2016;**220**:168-174

[70] Yang X, Tsibart A, Nam H, Hur J, El-Naggar A, Tack FM, et al. Effect of gasification biochar application on soil quality: Trace metal behavior, microbial community, and soil dissolved organic matter. Journal of Hazardous Materials. 2019;**365**:684-694

[71] Wang J, Wang S. Preparation, modification and environmental application of biochar: A review. Journal of Cleaner Production. 2019;**227**: 1002-1022

[72] Igalavithana AD, Mandal S, Niazi NK, Vithanage M, Parikh SJ, Mukome FN, et al. Advances and future directions of biochar characterization methods and applications. Critical Reviews in Environmental Science and Technology. 2017;**47**:2275-2330

[73] Wang T, Zhai Y, Zhu Y, Li C, Zeng G. A review of the hydrothermal carbonization of biomass waste for hydrochar formation: Process conditions, fundamentals, and physicochemical properties. Renewable and Sustainable Energy Reviews. 2018; **90**:223-247

[74] You S, Ok YS, Chen SS, Tsang DC, Kwon EE, Lee J, et al. A critical review on sustainable biochar system through

gasification: Energy and environmental applications. Bioresource Technology. 2017;**246**:242-253

[75] Fryda L, Visser R. Biochar for soil improvement: Evaluation of biochar from gasification and slow pyrolysis. Agriculture. 2015;**5**:1076-1115

[76] Ndirangu SM, Liu Y, Xu K, Song S. Risk evaluation of pyrolyzed biochar from multiple wastes. Journal of Chemistry. 2019;**2019**:1-28

[77] Meyer S, Genesio L, Vogel I, Schmidt H-P, Soja G, Someus E, et al. Biochar standardization and legislation harmonization. Journal of Environmental Engineering and Landscape Management. 2017;**25**: 175-191

[78] Llorach-Massana P, Lopez-Capel E, Peña J, Rieradevall J, Montero JI, Puy N. Technical feasibility and carbon footprint of biochar co-production with tomato plant residue. Waste Management. 2017;**67**:121-130

[79] Cowie AL, Downie AE, George BH, Singh B-P, Van Zwieten L, O'Connell D. Is sustainability certification for biochar the answer to environmental risks? Pesquisa Agropecuária Brasileira. 2012; **47**:637-648

[80] Gelardi DL, Li C, Parikh SJ. An emerging environmental concern: Biochar-induced dust emissions and their potentially properties. The Science of the Total Environment. 2019;**678**: 813-820

Chapter 3

Forest Trees for Biochar and Carbon Sequestration: Production and Benefits

Donald L. Rockwood, Martin F. Ellis, Ruliang Liu, Fengliang Zhao, Kyle W. Fabbro, Zhenli He and David R. Derbowka

Abstract

Many tree species worldwide are suitable for making biochar (BC), with planted eucalypts in particular being very productive and extensive. Above- and below-ground carbon sequestration by Eucalyptus plantations depends on plantation management options. An intensively managed cultivar could sequester over 100 mt of C/ha at a cost of $21–40/mt. BC production systems ranging in size from small mobile units to large centralized facilities and many kiln technologies influence the quality and price of the BC produced as well as the ability to control emissions. While BC from wood has many applications, its use as a soil amendment in forest plantations is appealing as a long-term sequestration strategy and opportunity to grow more robust trees and increase survival rates. Research in Florida USA and elsewhere addresses responses of forest and agronomic crops to wood BC soil amendments with and without other fertilizers. In combination with the carbon sequestered through tree growth, sequestration of 2.5 mt/ha of wood BC as a soil amendment in Eucalyptus plantations has estimated costs ranging from $3.30–5.49/ton of C.

Keywords: biochar, trees, Eucalyptus, production systems, carbon sequestration, soil amendment

1. Introduction

BC's multiple uses (www.biochar-international.org) and numerous benefits [1] include soil and crop improvement, carbon sequestration, retention of nutrients and water, reduced leaching, water purification as well as general and specialty industrial applications, and interest in and demand for BC are growing [2]. From a global BC market value of $1.04 billion in 2016, the market is projected to grow at ~13% annually to a value of $3.2 billion in 2025 [3].

Focusing on Florida USA, we previously published on eucalypts' suitability as a BC feedstock and assessed BC's potential for improving soil properties, tree nutrition, and tree growth [4].

In this chapter, we review (1) the advantages of forest trees for BC by documenting the availability and relative suitability of major tree species, particularly eucalypts (*Eucalyptus*) and related species, (2) carbon sequestration by planted

IntechOpen

Eucalyptus, (3) BC production systems ranging in size and the associated quality of BC, (4) promising BC applications, (5) recent and ongoing BC research, and (6) carbon sequestration potential and associated cost of *Eucalyptus* plantations using wood BC as a soil amendment.

2. Forest trees for biochar

Many tree species worldwide are suitable for making BC, with planted eucalypts in particular being very productive and extensive. Eucalypts are the world's most valuable and widely planted hardwoods (20 million ha in 2018 [5], up to 21.7 million ha in 61 countries by 2030 [6]) and have numerous potential applications [7, 8]. In Florida, several *Eucalyptus* species, including cultivars of *E. grandis* and *E. grandis* × *E. urophylla*, have promise as short rotation woody crops (SRWC, [9, 10]).

BCs from *E. grandis* × *E. urophylla* cultivar EH1, *Corymbia torelliana*, *E. grandis* cultivar G2, *E. amplifolia*, and *Quercus virginiana*, were similar and suitable for commercial BC production ([4], **Table 1**). Compared to high quality European *Quercus* spp., all five Florida trees were similar for recalcitrant carbon but higher in pH and water holding.

Other evaluations of BCs made from various woods and other feedstocks indicate that feedstock and pyrolysis condition influence properties important for using BC as a soil amendment [11, 12]. Since key objectives in BC production include minimizing the combustion of carbon, maximizing carbon content, and minimizing ash, consistency of feedstock and the production operating environment are imperative.

Property	Florida tree					Europe
	G2	CT	EH1	EA	QV	Qsp
Volatile matter (% of DW*)	83.3	85.0	85.9	82.5	83.3	
Fixed carbon (% of DW)	15.7	14.4	13.7	17.0	15.5	
Ash (% of green weight)	1.00	0.54	0.37	0.50	1.15	
Moisture content (% of DW)	36.4	48.0	43.1	30.1	33.1	
C (% of DW)	49.2	49.7	49.8	50.8	49.1	
O (% of DW)	43.0	43.1	43.1	42.0	43.1	
H (% of DW)	6.5	6.5	6.5	6.5	6.4	
N (% of DW)	0.21	0.17	0.17	0.26	0.29	
Cl (% of DW)	0.07	0.02	0.02	0.02	0.00	
S (% of DW)	0.01	0.00	0.00	0.01	0.00	
Recalcitrant carbon (%**)	76.0	71.6	74.0	70.8	71.8	67.6
pH	10.6	10.4	10.5	11.1	11.9	8.2
EC (mmhos/cm)	0.57	1.76	1.56	3.88	1.14	3.33
Water holding (ml/100 g)	75.9	78.8	79.8	69.0	68.5	43.4
Carbonate value (%)	2.6	2.5	5.6	16.7	2.5	—

Dry weight (DW).
Estimated at 80% of fixed carbon on a dry ash-free basis.

Table 1.
Properties of BC made from Florida E. grandis *cultivar G2, C. torelliana (CT),* E. grandis × *E. urophylla cultivar EH1,* E. amplifolia *(EA), and* Q. virginiana *(Qv), and European Q. sp. (Qsp) test trees (adapted from [4]).*

3. Carbon sequestration by planted *Eucalyptus*

Eucalyptus planting density trials have assessed the effect of stand density on biomass production. On former citrus lands and phosphate mined clay settling areas in central and south Florida, *E. grandis* cultivars can have maximum mean annual biomass increments (MAI_{max}) as high as 78.2 green mt/ha/year with associated internal rates of return greater than 10% [13]. Through 81 months, the intensively managed *E. grandis* × *E. urophylla* cultivar EH1 planted on former citrus beds at two planting densities yielded more at 2471 trees/ha than at 1181 trees/ha. Annual yield at 2471 trees/ha was over 58 green mt/ha/year in 3.7 years compared to 44 mt/ha/year at 5.0 years. However, planting density also inversely affected average tree Diameter Breast Height (DBH) as the higher planting density produced smaller trees.

To estimate carbon sequestration over a rotation in Florida, we applied carbon allocations for *E. grandis* in Brazil [14] and *E. grandis* × *E. urophylla* in China [15] to Florida tree data. The resulting total carbon sequestration estimates ranged from 38 to 95 mt/ha at the time of peak annual accumulation (**Table 2**), with longer-term totals over 100 mt/ha in 6 years, again depending on cultivar, site, planting density, and harvest age (**Figure 1**).

Sequestration estimates by *Eucalyptus* elsewhere vary. *Eucalyptus* plantations in southern China sequestered ~100 mt C/ha [16]. *E. urophylla* × *E. grandis* planted in southern China accumulated >70 mt C/ha in 6–8 years [15]. In South Africa, 10- and 25-year-old *E. grandis* plantations may store 47 and 270 mt C/ha [17]. *Eucalyptus tereticornis* plantations may accumulate up to 129 mt C/ha in 4 years [18].

Planting density (trees/ha)	Tree component				Rotation age at MAI_{max} (years)
	Stem (wood + bark)	Crown	Roots	Total	
G3 on clay settling areas					
2533	61.2	4.3	6.8	72.3	4.3
5066	80.5	5.6	8.9	95.0	4.2
8841	32.3	2.2	3.6	38.1	3.4
EH1 on former citrus beds with intensive culture					
1181	63.4	6.7	5.7	75.8	5.5
2471	64.5	6.9	5.8	77.2	4.7

Table 2.
Predicted carbon sequestration (mt/ha) by tree components at maximum mean annual increment (MAI_{max}) rotation age for cultivars G3 and EH1 at three and two planting densities, respectively.

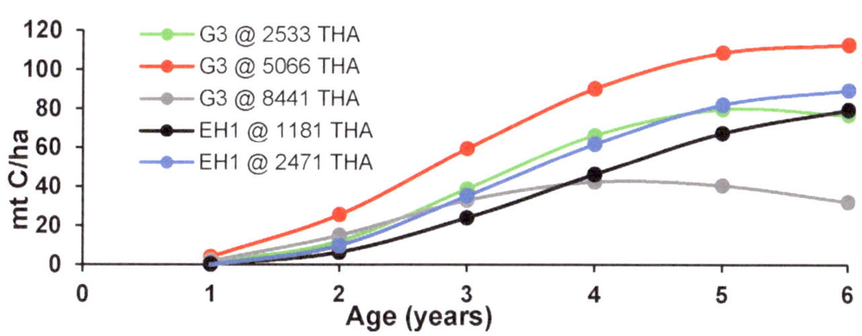

Figure 1.
Cumulative total (stem + crown + roots) carbon sequestration (C, mt/ha) for each genotype × planting density (trees/ha, THA) scenario by age (without BC application).

4. Biochar production systems

BC production needs to be considered in the context of charcoal production. Charcoal has been produced for millennia for various applications from art to soil amendments creating terra preta soils in the Amazon to metallurgical and other industrial applications. What differentiates high-quality BC from lower-quality BC or simple charcoal is the production system, i.e., the ability to control the process and operating conditions and ultimately BC's physical and chemical properties.

BC is produced via pyrolysis, which is the process of heating wood in a low oxygen environment (ideally close to zero oxygen) with the objective of removing all moisture and volatiles in the wood, maximizing carbon content, and minimizing ash content all while trying to increase porosity (pore structure) and maximize surface area.

BC from trees may be produced in systems ranging from small, simple kilns (e.g., mound and brick kilns) to large centralized, custom designed facilities (e.g., retort technologies). As with all production technologies, there are tradeoffs which impact cost, efficiency, quality, emissions, and product applications. BC production techniques are no different, and as the market and applications advance, these differences will become even more relevant.

Batch systems require less technical expertise, are easy to set up, and have low or very low capital requirements. Consequently, there are numerous batch production technologies used around the world and available for purchase. At the other end of the spectrum are continuous production technologies that are typically custom designed, require greater technical expertise, and require significant capital investment. However, the quality and consistency of the BC as well as the economies of scale are significantly enhanced, and a well-designed continuous process captures all components of the value-chain. As more sustainable and environmentally friendly production is sought, these issues will become increasingly important.

For perspective, we specifically review seven batch technologies (Pit, mound, and brick kilns, Metal kiln, Missouri-type kiln, Kon-Tiki kiln, and Rotary kiln), three mobile BC production units (Carbonator 6050, FireBoxes, and "cooker"), and two continuous production technologies (Polchar and GCS).

4.1 Pit, mound and brick kilns

Simple kilns—pit, mound and brick—are still widely used worldwide, and all operate as a batch process (**Figure 2**: 1–3). The earliest kilns were temporary in nature—either pits or mounds. They are low cost, simple, and are still widely used in developing countries. While there is value in simple and low-cost production, this simple "pyrolysis" technique has low yield with inconsistent quality and very limited ability to control the process.

Pit kilns (**Figure 2**: 1) are the lowest-tech "pyrolysis" technique. At their simplest, pit kilns are open pits with more advanced pits being covered with leaves and earth or mud to create a partially closed carbonization environment. An open pit is dug, wood is added to the pit and set alight with the goal of combusting the volatile material inherent in the wood without fully burning the wood to ash, in essence making charcoal. Control over oxygen in the carbonization process is limited. Pits require significant oversight, quality is poor, and BC yields (kilogram of BC produced per kilogram of wood) are low. Yields are independent of the size of the pit, but larger pits increase labor efficiency.

The mound kiln (**Figure 2**: 2) is an aboveground version of the pit kiln. Wood is stacked vertically into a mound with a built-in wood chimney in the center of the mound, and the mound is covered with twigs and leaves and then earth. This

Figure 2.
Representative kilns: (1) Open pit (source—Pacific biochar), (2) mound (courtesy register of the intangible cultural heritage, Slovenia), (3) brick (courtesy Kamado Joe Europe BV, Netherlands), (4) metal kiln (courtesy four seasons fuel ltd., UK), (5) Missouri-type kiln (courtesy the biomass project).

technique allows BC producers to better control heat and air during carbonization of the wood. Mound kilns are typically 4–5 m wide, 1.5–2.0 m high with a number of vents at the bottom of the mound to control air flow into the kiln.

The brick kiln (**Figure 2**: 3) is a step up from pit and mound kilns. With a relatively low capital cost, creates a better carbonization "chamber," produces better quality BC and generates better BC yields. Since the entire kiln is constructed from bricks it works similarly to a brick refractory by providing better heat insulation. Brick kilns are typically larger than mound kilns with diameters up to 7 m. However, production time is still relatively lengthy with carbonization and cooling taking up to 10 days.

4.2 Metal kilns

Advancement in kiln technology led to metal kilns a little over a century ago. Metal kilns (**Figure 2**: 4) have many benefits over brick and mound kilns; (1) they require less oversight and attention, (2) process wood faster (reduced residence time), and (3) have improved airflow all of which lead to improved yield and better-quality BC. With the reduced residence time, BC can be produced in as little at 3 days but as with all kilns, there is no control of air pollution/emissions.

4.3 Missouri-style kiln

In the quest to improve quality, combustion dynamics, and economies of scale, the Missouri Kiln (**Figure 2**: 5) was developed in the early 1900s. A rectangular kiln, with concrete or concrete block walls to improve thermal insulation and mechanized loading and unloading, allows producers to increase volume resulting in better economies of scale. Missouri-type kilns have approximately three times the capacity of a brick kiln and with the ability to modify and add chimneys, yields are better than metal kilns, and quality is improved.

While Missouri-type kilns can be designed to have reduced emissions compared to other kilns—chimneys can be connected to afterburners to reduce CO and volatile organic emissions—the batch production makes this difficult since there is no continuous process and no steady state.

4.4 Kon-Tiki kiln

Kon-Tiki kiln (**Figure 3**: 1, 2) was developed in Switzerland [19]. As opposed to high volume BC production, Kon-Tiki focused on the democratization of BC by developing a simple but ingenious invention that produces reasonably high-quality BC. Unlike earthen or brick kilns, steel walls reflect the pyrolysis and combustion heat back into the kiln, resulting in improved combustion dynamics and more uniform temperature distribution, thus ensuring more homogeneous charring conditions.

A steep cone shape is used. Air is drawn in over the hot outer wall of the kiln and swirls above the fuel bed creating a vortex that ensures good mixing of pyrolysis and combustion air. Once the kiln reaches its working temperature of 650–700°C, hardly any smoke is visible. The combustion air rolls in over the metal edge of the outer wall and into the kiln. But at the same time, the burning gases must escape upwards and so, a counter-rotating vortex is established in the center of the kiln. The wood gas, which is heavier than air, is kept in the vortex until it is burned. The garden scale Kon-Tiki kiln in **Figure 3**: 2 allows anyone to carbonize biomass quickly and cleanly. In approximately 2 hours, 0.2 m^3 of BC can be produced.

4.5 Rotary kiln

Rotary kiln technologies can be applied to powders, granules, or sized feedstock. A rotary kiln (**Figure 4**) consists of a cylindrical, rotating tube mounted between stationary material feed and outlet housings. The rotating tube mixes the material,

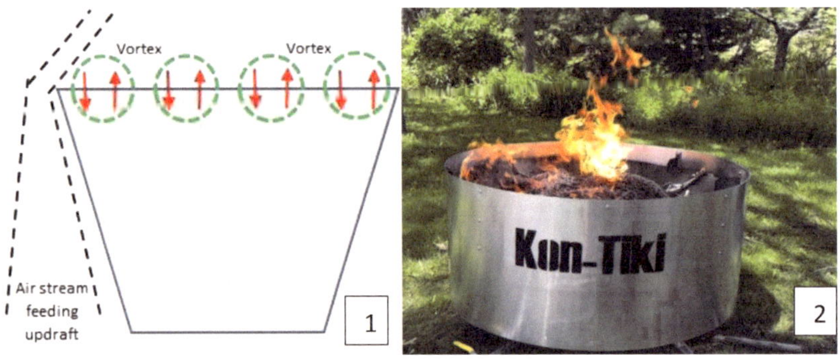

Figure 3.
Kon-Tiki kiln schematics (1, courtesy the democratization of biochar) and in operation (2, courtesy garden scale Kon-Tiki, Finger Lakes biochar, New York).

Figure 4.
Rotary kiln (1) and its schematics (2, courtesy IBU-tec AG, Germany).

ensuring that it is processed homogeneously. Rotary kilns can be heated either directly (inside the kiln) or indirectly (from outside).

In a direct-fired rotary kiln, the burner is situated inside the kiln body, that is, inside the reaction chamber. The material is heated directly by the burner flame and the stream of hot gas produced by the burner. These kilns are usually lined with a refractory (heat-resistant) material so that they can be operated at higher temperatures, as high as 1500° C. Directly fired kilns are generally robust and scalable.

4.6 Mobile biochar production

Tigercat's Carbonator 6050 (**Figure 5**: 1) is a mobile, carbon negative BC production system designed to cost effectively reduce logging, land clearing, etc., residues by 90% [20]. The single operator, remotely controlled, continuously running Carbonator 6050 converts low value, bulky woody biomass without chipping or grinding into high quality, variable sized BC (e.g., 71% recalcitrant carbon, 91% void space, 8% ash) that can be used on-site or easily transported.

Air Burners' FireBoxes (**Figure 5**: 2, airburners.com/products/) are cost-effective, self-contained, above ground air curtain burners (ACB) with thermal ceramic refractory walls designed to eliminate wood and vegetative waste by up to 98% with the lowest environmental impact, while producing clean carbon ash and BC. Burn rates are from 1 to 13+ tons/hour. Only Air Burners' Fireboxes have been tested by the U.S. EPA and other international government environmental agencies and proven to meet and exceed U.S. EPA regulations for ACBs.

A batch retort BC cooker (**Figure 5**: 3) built by Passive Remediation Systems Ltd. (www.prsi.ca) illustrates a cost effective small scale BC production system. Assembled from a 3785 l propane tank, a small propane fuel tank, a boat trailer, and upcycled pipes and valves, this 0.14 m^3 machine makes BC from any dried biomass. After sealing the bolted oven door with water-based caulking, pallets supply startup heat and spark to initiate pyrolysis. A thermocouple indicates internal conditions for BC production and facilitates the testing of biomass materials for improving organic

Figure 5.
Representative mobile BC production systems: (1) carbonator 6050, (2) air burner fireboxes, and (3) cooker.

farm growing operations and sequestering carbon. It has produced high quality hemp BC (82% pure, A&L Canada Laboratories, London, Ontario). With minimal operating impact, it can also produce valuable volatile organic compounds collectively called wood vinegar, which is rich in chemicals used for natural health products, fertilizer, insecticides, industrial chemicals, and manufacturing feedstocks [21].

4.7 Continuous production technologies

There are few "off-the-shelf" continuous production technologies. As a result, most continuous processes are custom designed around a specific pyrolysis technology. The benefit of a custom design continuous-process is that quality and consistency of the BC as well as economies of scale are significantly enhanced. One continuous production technology is vertical retort pyrolysis. Retorts may have a high capital cost, but the labor cost per unit of production is low.

GCS' sister company, Polchar (polchar.p) in Poland, is a large facility (**Figure 6**: 1) that specializes in pyrolysis and carbonization of different feedstocks using horizontal moving grate and vertical retort pyrolysis technologies. GCS and Polchar have developed a custom retort BC production process that is fully integrated.

Wood feedstock is cut to size, pre-dried and processed in a vertical, gravity-fed retort process. Pyrolysis temperatures can be modified to run the process hotter or cooler with typical temperatures between 600 and 700°C. With control over the process, high quality BCs are produced, and the volatiles emitted in the pyrolysis process are either fully combusted in a combustion chamber and the associated heat used in other applications, including the pyrolysis process, or the volatiles can be condensed into wood vinegars and bio-oils for further refinement. With the ability to control temperature and residence time, the retort technology allows production of a high quality BC in volume. For example, GCS' general BC specs are: fixed carbon: 93–95% DW, volatiles: 3–4% DW, ash: 2–3% DW, and surface area: 585–630 m^2/g.

High-end BCs can offer specs not achievable with most production techniques, including custom sizing and moisture content, enhanced pore structure—micro, meso, and macro pores, high surface area, and high-carbon content with low-ash.

GCS runs its process 24 hours a day, 7 days a week. The process quickly achieves steady state which allows GCS to control product quality, capture all components of the value-chain including excess heat produced from the pyrolysis process, condense vaporized volatiles into bio-oils and wood vinegars, and significantly reduce

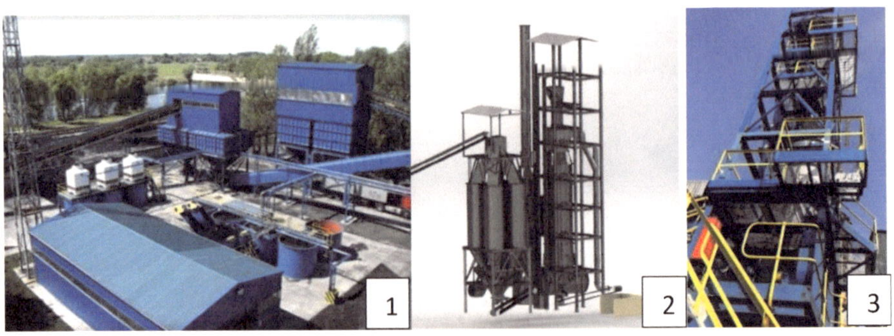

Figure 6.
Representative continuous BC production systems: (1) Polchar production facility, GCS's (2) vertical retort process and (3) vertical retort.

Technology	Technology cost	Technical expertise	Quality and consistency	Emission control
Pit (open and closed)	Very low	Very low	Low	None
Mound kiln	Very low	Low	Low	None
Brick kiln	Low	Low	Low/moderate	None
Metal kiln	Low	Modest	Moderate	None
Missouri-type kiln	Moderate	Modest	Moderate	Limited
Rotary kiln	Large/high	High	High	High
Kon-Tiki	Low	Low	Moderate	Low
Advanced mobile	Moderate	Modest	Moderate	None/modest
Custom continuous	Very high	High	High	Very high

Table 3.
Comparison of BC production technologies for cost, technical expertise, BC quality and consistency, and ability to control emissions.

the environmental footprint by reducing carbon monoxide and volatile organic emissions.

The BC production technologies reviewed, while varying in cost, expertise, consistency, and emissions (**Table 3**), can all process different wood feedstocks. The combination of varying feedstocks and production technologies results in significant variation of BC specs. *Eucalyptus* biomass components also influence BC properties such as grindability, slagging, and ash content [22]. GCS' new BC production facility in southern Florida will likely use eucalypts grown in nearby plantations.

5. Biochar applications

While interest in and demand for BC are growing [2], BC's multiple applications range widely in potential market size, timing, competitiveness, and pricing compared to alternative products, all of which influence the value of wood grown for BC (**Table 4**).

Historically BC has been viewed through the lens of a soil amendment to help build modern-day terra preta type soils, a type of very dark, fertile artificial (anthropogenic) soil found in the Amazon Basin. But with improved and more sophisticated BC production techniques, there is a vastly expanded market for BC. With the need to replace the substantial loss of soil carbon due to modern agricultural practices [23] and considering the emerging carbon cascades (utilizing BC for one application before it is used for a second) [24], the applications and future potential markets become quite large (**Table 4**).

Focusing on the soil amendment applications, there are growing opportunities to utilize BC to increase nutrient and water retention. However, more than this, many soils have been compromised with heavy metal buildup and other environmental toxicity. BC is now being utilized to help remediate these soils and similarly remediate contaminated water.

In general industrial markets, there are emerging BC applications in many large markets including concrete where BC is starting to be used as a partial substitute for sand and cement. While there are a number of benefits to including BC in concrete, some include reducing weight and increasing strength since BC's water retention

Application	Market size	Timing	Competition	Pricing
Soil carbon	Large	Current	Low	Low
Soil nutrients	Limited	Emerging	High	Low
Crop yield	Moderate	Current	High	Low/moderate
Carbon sequestration	Very large	Emerging	Moderate	Moderate
Nutrient retention	Large	Current	Moderate	Moderate
Water retention	Large	Current	Moderate	Moderate
Water purification	Large	Emerging	Low	High
General industrial	Large	Current	Moderate	Moderate
Specialty industrial	Moderate	Emerging	Low	High

Table 4.
Comparison of market size, timing, competition, and pricing for nine BC applications.

capacity improves the curing process. Other industrial applications include the metallurgical industry and as a filler in asphalt. With 1.8 billion tons of asphalt poured every year, if only 0.5% BC were added this would result in demand for 8 million tons of BC a year; increase the BC component and this market could be very large.

BC applications for specialty industrial markets are growing quickly too. There are opportunities in applications such as acoustic and thermal insulation in walls, ceilings and floors. There are also applications for BC in carbon fiber and polymer space where BCs increase strength, reduce weight, and improve thermal properties. Other emerging specialty industrial applications include protecting against electrosmog, filtration media, heavy metal adsorption, and as partial filler in car and truck tires. Growing trends in developing sustainable supply chains and reducing societal carbon footprint will help accelerate growth of many of these markets.

6. Biochar research

Based on results elsewhere, forest and agronomic crops in Florida USA are likely to respond to organic soil amendments such as BC. Applying BC improves soil physicochemical properties, including bulk density, porosity, cation exchange capacity, and pH. BC also increases soil water and nutrient retention and consequently influences crop production while reducing leaching [25]. Productivity of many crops significantly increased after soils were amended with BC [26, 27]. Sandy soils are more responsive to BC than clayey soils [28] due to their low water and nutrient holding capacities [29]. Most soils in Florida are sandy (>90% of soil particles) and have low nutrient and moisture holding capacities. Fertilizers necessary for crop yield and quality are readily leached, causing environmental pollution. The following recent and ongoing studies in Florida illustrate the response of a few crops to organic soil amendments with and without wood BC.

Green Edge (GE), an organic, slow-release fertilizer (6-4-0), and planting density both influenced the productivity of cultivar EH1 in a study at the IRREC near Ft Pierce, FL ([4], **Table 5**). While the differences among five fertilizers [GE equivalent to 0, 112, 224, and 336 kg N/ha and diammonium phosphate (DAP) at 336 kg N/ha] initially favored the higher GE rates, subsequent differences were inconsistent. The highest planting density had the smallest tree DBH through harvesting at 47 months.

Treatment	Trt. level	Tree DBH at age			Coppice trait		
		36	41	47	Ht	DBH	No.
Planting density (trees/ha)	3588	9.3 b*	10.3 b	11.4 b	2.4	2.6	3.9
	1794	10.9 ab	12.3 ab	13.5 ab	2.2	3.2	3.4
	1196	13.2 ab	14.2 a	15.4 a	2.3	3.3	4.3
Fertilizer (kg N/ha)	0	11.8	13.7 a	14.9 a	2.3	3.4	3.9
	GE 112	10.7	11.2 b	12.4 b	2.3	3.1	3.8
	GE 224	12.6	13.2 ab	14.5 a	2.3	3.0	3.7
	GE 336	11.3	12.6 ab	14.1 ab	2.3	3.0	4.3
	DAP 336	11.2	12.0 ab	13.3 ab	2.2	3.1	3.8

Table 5.
Effects of three planting densities and five fertilizers on EH1 tree (DBH in cm at 36-, 41-, and 47-months of age) and coppice [height (Ht) at 3 months and DBH and number of stems at 8 months of age] traits.

Early coppice responses to planting density and fertilization were nonsignificant, although some earlier trends remained (**Table 5**). The lower planting densities had somewhat larger coppice stem DBHs. The early fertilizer effects were no longer evident.

In a BC-GE study at the IRREC, BC at 11.2 tons/ha enhanced the soil nutrients and tree leaf nutrient levels [4]. GE at 336 kg/ha generally increased available soil nitrogen, and GE + BC further increased NH_4-N. Soil available P significantly increased for GE + BC 5 months after amendment. However, both available N and P in the soil decreased 11 months after amendment, likely due to uptake by the trees. At 11 months, soil NH_4-N was significantly higher with GE and GE + BC. BC also generally enhanced tree leaf nutrient levels, with GE increasing leaf concentrations of Ca, K, Mg, P, Fe, and Mn, and GE + BC significantly increasing 5-month Zn. GE + BC further improved tree leaf Ca, Mg, Zn and Mn, and such improvement was also observed 11 months after amendment. However, a general decrease in leaf Ca, K, Mg, P, and Zn occurred over time, likely due to decreased availability in the soil and dilution by rapidly increased tree biomass.

Further, GE and GE + BC gradually enhanced tree growth [4]. Six months after treatment applications, the cultivars receiving GE only and GE + BC had doubled in height, approximately twice the increase with no treatment. Eleven months after application, cultivars receiving GE + BC were bigger than those receiving GE only and no GE.

Longer term, GE and GE + BC further enhanced tree growth (**Table 6**). At 16 months, the cultivars receiving GE only and GE + BC had nearly tripled tree DBH with no GE and BC. Subsequently, cultivars receiving GE + BC continued to surpass

Age	Treatment*		
	Control	GE	GE + BC
16	2.0 b	5.1 ab	6.5 a
27	2.6	6.3	8.6
31	3.9	8.3	11.2

**Treatment means not sharing the same letter in a trait differ at the 5% level.*

Table 6.
Tree DBH (cm) at ages 16-, 27-, and 31- (11, 22, and 26 months after treatment applications) months of E. grandis *cultivar G5 receiving three treatments: Control, GE, GE + BC.*

those receiving GE only and no GE. BC plus fertilizer greatly improved reforestation performance of two tropical species [30].

The fertilizer and planting density differences observed in the IRREC fertilizer × planting density and BC-GE studies are consistent with previously observed influences of fertilizer and planting density on eucalypt productivity in Florida [31–34] and worldwide [35–38]. While inorganic fertilizers have been necessary for rapid growth of eucalypts on Florida's infertile sandy soils, the observed response here to the slow release organic GE, and its apparently beneficial coupling with BC, is encouraging for sustainable eucalypt management. Planting density effects were evident early, with, for example, the 3588 trees/ha in the fertilizer × planting density having the tallest trees at 9 months and the largest stand basal area but smallest tree DBH at subsequent ages. Similar effects of planting density have been noted for *E. dunnii* seedlings and clones [39].

BC enhanced the soil properties of infertile sandy Florida soils as well as the nutrient status of *E. grandis*, especially when applied together with GE and/or chemical fertilizers. BC's large cation exchange capacity facilitates retention of nutrients, particularly Ca, Mg, K, Fe and Mn against leaching. BC's large water holding capacity improves water availability, which is especially important during the dry season. Due to high temperature and humidity, decomposition of organic materials in sandy soils is very rapid, leading to low organic matter contents. BC is a good organic amendment for sandy soils because it stays in soil much longer than other organic materials.

Outside Florida, BC applications on forest trees have given mostly positive results. When broadcast on temperate hardwoods, the major short-term BC impact was an increase in limiting soil P and Ca [40]. BC application in forest ecosystems generally improved soil physical, chemical, and microbial properties [41]. BC from *E. marginata* decreased soil microbial carbon in a coarse soil [42], and BC added to a sandy desert soil did not significantly change soil physical properties [43]. Two BC types had different impacts on growth of young *Pinus elliottii* in subtropical China [44]. Varying doses of macadamia BC combined with two fertilizer rates had contrasting results on soil nutrients and ambiguous trends in the growth of young *E. nitens* [45]. BC did not enhance *Eucalyptus* hybrid survival or growth on degraded soils in southern Amazonia [46]. Compost and BC-compost mixes did not improve the performance of poplar, willow, and alder [47]. BC made from forest thinnings, when applied to temperate managed forests, had no detrimental effects, suggesting that BC can be used for carbon sequestration [48]. A meta-analysis of wood-based BCs indicates a large potential for early tree growth responses to soil amendment in reforestation of boreal and tropical systems [49]. An increase in carbon accumulation in planted loblolly pine due to fertilization [50] suggests that eucalypt plantations receiving BC with fertilizer will also experience an increase in soil carbon.

In Florida, BC has also been recently tested on agronomic crops. Oak-derived BC combined with standard fertilizers enhanced lettuce (*Lactuca sativa*) productivity. Superphosphate (SP) derived from dolomite phosphate rock (DPR) alone generally resulted in less biomass compared to DAP, likely due to lower P availability (**Table** 7). However, lettuce growth was significantly enhanced by P fertilizer plus BC, as indicated by a significant increase in lettuce height and leaf chlorophyll content (by 19.3–138%). Lettuce dry biomass on average is increased by 61.7–76.8%. The maximum biomass yield occurred with DAP and BC combined.

BC application increased soil pH by 1.2–1.7, which was significantly higher than the treatments without BC ($P < 0.05$). As compared with DAP fertilizer, SP application increased soil electrical conductivity (EC) ($P < 0.05$), but this effect was mitigated with BC ($P > 0.05$).

Treatment	Biomass yield	Height	LeafC	Soil pH	Soil EC
	g/pot	cm	SPAD	—	μS/cm
CK	1.1 ± 0.5 a*	7.7 ± 1.5 a	15.2 ± 1.5 a	5.65 ±0.04 b	1215 ±288 b
DAP	24.6 ± 2.3 c	24.3 ± 0.8 cd	28.1 ± 0.5 d	4.94 ± 0.08 a	833 ± 41 ab
SP1	2.6 ± 0.9 a	10.7 ±2.2 a	26.7 ± 1.5 cd	5.06 ± 0.08 a	2526 ±57 cd
SP5	11.4 ± 6.7 ab	17.7 ± 4.9 b	29.7 ± 0.9 d	5.07 ± 0.01 a	1950 ± 291 c
SP7	2.7 ± 1.2 a	8.4 ± 1.8 a	29.5 ± 0.7 d	5.05 ± 0.07 a	2643 ± 298 d
BC-DAP	39.8 ± 4.5 d	29.0 ± 0.7 d	22.0 ± 1.5 b	6.60 ± 0.07 d	410 ± 107 a
BC-SP1	22.4 ± 4.1 bc	22.6 ±1.5 bc	23.5 ± 1.4 bc	6.34 ± 0.05 c	2148 ± 300 cd
BC-SP5	26.1 ± 8.4 c	23.7 ± 3.2 bcd	23.4 ± 1.7 bc	6.44 ± 0.04 c	901 ± 44 cd
BC-SP7	17.4 ± 5.9 bc	19.9 ± 1.9 bc	26.6 ± 2.2 cd	6.28 ± 0.09 c	1964 ± 314 c

**The same trait means not sharing the same letter differ at the 5% level.*

Table 7.
Effects of P fertilizer and BC treatments (CK, control; DAP; SP1, SP2 and SP3: Superphosphate derived from DPR1, DPR2 and DPR3; and BC at 1% application rate) on lettuce biomass yield, height, leaf chlorophyll (LeafC), and soil properties.

Other recent agronomic BC studies are summarized in **Table 8**. As suggested most by the cauliflower response in Gainesville, notable soil and plant responses to BC may take up to 2 years, although BC immediately increased soil organic matter

Location—Crop	BC treatments	Notable responses
Gainesville—Cauliflower	(1) 11.2 mt/ha in 3/2018, (2) 11.2 mt/ha in 9/2018, (3) 0 mt/ha; 1961 kg/ha of 10-10-10 in 10/2019, 78 kg/ha of liquid 5–1-1 every 3 weeks	In 1/2020, (1), (2), (3)—Soil NO3-N: 3.44, 2.19, 2.45 ppm Leaf N: 5.29, 4.70, 4.68%
Gainesville—Perennial peanut	(1) 11.2 mt/ha in 1/2020, (2) 0 mt/ha in 1/2020, (3) 112 kg/ha of N as GE in 2/2020	In 1/2020, (1), (2)—OM: 1.93, 1.42% In 2/2020, (1), (3)—pH: 5.5, 5.2
Old Town—Vegetables	(1) 16.8 mt/ha in 1/2019, (2) 11.2 mt/ha, (3) 5.6 mt/ha, (4) 0 mt/ha; 672 kg/ha of DAP, 448 kg/ha of potassium sulfate, and 15.7 mt/ha of lime in 5/2019	In 5/2019, (1), (2), (3), (4)—No evident trends
Old Town—Sorghum	(1) 22.4 mt/ha in 1/2019, (2) 16.8 mt/ha, (3) 11.2 mt/ha, (4) 5.6 mt/ha, (5) 0 mt/ha	In 5/2019, (1), (2), (3), (4)—Soil NO3-N: 2.25, 2.05, 2.16, 3.69, 1.47 kg/ha Soil Ca: 3525, 3255, 3670, 3094, 3015 kg/ha Soil CEC: 8.8, 8.2, 9.0, 8.0, 7.6 meq/100 g
Old Town—Bahiagrass	(1) 11.2 mt/ha in 1/2019, (2) 0 mt/ha	In 2/2020, (1), (2)—Soil K: 166, 21 kg/ha Soil OM: 1.3, 0.8% Soil CEC: 9.1, 5.0 meq/100 g

Table 8.
Summary of notable results from recent wood BC agronomic studies in Florida.

in a new perennial peanut-based study in Gainesville. The studies near Old Town had varied responses after several BC rates were applied.

Outside Florida, BC from forest trees has benefited agronomic crops. Soil incorporation of BC produced from *E. camaldulensis* increased critical soil properties and the yield of groundnut in south Senegal [51]. BC-blended compost significantly improved quantity and quality of four crops in Europe [52].

7. Carbon sequestration potential of *Eucalyptus* plantations

Given eucalypts high productivity and their use for traditional forest products and because economic feasibility is one of several conditions for a sustainable BC system [53], our financial analysis goal here is to estimate the cost of potential carbon sequestration by *Eucalyptus* plantations using BC as a soil amendment. The following scenario for growing cultivar EH1 on sandy former citrus beds assumed original and two coppice rotations of 5.5 and 4.7 years for 1187 and 2471 trees/ha, respectively, with the first and second coppice growing at 90 and 80% of the original planting. The application of 2.5 tons/ha of BC priced at $750 and $1000/ton assumed a 7% growth increase per ton of BC.

The price of adding BC to the soil was approximately $3 and $5/ton for BC priced at $750 and $1000/ton, respectively, across two planting densities and three discount rates (**Table 9**). Increasing planting density from 1181 to 2471 trees/ha typically increased BC price by $2 per ton. Given the at least 91.5% C content of GCS's BC, the resulting cost for sequestering C in plantation soils ranges from $3.30–5.49/ton added to the soil. This cost is considerably less than the $30–50/ton estimated in 2005 for US forestry sequestering up to 500 million tons of C/year [54]. In 2015, the California Air Resources Board listed C sequestration credits at $12–13/ton [55]. BC produced from hardwoods has a soil residence time in excess of 1000 years [56].

Besides economic feasibility, other conditions of a sustainable BC system include producing/deploying BC safely and not competing with other wood uses; initiatives are necessary in research, policies, and implementation to meet these standards [53]. BC application to soil in Poland is viewed as an important component of the region's circular economy and means of counteracting climate change [57]. In South Africa, carbon storage by *Eucalyptus* and pine plantations and by their long-lived

Planting density (trees/ha)	Discount rate (%)		
	4	6	8
No BC added to soil			
1181	5.6	6.7	7.9
2471	7.7	9.0	10.4
2.5 mt/ha of BC added to soil at $750/mt			
1181	8.5	10.2	12.0
2471	10.3	12.0	13.9
2.5 mt/ha of BC added to soil at $1000/mt			
1181	9.6	11.5	13.6
2471	11.3	13.2	15.3

Table 9.
Effects of planting density, discount rate, and BC price on carbon price ($/mt) for E. urophylla × E. grandis *cultivar EH1 grown on former citrus land.*

Metabolite	*Eucalyptus* genotype					
	G2	G3	G4	G5	5408	EH1
α-Pinene	32.2	28.7	59.3	12.2	38.3	11.6
1,8-Cineole	41.8	47.2	0.1	1.5	1.0	32.5
g-Terpinene	0.1	0.2	0.1	0.6	9.9	17.6
15.81195238	0	0	0	27.9	0.1	5.9
Squalene	5.9	5.7	16.3	9.9	10.4	2.1
Vitamin E	7.6	5.9	7.3	4.0	11.4	1.3

Table 10.
Percent of total hexane extractable metabolites in the crown biomass of six Eucalyptus *genotypes.*

forest products may equally contribute to offsetting almost a total of 4% of the country's carbon emissions [17]. However, because soil C may decrease as *Eucalyptus* plantations mature [15], BC incorporation into plantation soil can be beneficial.

Other bioproducts may also enhance *Eucalyptus* value [4]. Classified as naturally occurring, generated by biochemical processes, or by thermochemical processes [8, 58], many bioproducts have higher value than or could augment traditional wood products, thereby reducing the cost of carbon sequestration by planted eucalypts. However, genetic variation in *Eucalyptus* influences which genotype is best for a particular bioproduct. For example, Florida eucalypts that have been evaluated as jet fuel feedstocks vary widely in percentage of six important metabolites (**Table 10**), with cultivar EH1 not only having a relatively high proportion of 1,8-cineole, but also a much higher absolute amount of this important chemical.

8. Conclusions

While planted eucalypts are very productive worldwide, their above- and below-ground carbon sequestration depends on plantation management options such as cultural intensity, planting density, and rotation length. In Florida USA, *E. grandis* × *E. urophylla* cultivar EH1 planted on former citrus beds and managed at relatively low intensity could sequester over ~20 mt of C/ha/year at a cost of $30–40/mt. BC production systems ranging from small mobile units to large centralized facilities influence the quality and price of the BC produced, and high-quality feedstocks are critical to producing consistently high-quality BC with uniform quality and specifications for many promising applications. Research in Florida USA and elsewhere addresses responses of forest and agronomic crops to wood BC soil amendments with and without other fertilizers. While BC from wood has many applications, its use as a soil amendment in forest plantations is appealing as a long-term sequestration strategy. In combination with the carbon sequestered in trees, cost estimates of sequestration in *Eucalyptus* plantations by using wood BC as a soil amendment for those plantations are around $5 per mt of BC added per ha.

Acknowledgements

The authors gratefully acknowledge the direct and/or indirect support provided by the IRREC, GCS, GreenTechnologies, Evans Properties, Becker Tree Farm, and ArborGen, the provision of research sites by R Hodges, Evergreen Polled Herefords LLC, G Kunkel, and J Rodriguez, the analyses by S Murch of the University of

British Columbia and by G Tuskan, T Tschaplinski, and N Engle of the Biosciences Division, Oak Ridge National Laboratory, and the field assistance of R Cave and J Rockwood.

Conflict of interest

The authors declare no conflicts of interest.

Author details

Donald L. Rockwood[1,2]*, Martin F. Ellis[3], Ruliang Liu[4], Fengliang Zhao[5], Kyle W. Fabbro[2], Zhenli He[5] and David R. Derbowka[6]

1 Florida FGT LLC, Gainesville, USA

2 UF/IFAS School of Forest Resources and Conservation, Gainesville, USA

3 Green Carbon Solutions (GCS), Pepper Pike, USA

4 UF/IFAS Indian River Research and Education Center (IRREC), Ft Pierce, USA

5 IRREC, Ft Pierce, USA

6 Passive Remediation Systems Ltd., Armstrong, USA

*Address all correspondence to: dlrock@ufl.edu

References

[1] Hussein H, Farooq M, Nawaz A, Al-Sadi AM, Solaiman ZM, Alghamdi SS, et al. Biochar for crop production: Potential benefits and risks. Journal of Soils and Sediments. 2017;**17**(3):685-716

[2] Grand View Research. Global Biochar Market Estimates and Forecast, 2012–2025; Report issued April 2017

[3] Business Wire. Global Biochar Market Outlook Report, 2019-2024; 30 January, 2020

[4] Rockwood D, Ellis M, Liu R, Zhao F, Ji P, Zhu Z, et al. Short rotation eucalypts: Opportunities for biochar. Forests. 2019;**10**:314. Available from: https://www.mdpi.com/1999-4907/10/4/314/html

[5] CIRAD - FRA, IUFRO - AUT, MUSE - FRA. Eucalyptus 2018: Managing Eucalyptus Plantation under Global Changes. Montpellier, France; 17 September 2018/21. DOI: 10.19182/agritrop/00023

[6] Carle J, Homgren P. Wood from planted forests: A global outlook 2005-2030. Forest Products Journal. 2008;**58**(12):6-18

[7] Sims REH, Hastings A, Schlamadinger B, Taylor G, Smith P. Energy crops: Current status and future prospects. Global Change Biology. 2006;**12**(11):2054-2076

[8] Rockwood DL, Rudie AW, Ralph SA, Zhu J, Winandy JE. Energy product options for *Eucalyptus* species grown as short rotation woody crops. International Journal of Molecular Sciences. 2008;**9**:1361-1378. Available from: http://www.mdpi.org/ijms/papers/i9081361.pdf

[9] Rockwood DL. History and status of Eucalyptus improvement in Florida. International Journal of Forest Research. 2012;**2012**:607879. 10 pages. Available from: http://www.hindawi.com/journals/ijfr/2012/607879/

[10] Rockwood DL, Peter GF. Eucalyptus and Corymbia species for mulchwood, pulpwood, energywood, bioproducts, windbreaks, and/or phytoremediation. In: Florida Cooperative Extension Service Circular 1194. Gainesville, FL: Institute of Food and Agricultural Sciences; 2018. p. 6. Available from: http://edis.ifas.ufl.edu/FR013

[11] Singh B, Singh BP, Cowie AL. Characterisation and evaluation of biochars for their application as a soil amendment. Australian Journal of Soil Research. 2010;**48**(7):516-525. DOI: 10.1071/SR10058

[12] Kloss S, Zehetner F, Dellantonio A, Hamid R, Ottner F, Liedtke V, et al. Characterization of slow pyrolysis biochars: Effects of feedstocks and pyrolysis temperature on biochar properties. Journal of Environmental Quality. 2012;**41**:990-1000

[13] Fabbro KW, Rockwood DL. Optimal management and productivity of *Eucalyptus grandis* on former phosphate mined and citrus lands in central and southern Florida: Influence of genetics and spacing. In: Proceedings 18th Biennial Southern Silvicultural Research Conference, March 2–5, 2015, Knoxville, TN. *e-Gen. Tech. Rpt. SRS-212*. 2016. pp. 510-517. Available from: http://www.srs.fs.usda.gov/pubs/gtr/gtr_srs212.pdf

[14] Campoe OC, Stape JL, Laclau JP, Marsden C, Nouvellon Y. Stand-level patterns of carbon fluxes and partitioning in a *Eucalyptus grandis* plantation across a gradient of productivity, in Sao Paulo State, Brazil. Tree Physiology. 2012;**32**:696-706. DOI: 10.1093/treephys/tps038

[15] Du H, Zeng F, Peng W, Wang K, Zhang H, Liu L, et al. Carbon storage in a *Eucalyptus* plantation chronosequence in southern China. Forests. 2015;**6**: 1763-1778. DOI: 10.3390/f6061763

[16] Zhang H, Guan D, Song M. Biomass and carbon storage of *Eucalyptus* and *Acacia* plantations in the Pearl River Delta, South China. Forest Ecology and Management. 2012;**277**:90-97. DOI: 10.1016/j.foreco.2012.04.016

[17] Christie S, Scholes R. Carbon storage in Eucalyptus and pine plantations in South Africa. Environmental Monitoring and Assessment. 1995;**38**:231-241

[18] Ulman Y, Avudainayagam S. Carbon storage potential of *Eucalyptus tereticornis* plantations. Indian Forester. 2014;**140**(1):53-58

[19] Schmidt HP, Taylor P, Eglise A, Arbaz C. Kon-Tiki - the democratization of biochar production. Biochar Journal. 2014;**1**:14-24

[20] Tigercat. 2020. Available from: www.tigercat.com/product/6050-carbonator/

[21] Murch S. Report for Passive Remediation Systems Biochar Project; 2016. Unpublished. Contact david.derbowka@prsi.ca

[22] Abdullah H, Mediaswanti K, Wu H. Biochar as a fuel: 2. Significant differences in fuel quality and ash properties of biochars from various biomass components of Mallee trees. Energy & Fuels. 2010;**24**(3):1972-1979. DOI: 10.1021/ef901435f

[23] Montgomery D. Growing a Revolution; Bringing our Soil Back to Life. New York: W Norton and Company; 2017

[24] Bales A, Draper K. BURN, Using Fire to Cool the Earth. London: Chelsea Green Publishing; 2018

[25] Ahmed A, Kurian J, Raghavan V. Biochar influences on agricultural soils, crop production, and the environment: A review. Environmental Reviews. 2016;**24**(4):495-502

[26] Jeffery S, Abalos D, Prodana M, Bastos AC, van Groenigen JW, Hungate BA, et al. Biochar boosts tropical but not temperate crop yields. Environmental Research Letters. 2017;**12**:5

[27] Ahmed F, Arthur E, Plauborg F, Razzaghi F, Korup K, Andersen MN. Biochar amendment of fluvio-glacial temperate sandy subsoil: Effects on maize water uptake, growth and physiology. Journal of Agronomy and Crop Science. 2018;**204**(2):123-136

[28] Blanco-Canqui H. Biochar and soil physical properties. Soil Science Society of America Journal. 2017;**81**(4):687-711

[29] Bruun EW, Petersen CT, Hansen E, Holm JK, Hauggaard-Nielsen H. Biochar amendment to coarse sandy subsoil improves root growth and increases water retention. Soil Use and Management. 2014;**30**(1):109-118

[30] Lefebvre D, Román-Dañobeytia F, Soete J, Cabanillas F, Corvera R, Ascorra C, et al. Biochar effects on two tropical tree species and its potential as a tool for reforestation. Forests. 2019;**10**: 678

[31] Langholtz M, Carter DR, Rockwood DL. Assessing the economic feasibility of short-rotation woody crops in Florida. In: Florida Cooperative Extension Service Circular 1516. Gainesville, FL: Institute of Food and Agricultural Sciences; 2007. p. 5. Available from: http://edis.ifas.ufl.edu/FR169

[32] Rockwood DL, Dippon DR, Lesney MS. Woody species for biomass production in Florida. In: Final Report 1983-1988. Oak Ridge National

Laboratory; 1988. p. 153. ORNL/Sub/81-9050/7. Available from: http://web.ornl.gov/info/reports/1988/3445602753456.pdf

[33] Rockwood DL, Carter DR, Stricker JA. Commercial tree crops on phosphate mined lands. Florida Institute of Phosphate Research. FIPR Publication#03-141-225; 2008. p. 123. Available from: http://fipr.state.fl.us/wp-content/uploads/2014/12/03-141-225Final.pdf

[34] Zalesny RS Jr, Cunningham MW, Hall RB, Mirck J, Rockwood DL, Stanturf JA, et al. Chapter 2. Woody biomass from short rotation energy crops. In: In ACS Symposium Book: Sustainable Production of Fuels, Chemicals, and Fibers from Forest Biomass (Zhu J et al.). 2011. pp. 27-63. Available from: http://pubs.acs.org/doi/abs/10.1021/bk-2011-1067.ch002

[35] Perez-Cruzado C, Merino A, Rodriguez-Soalleiro R. A management tool for estimating bioenergy production and carbon sequestration in *Eucalyptus globulus* and *Eucalyptus nitens* grown as short rotation woody crops in north-west Spain. Biomass and Bioenergy. 2011;**35**:2839-2851

[36] Morales M, Aroca G, Rubilar R, Acuna E, Mola-Yudego B, Gonzalez-Garcia S. Cradle-to-gate life cycle assessment of *Eucalyptus globulus* short rotation plantations in Chile. Journal of Cleaner Production. 2015;**99**:239-249

[37] Hinchee M, Rottman W, Mullinax L, Zhang C, Chang S, Cunningham M, et al. Short-rotation woody crops for bioenergy and biofuels applications. In Vitro Cellular & Developmental Biology: Plant. 2009;**45**:619-629

[38] Harper RJ, Sochacki SJ, Smettem RJ, Robinson N. Bioenergy feedstock potential from short-rotation woody crops in a dryland environment. Energy & Fuels. 2010;**24**:225-231

[39] Stape JL, Binkley D. Insights from full-rotation Nelder spacing trials with *Eucalyptus* in Sao Paulo, Brazil. Southern Forests: A Journal of Forest Science. 2010;**72**:91-98

[40] Sackett TE, Basiliko N, Noyce GL, Winsborough C, Schurman J, Ikeda C, et al. Soil and greenhouse gas responses to biochar additions in a temperate hardwood forest. GCB Bioenergy. 2015; 7:1062-1074. DOI: 10.1111/gcbb.12211

[41] Li Y, Hu S, Chen J, Muller K, Li Y, Fu W, et al. Effects of biochar application in forest ecosystems on soil properties and greenhouse gas emissions: A review. Journal of Soils and Sediments. 2018;**18**:546. DOI: 10.1007/s11368-017-1906-y

[42] Dempster D, Gleeson D, Solaiman Z, Jones D, Murphy D. Decreased soil microbial biomass and nitrogen mineralisation with *Eucalyptus* biochar addition to a coarse textured soil. Plant and Soil. 2012;**354**:311-324

[43] Mertens J, Germer J, de Araújo Filho JC, Sauerborn J. Effect of biochar, clay substrate and manure application on water availability and tree-seedling performance in a sandy soil. Archives of Agronomy and Soil Science. 2017;**63**: 969-983

[44] Lin ZB, Liu Q, Liu G, Cowie AL, Bei QC, Liu BJ, et al. Effects of different biochars on *Pinus elliottii* growth, N use efficiency, soil N_2O and CH_4 emissions and C storage in a subtropical area of China. Pedosphere. 2017;**27**:248-261

[45] Wrobel-Tobiszewska A, Boersma M, Adams P, Singh B, Franks S, Sargison JE. Biochar for eucalyptus forestry plantations. Acta Horticulturae. 2016;**1108**:55-62

[46] de Farias J, Marimon BS, de Carvalho Ramos Silva L, Petter FA, Andrade FR, Morandi PS, et al. Survival and growth of native *Tachigali vulgaris*

and exotic *Eucalyptus urophylla* × *Eucalyptus grandis* trees in degraded soils with biochar amendment in southern Amazonia. Forest Ecology and Manage ment. 2016;**368**:173-182

[47] Glisczynski FV, Pude R, Amelung W, Sandhage-Hofmann A. Biochar-compost substrates in short-rotation coppice: Effects on soil and trees in a three-year field experiment. Journal of Plant Nutrition and Soil Science. 2016;**179**:574-583

[48] Sarauer J, Page-Dumroese D, Coleman M. Soil greenhouse gas, carbon content, and tree growth response to biochar amendment in western United States forests. GCB Bioenergy. 2019;**11**: 660-671

[49] Thomas S, Gale N. Biochar and forest restoration: A review and meta-analysis of tree growth responses. New Forests. 2015;**46**:931-946. DOI: 10.1007/s11056-015-9491-7

[50] Bracho R, Vogel J, Will R, Noormets A, Samuelson L, Jokela E, et al. Carbon accumulation in loblolly pine plantations is increased by fertilization across a soil moisture availability gradient. Forest Ecology and Management. 2019;**424**:39-52

[51] Goudiaby A, Diedhiou S, Diatta Y, Adiane A, Diouf P, Fall S, et al. Soil properties and groundnut (*Arachis hypogea* L.) responses to intercropping with *Eucalyptus camaldulensis* Dehn and amendment with its biochar. Journal of Materials and Environmental Science. 2020;**11**(2):220-230

[52] Sánchez-Monedero M, Cayuela M, Sánchez-García M, Vandecasteele B, D'Hose T, López G, et al. Agronomic evaluation of biochar, compost and biochar-blended compost across different cropping systems: Perspective from the European project FERTIPLUS. Agronomy. 2019;**9**:225. DOI: 10.3390/agronomy9050225

[53] Shackley S, Sohi S, Ibarrola R, Hammond J, Masek O, Brownsort P, et al. Biochar, tool for climate change mitigation and soil management. In: Meyers RA, editor. Encyclopedia of Sustainability Science and Technology. New York, NY: Springer; 2012. DOI: 10.1007/978-1-4419-0851-3

[54] Stavins RN, Richards KR. The cost of U.S. forest-based carbon sequestration. In: Pew Center on Global Climate Change. 2005. Available from: https://www.c2es.org/document/the-cost-of-u-s-forest-based-carbon-sequestration/

[55] California Air Resources Board. Compliance Offset Protocol U.S. Forest Offset Projects. 2015. Available from: arb.ca.gov/cc/capandtrade/protocols/usforest/usforestprojects_2015.htm

[56] Lehman J, Joseph S. Biochar for Environmental Management. New York: Earth Scan; 2006. pp. 188-200

[57] Bis Z, Kobyłecki R, Ścisłowska M, Zarzycki R. Biochar – Potential tool to combat climate change and drought. Ec ohydrology & Hydrobiology. 2018; **18**(4):441-453

[58] Gardiner ES, Ghezehei SB, Headlee WL, Richardson J, Soolanayakanahally RY, Stanton BJ, et al. The 2018 Woody Crops International Conference, Rhinelander, Wisconsin, USA, 22–27 July 2018. Forests. 2018;**9**:693. Available from: https://www.mdpi.com/1999-4907/9/11/693

Chapter 4

Mechanism of In-Situ Catalytic Cracking of Biomass Tar over Biochar with Multiple Active Sites

Dongdong Feng, Yu Zhang, Yijun Zhao and Shaozeng Sun

Abstract

Biomass tar is the bottleneck in the development of efficient utilization of biomass syngas. The in-situ catalytic cracking biomass tar with multi-active biochar is investigated in a two-stage fluidized bed-fixed bed reactor. It indicates that adding H_2O or CO_2 is found to improve the homogeneous and heterogeneous cracking of biomass tar. Activation of biochar by H_2O or CO_2 impacted the morphology of biochar surface and distribution of metal species. H_2O or CO_2 affects the creation and regeneration of pore structures, influencing the biochar structure and dynamical distribution of alkali and alkaline earth metal species (AAEMs), which ensure enough surface active sites to maintain the catalytic activity of biochar. The tar cracking into low-quality tar or small-molecule gas may be catalyzed by K, while the combination of tar with biochar would be promoted by Ca. The volatilizations of K and Ca, due to their reaction with volatiles, are to a large extent in accordance with their valences and boiling points. The subsequent transformation from the small aromatic ring systems to the larger ones occurs due to the volatile-biochar interaction. During tar cracking over biochar, K and Ca act as the active sites on biochar surface to promote the increase of active intermediates (C–O bonds and C–O–K/Ca).

Keywords: biochar, tar, catalytic cracking, AAEM species

1. Introduction

Tar is a generic term comprising all organic compounds present in syngas except for gaseous hydrocarbons. Tars can condense to more complex structures in pipes, filters, or heat exchangers of downstream equipment and processes, which may cause mechanical breakdown of the entire system [1]. For biomass gasification, the allowable limit for tar in the producer gas is less than 5 mg/Nm^3 for a direct-fired gas turbine [2], and for some fuel synthesis processes, the contents of tar and ammonia are required to be <0.1 mg/Nm^3 and <10 ppm, respectively [3–5], in order to protect the catalysts and downstream equipment and to improve the overall efficiency and economics. The tar mixture is classified into five classes by Padban [6]: undetectable, heterocyclic, light aromatic hydrocarbons (LAHs), light polyaromatic hydrocarbons (LPAHs), and heavy polyaromatic hydrocarbons (HPAHs). The removal of biomass tar is one of the main challenges for the biomass gasification industry [7, 8]. Catalytic cracking is a known method for the efficient removal of biomass tar [9–12]. Biochar, as a product of pyrolysis and gasification of biomass, is a relatively cheap catalyst with high activity in tar heterogeneous cracking [13–20]. During tar catalytic cracking over biochars, even after the loss of catalytic activity through coking, the biochar samples can still be directly combusted, so as to recover the chemical energy of catalyst, thus avoiding any reprocessing as a result of deactivation.

In addition to the analysis of model tar compounds [21–23], studies of biomass tar over biochar mainly discuss the reforming of real tar from raw materials [24, 25]. However, the AAEM species (e.g., Na, K, Mg, and Ca) in raw biomass play an important role as the "cross points" during tar formation. The chemical bonds between AAEM species and the carbon matrix are repeatedly breaking and reforming. This process promotes the production of gaseous products from the fatty acid tar and a degree of small aromatic compounds. Simultaneously, larger aromatic ring compounds (≥5 aromatic ring system) are formed within the biochar structure [26, 27]. The presence of AAEM species can inhibit the release of volatile matter (especially for biomass tar)—even the strong interaction between volatile materials and biochar will affect tar composition, leading to the catalytic conversion of the real tar components before contacting with the catalyst, which misleads mechanistic studies of subsequent heterogeneous reforming over biochar catalyst.

The formation (e.g., 500–700°C) and thermal decomposition (e.g., 700–900°C) of tar during the gasification process are an extremely complex multistep reaction [28–32], which involves not only homogeneous conversion, but also heterogeneous reforming. H_2O and CO_2 are two important reforming agents [33] in the biomass gasification industry. Studying the influence of H_2O and CO_2 on tar homogeneous transformation and heterogeneous reformation is valuable to better understand the analysis of the tar complex gas-solid phase reaction. However, there is still less research on separate discussion between homogeneous conversion and heterogeneous reforming of biomass tar over biochar. Although there are reports detailing the influence of H_2O and CO_2 on tar during the gasification process [34–39], they were mainly focused on the single concentration of reforming agent (15 vol.% H_2O or pure CO_2 atmosphere). There has yet to be detailed a complete understanding of the influence of H_2O and CO_2 on the homogeneous conversion and heterogeneous reforming over biochar as a function of biomass tar evolution.

The effects of reforming agent concentration and reaction temperature on the tar homogeneous conversion and heterogeneous reforming over biochar were investigated in a two-stage fluidized bed/fixed bed reactor. The H-form biomass samples (with little AAEM species) were used to provide the real tar components, which effectively inhibited the tar-AAEM interactions in gas phase during H_2O/CO_2 homogeneous conversion and prevented any secondary catalytic effects of AAEM species from the volatilization of raw materials on the biochar catalyst surface. The analysis of biochar structures examined with Raman spectroscopy to comprehensively elucidate the changes of biochar catalyst structure after the H_2O and CO_2 heterogeneous reforming of biomass tar. In addition to the measurement of tar yields, GC/MS spectroscopy was used to characterize the detailed structural features of tar [40], so as to understand the molecular biomass tar transformation pathway and the coupling mechanism (e.g., collaboration and interaction effects) between the biochar structure and the AAEM species during tar reforming.

2. Experiment

2.1 Material preparation

Biomass (rice husks) obtained from the Wu Chang area in Harbin, Heilongjiang Province, China, was used in the experiments. The samples were dried overnight at 105°C, pulverized, and sieved to obtain a fraction with particle sizes between 0.15 and 0.25 mm. The proximate and ultimate analyses data [41] for the rice husk samples are listed in **Table 1**, which could be used to characterize the composition of biomass, grasping its reaction characteristics and application value (*M*: moisture, *A*: ash, *V*: volatile, *FC*: fixed carbon; *C*: carbon, *H*: hydrogen, *O*: oxygen, *N*: nitrogen, *S*: sulfur).

The H-form rice husk was used as the raw material to supply real biomass tar. The raw pyrolysis biochar was mixed with an aqueous solution of 0.2 M H_2SO_4 in an acid solution:sample mass ratio of 30:1 and stirred in an argon atmosphere for 24 h. The slurry was filtered and washed with deionized water until the filtrate pH was constant (pH ≈ 7). After drying, the acid-washed sample is termed as the H-form char.

2.2 Biochar catalyst preparation

Pyrolysis biochar was used as the catalyst for biomass tar reforming. The set-up to pyrolysis biochar comprises a quartz reactor and a standard muffle furnace, as shown in **Figure 1**.

The quartz tray (red tray) with 5.0 g raw rice husk was placed into the quartz reactor. Along with the reactor cover, the quartz reactor was placed into the muffle furnace. At room temperature, the air in the reactor was displaced by Ar at a rate of 2.0 L/min for 30 min. Pyrolysis was performed at a slow-heating rate of 10°C/min up to a final pyrolysis temperature of 700°C with 70 min. Thereafter, with the temperature of turn-off furnace back to room temperature, the door of the muffle furnace was opened, and the reaction quenched by removing the reactor from the furnace. Ar gas was passed continuously through the reactor to prevent oxidation during cooling. The pyrolyzed biochar was removed from the reactor and stored at 4°C. Ar gas is supplied through a gas pipe (400 mm, long) into the porous distributor (with a diameter of 120 mm) and fed from the bottom of the quartz reactor filling the entire reactor. The upper cover acts as a partial seal under the action of its own gravity; however, with increasing internal gas volume produced as a function

Sample	Proximate analysis (wt.%)				Ultimate analysis (wt.%)				
	$M_{ad.}$	$A_{ad.}$	$V_{ad.}$	$FC_{ad.}$	$C_{ad.}$	$H_{ad.}$	$O_{ad.(diff)}$	$N_{ad.}$	$S_{t,ad.}$
Rice husk	6.86	17.00	60.92	15.22	37.35	4.40	34.05	0.20	0.14

Note: diff. = by difference, ad. = air dry basis.

Table 1.
Proximate and ultimate analyses of rice husk samples.

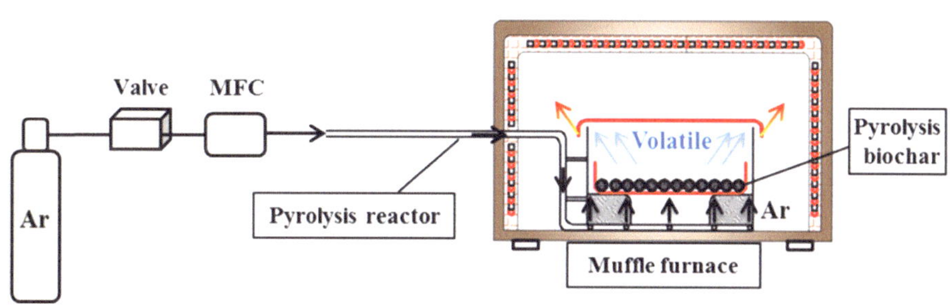

Figure 1.
Biochar catalyst preparation using a muffle furnace.

of muffle furnace temperature, some gaps between the upper cover and the reactor allow the release of gases under internal positive pressure. The volatile matters formed during the volatilization of biomass were rapidly dispersed away from the reactor, carried by Ar gas, so as to ensure an inert atmosphere inside the reactor. The metal contents of the origin and H-form biochar are listed in **Table 2**.

2.3 Homogeneous/heterogeneous reforming of biomass tar

As shown in **Figure 2**, a two-stage fluidized bed/fixed bed reactor was used for the investigation of the homogeneous conversion and heterogeneous reforming of biomass tar over biochar. The inner diameter of the reactor was 37 mm. The reactor is divided into two layers by four quartz frits. The upper layer is fixed bed, while the lower is fluidized bed. The heights of upper and lower layers are 30 and

Biochar	Primary metal contents (wt.%)					
	Na	K	Mg	Ca	Al	Fe
Origin biochar	0.03	1.44	0.09	0.14	0.03	0.05
H-form biochar	0.00	0.02	0.01	0.01	0.01	0.01

Table 2.
Primary metal contents of pyrolysis rice husk biochar samples.

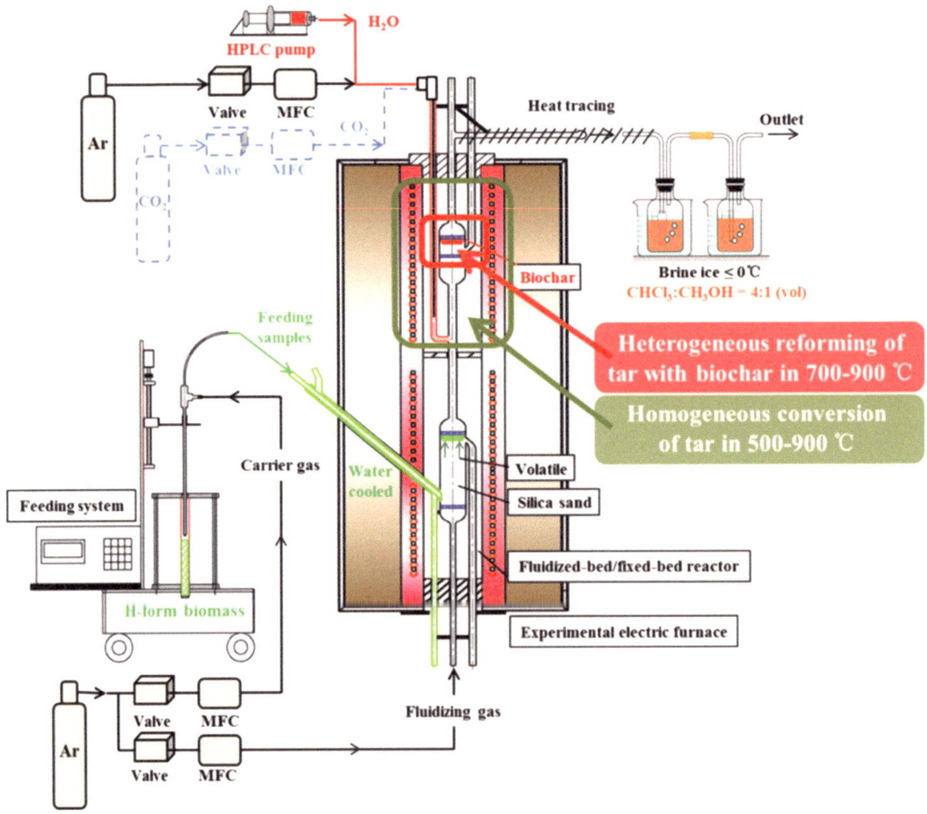

Figure 2.
Schematic diagram of a two-stage fluidized bed/fixed bed reactor for the homogeneous conversion and heterogeneous reforming of biomass tar.

130 mm, respectively. For the homogeneous conversion of tar, the lower fluidized bed reactor was heated to 500°C, with the temperature increased to 500–900°C for the upper fixed bed reactor (without catalyst). The silica sand with the weight of 60 g was pre-loaded into the bottom stage of the quartz reactor followed by Ar purging (1.0 L/min carrier gas and 1.5 L/min fluidizing gas) before heating the desired temperature. Once stabilization of the temperatures was achieved, the H-form rice husk was injected into the fluidized bed through the water-cooled pipes at a feeding rate of 100 mg/min. Simultaneously, for the CO_2 or H_2O separate treatment, the atmosphere was switched to CO_2 (29 vol.%) or H_2O (15 vol.%) through a dedicated gas injection system located in between the lower and upper reactors as shown in **Figure 2**. Pure Ar gas was also injected into the dedicated gas injection system, to balance the system, at rates of 1.03/0.92/0.82/0.75/0.68 L/min for reaction temperatures of 500/600/700/800/900°C, respectively, to maintain constant residence times for each reforming temperature. For 15 vol.% H_2O, steam injection was achieved by feeding a metered amount of water through a high-performance liquid chromatography (HPLC) pump into the heated zone of the reactor where the water was evaporated into steam. De-ionized H_2O was injected at rates of 0.34/0.30/0.28 ml/min along with 0.40/0.36/0.33 L/min of balanced Ar for the 700/800/900°C reaction temperatures, respectively. CO_2 was injected through the dedicated gas injection system at rates of 0.82/0.75/0.68 L/min to achieve 29 vol.% for 700/800/900°C reaction temperatures, respectively. The temperature was held for 10 min for each reaction. Reactions were terminated by switching the atmosphere to argon and removing the reactor out of the furnace.

For the heterogeneous reforming of biomass tar over biochar activated by H_2O or CO_2, the activation of biochar was carried out for 10 min in the fixed-bed zone in a 15 vol.% H_2O or a 29 vol.% CO_2 atmosphere with no supplemental H-form rice husk added to the fluidized-bed zone. This was followed by another 10 min at 800°C in an Ar atmosphere, to maintain the same total reaction time (20 min) as the tar-reforming conditions. Details of the five experimental conditions (A–E) are shown in **Table 3**. The experiments involved three pyrolysis experiments: tar reforming (A) in Ar with unactivated pyrolysis biochar; (B) in Ar over H_2O-activated biochar; and (C) in Ar over CO_2-activated biochar. In (B) and (C), a 10-min activation of the biochar was first carried out with the activated biochar then used for 10 min of tar reforming in an Ar atmosphere at 800°C, while H-form biomass was also fed to the reactor. In addition, two gasification experiments were carried out: tar reforming in (D) a 15 vol.% H_2O atmosphere over H_2O-activated biochar and in (E) a 29 vol.% CO_2 atmosphere over CO_2-activated biochar. In (D) and (E), both atmospheres were kept constant for 20 min even though the period was evenly divided into a biochar-activation stage, which was followed by tar reforming over biochar.

Conditions	No.	Conditions of biomass tar reforming over biochar at 800°C
Pyrolysis	A	Biomass tar reforming in Ar over pyrolysis biochar
	B	Biomass tar reforming in Ar over H_2O-activated biochar
	C	Biomass tar reforming in Ar over CO_2-activated biochar
Gasification	D	Biomass tar H_2O reforming over H_2O-activated biochar
	E	Biomass tar CO_2 reforming over CO_2-activated biochar

Table 3.
Experimental conditions investigated for tar reforming over biochar.

2.4 Sampling and analysis of biochar and tar

The biomass tar compounds were trapped in two gas bottles, connected in series, and filled with a mixture of HPLC-grade chloroform and methanol (4:1, v/v), as shown in **Figure 2**. The bottles were placed in a brine ice bath (≤0°C). After the reaction, the total solution was transferred to a 200 mL volumetric flask and made up to volume with a mixture of chloroform and methanol (4:1, v/v).

The tar yield was determined by evaporating the solvents and water at 35°C for 4 h. The tar is thus experimentally defined as the material soluble in the chloroform/methanol (4:1, v/v) solvent mixture not being evaporated (with the solvents) at 35°C within 4 h [24, 25, 42]. The residues in the solvents themselves (i.e., blank) and the biomass moisture content were considered in the tar yield calculation. The equation used for tar concentration in the solution is shown as follows Eq. (1):

$$C = \frac{C_2 - C_1}{1 - C_2} \tag{1}$$

where C is the concentration of tar; C_1 is the concentration of the mixed solution residue (blank experiment); and C_2 is the concentration of residue in tar solution.

The equation used to determine tar yield is shown as follows Eq. (2):

$$\text{Tar yield} = \frac{C \times M_{\text{Tar solution}}}{M_{\text{biomass}}} \tag{2}$$

where C is the concentration of tar in the solution; $M_{\text{Tar solution}}$ is the total mass of tar collected in the solution; and M_{biomass} is the feed quality of the H-form rice husk into the reactor.

The samples were analyzed using an Agilent Gas Chromatography Mass Spectrometer (GC-MS) instrument (6890 series GC with a 5973 MS detector) with a capillary column (DB-5 ms; length 30 m, internal diameter 0.25 mm, film thickness 0.5 μm). The sample solution (5 μL) was injected into the injection port, set at 260°C, with a split ratio of 80:1. The column was operated in constant-flow mode using 2.0 mL/min of helium as the carrier gas. The column temperature was initially maintained at 35°C for 3 min, then increased to 260°C at a heating rate of 10°C/min, and then maintained at 260°C for 5 min. Mass spectra were acquired after a 4-min solvent delay [21]. The chromatogram peaks were identified by comparison with the standard spectra of compounds in the National Institute of Standards and Technology library (NIST) and/or from the retention times/spectra of known injected species.

The total amount of metal species in the biochar samples was quantified by employing a previously established procedure [43]. Using a microwave system (Ethos 1, Milestone, Sorisole, Italy), the sample (0.1 g) was digested in a 1:3:8 (v/v/v) mixture of 40% HF, 30% H_2O_2, and 65% HNO_3 at 200°C for 60 min. The metal species content was then quantified by inductively coupled plasma-atomic emission spectroscopy (ICP-AES). Three measurements were conducted with the average values and then taken as the results.

The biochar's particle morphology and surface composition were measured by an EVO18 scanning electron microscope coupled to an energy dispersive X-ray spectrometer (SEM-EDX, Carl Zeiss, Germany).

Biochar samples were set for at least 24 h to displace the reaction gas within the pore structure with air. N_2-adsorption isotherms were then obtained at −196°C (ASAP 2020M, Micromeritics Instrument Crop, USA) and analyzed by the BET model to determine the sample's surface area and pore volume.

XPS analysis was used to evaluate the characteristics of surface elements in biochar. This was performed using a K-Alpha spectrometer (Thermo Fisher Scientific) equipped with monochromatic Al Kα X-rays at 1486.6 eV. To exclude effects on the binding energies caused by changing the sample during measurements, the data were corrected by a linear shift with the maximum peak of the C1s binding energy of the adventitious carbon corresponding to 284.6 eV. The surface's elemental condition was analyzed using the number of escaped electrons from the char surface at a depth of 1–10 nm, according to the findings in our previous investigations [44, 45].

3. Results and discussion

3.1 Homogeneous conversion of biomass tar

The homogeneous conversion of tar mainly refers to the initial pyrolysis tar experiences during a series of decomposition and polymerization processes under ambient conditions (heat and atmosphere). Tar yield during the Ar, H_2O, and CO_2 homogeneous conversion experiments performed at 500–900°C can be seen in **Figure 3**. In the presence of an Ar-only atmosphere, the tar yield decreased gradually as a function of increasing temperature from 500 (26.18%) to 900°C (6.38%). Temperature in the Ar-only experiments has a greater influence between 500 and 700°C. Further increasing the temperature to 700–900°C results in increased biomass decomposition, thus lowering tar yields. Thermal decomposition is considered to be the main factor in the conversion of tar [46, 47]. As shown in **Figure 3**, in 15 vol.% H_2O and 29 vol.% CO_2, the effects of H_2O and CO_2 on the homogeneous transformation of biomass tar over biochar are significant. At the same temperature, the tar reforming effect of 15 vol.% H_2O is significantly higher than that of 29 vol.% CO_2. In 15 vol.% H_2O, the tar yield decreased from 6.95% at 700°C to 3.56% at 900°C. In 29 vol.% CO_2, the tar yield decreased from 7.99% at 700°C to 5.01% at 900°C. For the higher temperatures, 700–900°C are required for H_2O and CO_2 to influence tar homogeneous transformations, while for the lower temperatures, 500 and 600°C are not in the gasification thermal range.

As shown in **Figure 4**, it can be seen that at lower temperatures (500–600°C), the majority of the biomass tar still comprises components based on the primary biomass tar containing oxygen and substituent compounds, such as levoglucosan and dimethoxymethane. However, when subjecting the biomass to higher temperatures (700–900°C), most of the primary pyrolyzed tar gradually transforms [48]. The tar composition seems to be mainly composed of aromatic compounds having

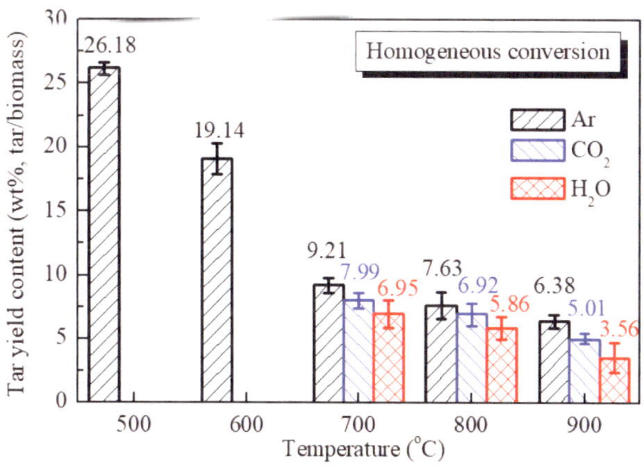

Figure 3.
Tar yield during homogeneous conversion at 500–900°C.

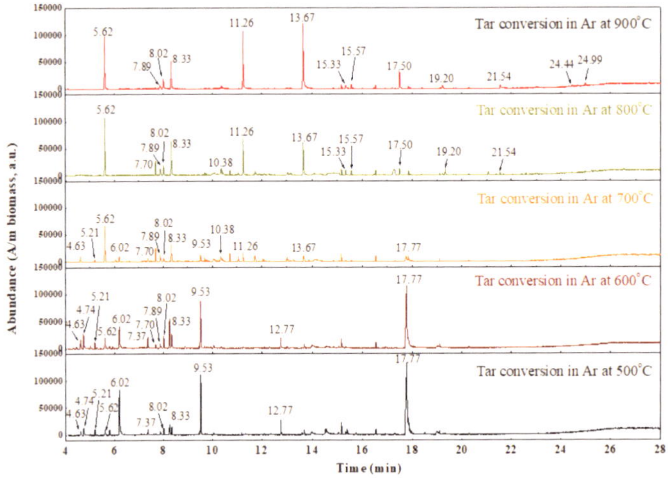

Figure 4.
GC-MS analysis of biomass tar in Ar at 500–900°C.

good thermal stability, such as toluene, indene, and naphthalene, among others. Increasing the temperature resulted in either a gradual reduction or a complete removal of tars containing branched or heteroatom compounds, and polycyclic aromatic hydrocarbons (PAHs) were gradually formed. For the biomass tar homogeneous conversion, the aromatic ring structure has higher thermal stability than that of the non-aromatic structures. Specific tar components decompose into small molecular gases and C_1–C_5 hydrocarbons, while there is evidence for the promotion of aromatic rings as a function of increasing temperature. H-abstraction, C_2H_2-addition (HACA), and cyclodehydrogenation are the mechanisms typically responsible for such processes [49, 50]. Performing the reactions at the mid-temperature range (700–800°C) results in aromatic conversion with oxygen and substituents. Thermal decomposition [51] and additional reactions convert short-chain hydrocarbons (C_1–C_5) into compounds containing unsaturated double and triple bonds that gradually increase in concentration by the acetylene addition reaction. The aromatic components can also be polymerized by dehydrogenation. Further increasing the temperature to 900°C results in the relative content of PAHs, such as naphthalene, phenanthrene, and anthracene, to increase the above conversion pathway yielding highly stable aromatic hydrocarbons.

GC-MS analysis during H_2O and CO_2 homogeneous conversion at 700–900°C can be seen in **Figure 5**. At 700–900°C, H_2O and CO_2 have a degree of influence on the conversion of tar. The degree of tar homogeneous conversion in the presence of either a H_2O or CO_2 atmosphere was significantly higher than that of the thermal decomposition in Ar. PAH concentration was low. The results show that H_2O and CO_2 have obvious effects on the transformation of aromatics, especially PAHs [52]. The free radical theory is used to explain the homogeneous transformation of tar. The formation of aromatic radicals in the polymerization of aromatic hydrocarbons is considered to be the key to the reaction. The continuous polymerization process is considered to be the main pathway [53–56]. Thermal decomposition is a method of generating free radicals through thermal breaking of bonds. The free radicals generated by the original tar form different final products by reacting with different free radicals produced as a function of the atmosphere. The presence of H_2O and CO_2 promoted the formation of free radicals with H/O/OH moieties. The influence of temperature is mainly reflected in the promotion of the decomposition reaction caused by free radicals [57]. CO_2 is a pure oxygen donor. **Figure 6** shows that the active oxygen atoms used for oxidative decomposition of hydrocarbons and intermediate

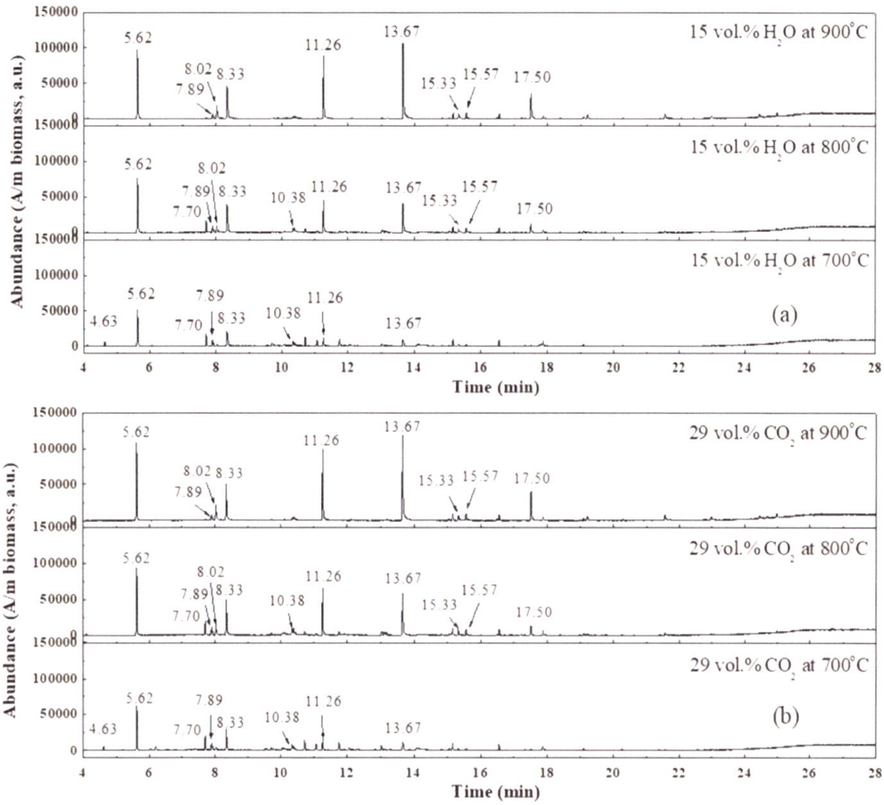

Figure 5.
GC-MS tar analysis during (a) 15 vol.% H_2O and (b) 29 vol.% CO_2 homogeneous conversion at 700–900°C.

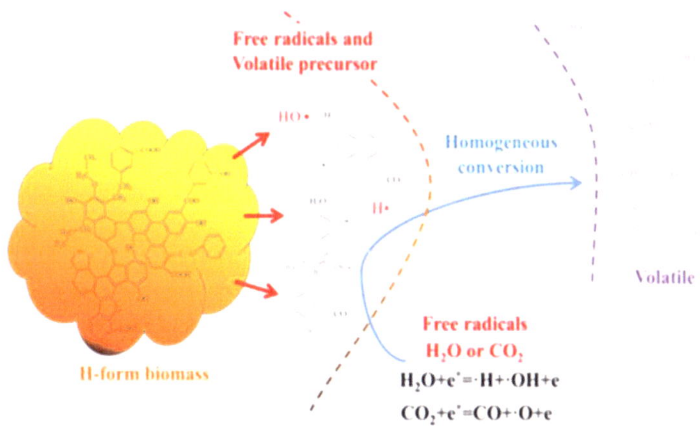

Figure 6.
Homogeneous conversion of tar in Ar, H_2O, and CO_2 at 500–900°C [52].

products are mainly produced by the reaction CO_2 + e× → CO + O·+ e. Active OH free radicals can be formed by replacing the hydrogen atoms in the hydrocarbons with oxygen atoms. Increasing the content of CO_2 is helpful to inhibit the cyclization of aromatics. The addition of CO_2 promotes the formation of free radicals such as O, which can further react with hydrocarbon groups. The oxidation reaction of active oxygen atoms with hydrocarbons forms CO, H_2O, and other products. The oxidative cracking process of tar is initiated, and the polymerization process of aromatic hydrocarbons is also inhibited. H_2O not only promotes tar cracking conversion but also inhibits the polymerization reaction. This is related to the higher activity of free radical formation being a more active reformer in the conversion of tar. H_2O and CO_2 have similar oxidation capacities. The difference between the two is mainly reflected in the product—H_2O produces higher numbers of H free radicals than CO_2 [58]. O and OH free radicals can be formed by ionization of H_2O ($H_2O \rightarrow H + OH$) in the presence of steam. The fracture of OH can form new H and O free radicals. The H/O/OH atoms in the gas phase exist in radical form. According to the free radical mechanism, the primary constituents of the biomass are broken into activated tar fragments at high temperatures. A large number of H/O/OH free radicals will combine with activated tar fragments before tar polymerization.

As shown in **Figure 6**, the conversion of the tar homogeneous transformation process is considered to be a two-stage process. The first stage involves the decomposition and transformation of the active heteroatom-containing groups in the tar, along with the decomposition of dealkylated side chains, hydrocarbon molecular cyclization, and aromatization reactions. The products include low-chain aliphatic hydrocarbons, oxygen-containing small molecular gases, and single-ring aromatic hydrocarbons. The second stage is the reforming of tar components; the dehydrogenation of cyclization products; the addition of acetylene; and the growth, recombination, and isomerization of aromatics. The two processes constitute the basis of the biomass tar homogeneous reaction. In the presence of different reforming agents (H_2O or CO_2), the atmosphere can promote or inhibit tar pyrolysis conversion, thus influencing the composition of the final tar. The addition of H_2O and CO_2 can promote the generation of active free radicals such as O, OH, H, and so on. These free radicals can react with the active free tar fragments generated from the first stage of thermal decomposition demonstrating the importance of the H_2O and CO_2 reforming agents in the homogeneous conversion of biomass tar.

3.2 Heterogeneous reforming of biomass tar over biochar

3.2.1 Biomass tar reforming

As shown in **Figure 7**, the highest proportion of bio-tar was reformed (including homogeneous and heterogeneous phases) in the 15 vol.% H_2O atmosphere over H_2O-activated biochar (D). The proportion of tar reformed in the 29 vol.% CO_2 atmosphere over CO_2-activated biochar (E) was also considerably higher than results for reforming in an Ar atmosphere (A, B, and C). This illustrates that the presence of a gasification agent (H_2O/CO_2) greatly promotes in-situ reforming of nascent bio-tar over biochar. Under pyrolysis conditions, the homogeneous transformation of biomass tar was mainly based on secondary reactions (i.e., tar thermal cracking at 800°C), yielding a conversion efficiency of 70.86%. The ability of H_2O/CO_2 activation to improve biochar reactivity was also clearly observed. In the Ar atmosphere, the highest proportion of tar was reformed over the H_2O-activated biochar (B, 20.08%), followed by that over the CO_2-activated biochar (C, 19.01%), while the lowest conversion was for the (unactivated) pyrolysis biochar (A, 17.41%). El-Rub and Kamel [59] and Chen et al. [60] studied biochar's catalytic activity for tar reforming using a fixed char bed. They concluded that in an inert atmosphere, the tar molecules were mainly adsorbed on biochar active sites and converted into larger polyaromatic molecules and coke via a series of dehydrogenation, cyclization, and condensation reactions. Differences between the unactivated and H_2O/CO_2-activated biochars may be attributed to differences in inherent catalytic AAEM species (such as K and Ca) and the biochars' physiochemical structures [61], as discussed later.

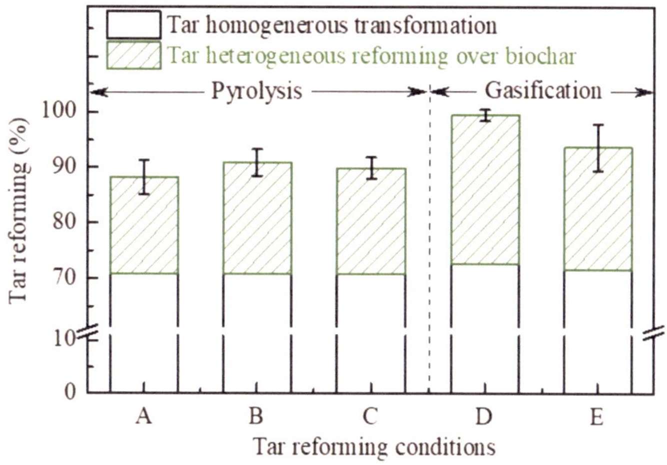

Figure 7.
Proportion of tar reformed under different conditions: (A) in Ar over pyrolysis biochar, (B) in Ar over H_2O-activated biochar, (C) in Ar over CO_2-activated biochar, (D) in 15 vol.% H_2O over H_2O-activated biochar, and (E) in 29 vol.% CO_2 over CO_2-activated biochar.

The presence of a gasification agent further improved the homogeneous reforming of the bio-tar (72.62 and 71.57% with H_2O and CO_2 present, respectively). The homogeneous transformation of biomass tar in the gas phase was in a certain extent in H_2O/CO_2 gasification condition, which was in broad agreement with the results obtained by Wang et al. [17] and Min et al. [24]. Besides, in the gasification conditions, H_2O and CO_2 further improved the biochar's catalytic reactivity for the heterogeneous reforming process with the proportion reformed increasing from 20.08% (B) to 26.85% (D) with H_2O and from 19.01% (C) to 22.17% (E) with CO_2. According to Min et al. [24], the reforming of tar molecules over biochar may be activated in two ways. First, in the gas phase, nascent tar (volatiles) contains abundant free radicals that would react with extant tar molecules to form activated tar fragments. Second, tar molecules and biochar may be activated by H_2O/CO_2 during the chemisorption of tar on the biochar surface with reactions between tar and H_2O/CO_2 adsorbed on the biochar's active sites then leading to further reforming reactions. Overall, gasification agents (H_2O and CO_2) improved and maintained the system's ability to carry out homogeneous and heterogeneous reforming of biomass tar.

3.2.2 H_2O/CO_2 activation of biochar

The metal contents of the biochar samples are shown in **Table 4**. Apart from K, there was little difference (±0.02 wt.%) observed for the metals between the biochar samples. During the H_2O/CO_2 activation of biochar, K appears to have been released from the biochar, decreasing from 1.12 wt.% in the pyrolysis biochar to 1.06 wt.% when activated by H_2O and 1.09 wt.% in the CO_2-activated biochar.

Biochar samples	**Metal species content (wt.% in biochar)**				
	K	**Mg**	**Ca**	**Fe**	**Al**
H_2O-activated biochar	1.06	0.10	0.16	0.06	0.02
CO_2-activated biochar	1.09	0.09	0.15	0.05	0.03

Table 4.
Biochar samples' metal-content analysis.

Because of their valence states [43], other metal species like Ca bond to the biochar more strongly (i.e., at two or more sites) than K (which only bonds at one site).

3.2.3 SEM-EDX analysis of biochar

As shown in **Figure 8(a)**, the unactivated biochar particles' surfaces showed more, larger (40 × 60 μm) hill-like structures than the surfaces of the activated biochars. H_2O and CO_2 activate the biochar via $C + H_2O \rightarrow CO + H_2$ and $C + CO_2 \rightarrow 2CO$, respectively. However, its larger size meant activation by CO_2 was limited to the surface of biochar, resulting in small structures (15 × 15 μm), which can be seen in **Figure 8(c)**. However, as illustrated by the structure shown in **Figure 8(b)**, H_2O, as well as H/O/OH radicals, was able to alter the surface morphology (creating structures of 20 × 20 μm) and infiltrate into the particle's carbon matrix to produce new larger pore structures from the inside out. According to Wu et al. [62], interactions between radicals and metal species take place on the surface of internal pores or inside the char matrix. **Table 4** shows little change in the biochar's internal metal (K, Ca, Mg, Fe, and Al) contents before and after H_2O/CO_2 activation. Thus, the effect of H_2O/CO_2 may be more focused on changing the distribution of metal species within the biochar samples. As we reported previously [43], the effect of K in biochar on tar reforming is stronger than that of Ca and other species. Thus, the surface content and distribution of K were studied, as shown in **Figure 8(a)–(c)**. The surface content of K significantly increased from 0.18% in the unactivated biochar to 0.35% in the H_2O-activated biochar and 0.21% in CO_2-activated biochar. In addition, an obvious enrichment occurred on the surface of H_2O-activated biochar. Klinghoffer et al. [63] reported that during thermal treatment the metal species migrated to the biochar surface, some of which formed clusters that then acted as an active site for catalytic reactions. During H_2O/CO_2 activation, an increase in surface O content occurred alongside the migration of AAEM species from the interior of the particles to the surface, forming metal-carbon complexes. The redox properties of these metal-carbon complexes may have had implications for the biochar's catalytic properties. Also, highly dispersed metal species in a highly porous carbon matrix could have effectively acted as active adsorption sites that

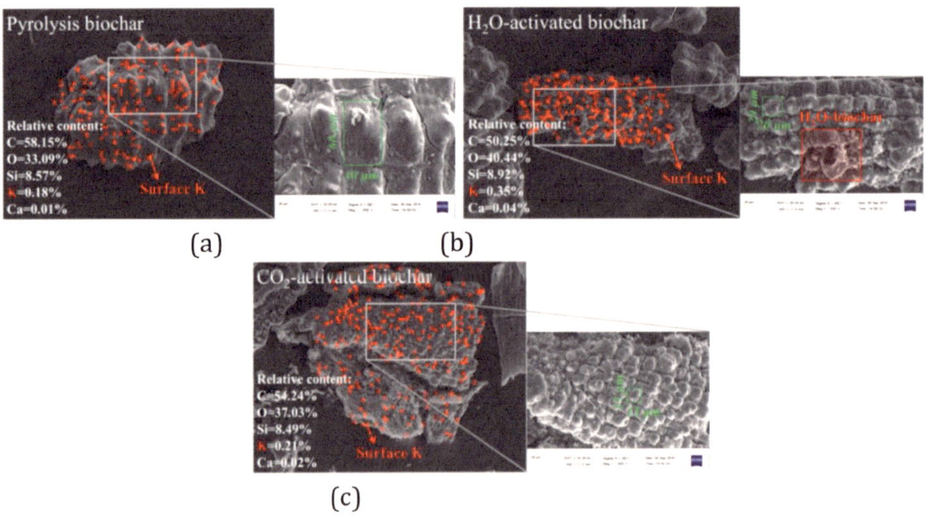

Figure 8.
SEM-EDX analysis of biochar samples. (a) Pyrolysis biochar, (b) H_2O activated biochar, and (c) CO_2 activated biochar.

promote volatile hydrocarbon condensation reactions to form coke [25], which were caused by hydrodeasphalting (HDA) reactions. Thus, the H_2O/CO_2 activation of biochar impacted the biochar surface's morphology and metal content, both of which influenced the reforming of biomass tar over biochar.

3.2.4 BET analysis of biochar during H_2O/CO_2 activation and tar reforming

The biochar samples' N_2-absorption/desorption isotherms at 77 K during H_2O/CO_2 activation and biomass tar reforming are shown in **Figure 9**. Compared with those of the original pyrolysis biochar, the pore systems of H_2O/CO_2-activated biochar samples and that from tar H_2O reforming over H_2O-activated biochar were better developed. Conversely, the other conditions exhibited pore structures that were somewhat blocked, especially for reforming over pyrolysis biochar in Ar and for the CO_2-activated biochar.

To further investigate the microphysical structures of the biochar samples, their BET surface properties were evaluated and are presented in **Table 4**. The unactivated pyrolysis biochar presented a BET surface area of 195.35 m^2/g and a pore volume of 0.0999 cm^3/g. Activation by H_2O and CO_2 increased the BET surface area to 307.45 and 237.71 m^2/g, respectively. The biochar's porous structure enabled efficient tar adsorption, resulting in a good residence time of the tar reacting with the catalyst [46]. **Table 5** also shows that the concentration of micropores (<2 nm), mesopores (2–50 nm), and macropores (>50 nm) varied between the samples. Thus, the ratio of micropores (<2 nm) to mesopores and macropores (>2 nm) ($S_{Mic.}/S_{Ext.}$) was employed. The H_2O-activated biochar showed a lower value of this ratio (2.28) than that of the CO_2-activated biochar (4.57) indicating that activation/gasification under a CO_2 atmosphere produced a higher relative micropore content, whereas under an H_2O atmosphere mesopores were favored. This may be explained by considering that H_2O removes carbon atoms from the particle's interior, enlarging open micropores and opening closed micropores, promoting the formation of mesopores. Meanwhile, CO_2 causes changes in the biochar surface that create more micropores. According to Klinghoffer et al. [63], the higher biochar surface area was the main reason for better catalyst performance, but pore size distribution also affected its activity, and evidence of diffusion limitations in microporous biochar was observed. Elsewhere, it has been confirmed that mesopores significantly

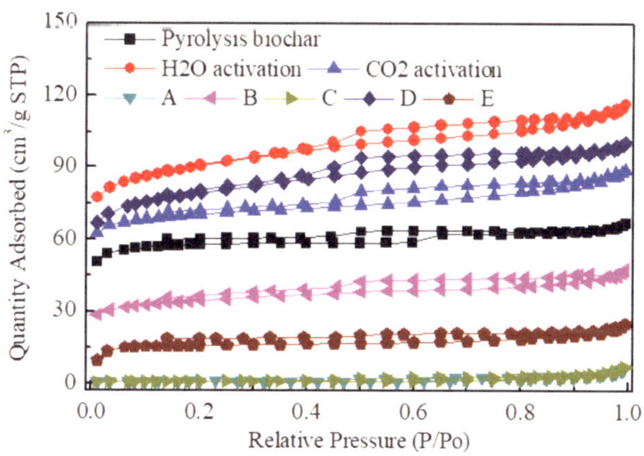

Figure 9.
N_2-absorption/desorption isotherms at 77 K for biochar obtained from different conditions: (A) in Ar over pyrolysis biochar; (B) in Ar over H_2O-activated biochar; (C) in Ar over CO_2-activated biochar; (D) in 15 vol.% H_2O over H_2O-activated biochar; and (E) in 29 vol.% CO_2 over CO_2-activated biochar.

enhance catalytic activity by allowing the penetration of macromolecules, facilitating their adsorption on the catalyst surface [64–66]. Thus, biochar used for catalytic tar reforming should ideally possess a high surface area and high mesoporosity (i.e., a small $S_{Mic.}/S_{Ext.}$ value).

After tar reforming in the Ar atmosphere, the biochar samples' BET surface area and pore volume markedly decreased. This was especially the case for tar reforming in Ar over the pyrolysis biochar and the CO_2-activated biochar where the BET surface areas fell to 3.78 and 4.07 m_2/g, respectively. According to the findings of Hosokai et al. [19], the decrease in surface area was attributed to tar forming coke deposits on the biochar's surface. In the Ar atmosphere, tar was mainly decomposed via coking [C_mH_n (aromatic compounds) = C_mH_x(coke) + $(n - x)/2$ H_2]. Thus, the biochar's activity could have fallen with a decrease in the biochar's surface area and/or pore volume caused by coke deposition. This implies that when some tar molecules reacted with the biochar they were absorbed in a way that yielded a condensed-phase product (coke) that remained on the biochar surface.

However, with the gasification agents, especially H_2O, the relatively high BET surface area and pore volume of biochar were maintained following the tar-reforming reactions (see **Table 5**). This indicated that the tar was not reformed directly to give gaseous products but instead involved the intermediate formation of coke, which was subsequently gasified by H_2O/CO_2. El-Rub and Kamel [59] suggested that tars can be adsorbed onto the active sites of biochar particles. Adsorbed tar and coke molecules can be catalytically reformed to give CO and H_2 by steam and dry CO_2 thermochemical reactions, regenerating the pore structure. Meanwhile, free radicals that enter polymerization reactions and coke on biochar surfaces can be formed from tar decomposition. For the H_2O-activated biochar, the BET surface area only decreased a little following reforming (to 268.52 m^2/g) and the $S_{Mic.}/S_{Ext.}$ value remained 2.30. Given that these conditions also reformed the greatest portion of the tar, the existence of gasification agents, especially H_2O, appeared to stimulate and maintain catalytic activity by continually creating and regenerating pore structures in the biochar.

Conditions	BET surface area (m^2/g)	Pore volume (cm^3/g)	Micro pore < 2 nm (m^2/g)	External pore > 2 nm (m^2/g)	$S_{Mic.}/S_{Ext.}$
Pyrolysis biochar	195.35	0.0999	170.46	24.89	6.85
H_2O-activated biochar	307.45	0.1745	213.83	93.62	2.28
CO_2-activated biochar	237.71	0.1330	195.02	42.69	4.57
(A) Pyrolysis and Ar reforming	3.78	0.0070	3.14	0.65	4.86
(B) H_2O activation and Ar reforming	117.53	0.0693	78.28	39.25	1.99
(C) CO_2 activation and Ar reforming	4.07	0.0074	2.12	1.95	1.09
(D) H_2O activation and H_2O reforming	268.52	0.1512	187.08	81.44	2.30
(E) CO_2 activation and CO_2 reforming	54.31	0.0346	42.98	11.33	3.79

Table 5.
BET properties of biochar samples during activation and tar reforming.

3.2.5 XPS analysis of biochar during biomass tar reforming

The elemental contents (C, O, K, and Ca) at the surface of the biochar samples are shown in **Figure 10**. Samples taken the following tar reforming in (A) Ar over pyrolysis biochar, (B) Ar over H_2O-activated biochar, and (C) Ar over CO_2-activated biochar showed that the H_2O/CO_2 activation of biochar played an important role in maintaining the biochar's active sites, such as surface O-containing functional groups and AAEM species (especially K and Ca) and improved its tar-reforming performance. According to Du et al. [67], XPS revealed the evolution of AAEM species and char structures, and concentrations of AAEM species agreed well with surface atomic O concentrations. Similar results were obtained in **Figure 10**, where Ar reforming over H_2O-activated biochar yielded a biochar with a higher surface content of O (16.25 atomic%), K (0.80 atomic%), and Ca (0.45 atomic%) than the samples from Ar reforming with the CO_2-activated biochar and the pyrolysis biochar. Abundant O-containing groups on the biochar surface can form acidic centers that can combine with biomass tar precursors, which have negatively charged π-electron systems and activate thermal cracking reactions [61]. For tar reforming in Ar, more carboxylic (O=C–O)/carbonyl (C=O) groups and fewer aromatic (C–C/C=C) groups were formed on the H_2O/CO_2-activated biochar surface. Franz et al. [68] investigated the effects of O-containing groups, particularly carboxylic and carbonyl groups, on the adsorption of dissolved aromatics on ash-free activated carbon. They found that the adsorption mechanism was influenced by the surface functional group's properties, especially its ability to hydrogen-bond, and through its activating/deactivating influence on the tar's aromatic ring. As shown in **Figure 5** for conditions (D) and (E), the existence of the gasification agents during tar reforming over H_2O/CO_2-activated biochar helped to limit coke formation on the biochar surface, likely by continually creating and regenerating surface active sites. This finding was consistent with that of a previous investigation [24].

The surface AAEM content remained high. For example, 2.12 atomic% K in H_2O and 1.83 atomic% in CO_2. This was similar for the surface O content (34.01 atomic% in H_2O and 32.07 atomic% in CO_2). A biochar with a higher O content appeared to favor the retention of AAEM species, with O likely serving as a link between the AAEM species and the char matrix [69]. In addition, the results of Wu et al. [70] suggest that adding H_2O was likely to have eliminated more tar, while the presence of CO_2 induced the formation of OH, H, and O radicals, which increase hydrocarbon conversion.

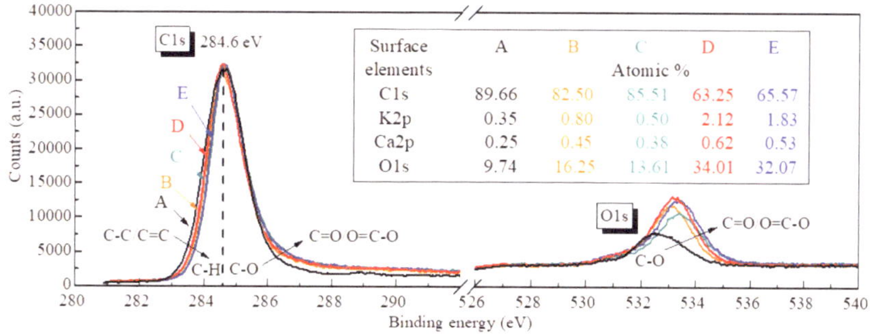

Figure 10.
XPS analysis of biochar samples from different conditions: (A) in Ar over pyrolysis biochar; (B) in Ar over H_2O-activated biochar; (C) in Ar over CO_2-activated biochar; (D) in 15 vol.% H_2O over H_2O-activated biochar; and (E) in 29 vol.% CO_2 over CO_2-activated biochar.

3.2.6 GC-MS analysis of biomass tar reforming over biochar

The biomass tar reformed in the gas phase (without involving biochar) in Ar at 800°C was mainly composed of aromatic tar compounds, owing to the secondary thermal cracking of in-situ biomass tar [52], compared with tar from H-form rice husk pyrolysis in fluidized bed at 500°C. Thermal cracking cannot completely convert tars [71]. Regarding the biochar catalyzed reactions, defined as the net tar loss owing to exposure to the biochar (i.e., the amount remaining after subtracting the amount of tar destroyed by vapor-phase cracking upstream and downstream of the biochar bed from the total change in tar amount during thermal treatment) [16], no new tar compounds were observed. According to Yao et al. [72], absent biochar, the gasification agent has a larger effect on the reforming of large aromatic ring systems (e.g., ≥2 fused benzene rings) than on smaller and isolated aromatics. However, here, biomass tars with a single aromatic ring or more than one ring structure were catalytically reformed over the various biochars. The conversion rates of tar compounds for the tar reformed without biochar in Ar at 800°C can be seen in **Figure 11(a)** and **(b)**. These figures also show that H_2O/CO_2 notably enhanced in-situ reforming of both large and small aromatic ring systems in biomass tar. For the experiments in Ar (conditions (A), (B), and (C)), individual tar conversion rates were improved by activation by H_2O/CO_2. For example, phenylethyne conversion increased from 42.27% over pyrolysis biochar to 77.43% over H_2O-activated biochar and to 49.93% over CO_2-activated biochar. However, the magnitude of the improvement was limited because of coke formation on the active sites, which deactivated the biochar. Thus, continuously supplying gasification agents (H_2O and CO_2 in conditions D and E, respectively) made more complete biomass tar conversion possible. For example, reforming in 15 vol.% H_2O over H_2O-activated biochar saw conversion rates of tars with both single aromatic ring structures (e.g., phenylethyne and benzofuran) and multiring structures (e.g., 1-methy-naphthalene, 2-methy-naphthalene, and phenanthrene) reach almost 100%. Although tar conversion was not completed in the 29 vol.% CO_2 atmosphere, it was also notably higher than for reforming in Ar over CO_2-activated biochar. As mentioned above, the gasification agent directly affected gas-phase tar reforming reactions [72], and it is likely that H_2O/CO_2 indirectly affected tar destruction by influencing the biochar structure and distribution of AAEM catalysts during the reaction by helping to ensure enough active sites on the biochar surface to maintain its catalytic activity.

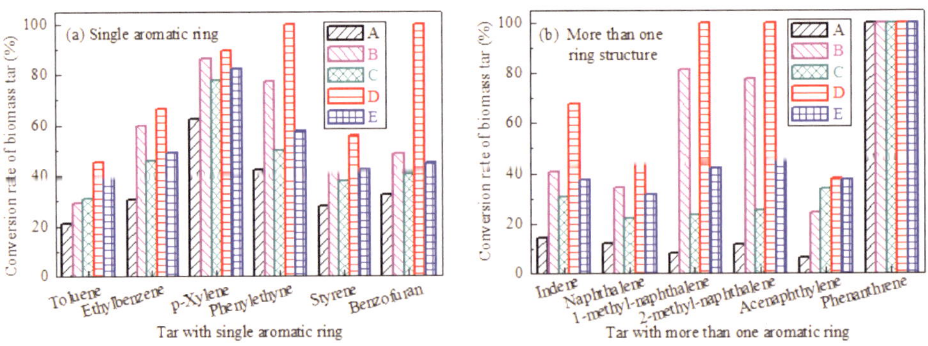

Figure 11.
Biomass tar conversion rates (based on tar observed following the treatment without biochar in Ar at 800°C) for different reforming conditions: (A) in Ar over pyrolysis biochar; (B) in Ar over H_2O-activated biochar; (C) in Ar over CO_2-activated biochar; (D) in 15 vol.% H_2O over H_2O-activated biochar; and (E) in 29 vol.% CO_2 over CO_2-activated biochar.

The heterogeneous reforming mechanism of the biomass tar over biochar and in the presence of the H_2O and CO_2 reforming agents at 800°C is shown in **Figure 12**. H_2O and CO_2 dissociate in space to form a large number of H/O/OH radicals, which play an important role in the tar-biochar reforming reaction. Biomass tar, through the biochar layer, is adsorbed onto the acid-base active sites (oxygen-containing functional groups and AAEM catalysts). The attraction effect of the carbon-rich biochar matrix invokes an electron pair shift in the tar molecules (relatively small mass), which promotes the tar molecules to break at high temperatures. According to the free radical theory [73], the tar adsorbed on the catalyst surface will catalytically crack to form the corresponding free radicals. The chemical reaction between these free radicals may permit new products. H_2O and CO_2 act as the reforming agents in the biochar carbon matrix, resulting in the fragmentation of the smaller aromatic rings. The empty active sites, formed by bond cleavage, were gradually occupied by H/O/OH radicals, forming active groups such as O-containing functional groups. In the presence of H_2O and CO_2, a significant amount of H/O/OH radicals in the vicinity ingress into the biochar carbon structure. The catalytic elements, such as AAEM species migrate at different rates and transformation from the carbon matrix onto the gas-solid interface or the gas phase undergoes, as shown in **Figure 12**. As the AAEM species are bonded with the C element on the biochar surface by the O element [21], H_2O and CO_2 react with these C elements on the biochar surface resulting in AAEM-O bond cleavage followed by precipitation. The valence state of Ca results in a stronger bonding interaction with the biochar when compared with K. Additionally, Ca migration and precipitation are more difficult than K. When tar adsorbs then cleaves the AAEM-O bond and functional group bond on the biochar surface, an aromatic fragmented radical is formed when other free radicals are encountered. At the same time, active AAEM species in the vicinity will continue to occupy active sites on the tar fragment groups, thereby inhibiting their secondary polymerization. At the same time, the H/O/OH radicals are exchanged to the AAEM species, which increases the possibility for the reforming of tar macromolecules. After the reaction, gas and light tars ($C_nH_y/CO/H_2$) were formed, thus realizing the H_2O or CO_2 heterogeneous reforming of biomass tar over the biochar catalyst.

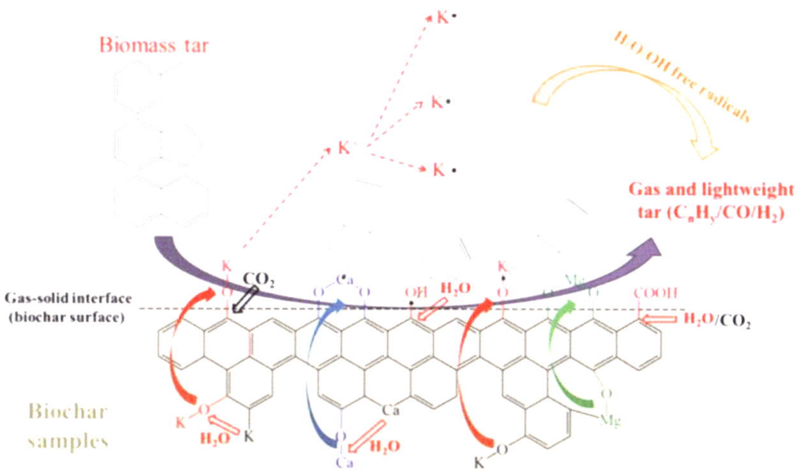

Figure 12.
Heterogeneous reforming mechanism of biomass tar over biochar in the presence of H_2O and CO_2 at 800°C [52].

4. Conclusions

The tar yield decreases as a function of increasing temperature from 26.18% at 500°C to 6.38% at 900°C. H_2O and CO_2 influence significantly the tar homogeneous transformations at 700–900°C, while the tar reforming effect of 15 vol.% H_2O is significantly higher than that of 29 vol.% CO_2. H_2O and CO_2 have obvious effects on the transformation of PAHs. H_2O and CO_2 not only directly affect the tar transformation on biochar but also indirectly influence the reforming of tar through changing the structure of biochar catalyst. The formation of additional oxygen-containing functional groups is strengthened with the concentration of H_2O and CO_2 increasing. During tar heterogeneous reforming over biochar, the transformation of small aromatic ring systems (3–5 fused rings) to larger aromatic ring systems (≥6 fused rings) in the biochar structure is promoted by the increasing concentration of H_2O and CO_2. The activation by H_2O/CO_2 of biochar impacted the biochar surface's morphology and distribution of metal species. Activation/gasification under a CO_2 in an Ar atmosphere produced more micropores, while adoption under a H_2O in an Ar atmosphere favored the formation of mesopores. With the existence of gasification agents, especially for H_2O, the simultaneous creation of pore structures is necessary to maintain biochar's catalytic activity during tar reforming. H_2O/CO_2 also indirectly affects tar destruction by influencing the biochar structure and distribution of AAEM catalysts, while the reaction is occurring to ensure enough active sites on the biochar surface to maintain its catalytic activity. The activation and/or activity-maintaining effects of H_2O/CO_2 can notably enhance the in-situ reforming of both large and small aromatic ring systems present in biomass tar.

Acknowledgements

This work is supported by the National Natural Science Foundation of China (51906052), the National Postdoctoral Program for Innovative Talents of China (BX20180086), China Postdoctoral Science Foundation Funded Project (2018M641826), Heilongjiang Provincial Postdoctoral Science Foundation, Foundation of State Key Laboratory of High-efficiency Utilization of Coal and Green Chemical Engineering (2019-KF-14), and Fundamental Research Funds for the Central Universities (Grant no. HIT. NSRIF. 2020052).

Conflict of interest

The authors declare no conflict of interest.

Author details

Dongdong Feng*, Yu Zhang, Yijun Zhao and Shaozeng Sun
School of Energy Science and Engineering, Harbin Institute of Technology, Harbin, China

*Address all correspondence to: 08031175@163.com

References

[1] Shen Y, Zhao P, Shao Q, Ma D, Takahashi F, Yoshikawa K. In-situ catalytic conversion of tar using rice husk char-supported nickel-iron catalysts for biomass pyrolysis/gasification. Applied Catalysis B: Environmental. 2014;**152-153**:140-151

[2] Milne T, Evans R. Biomass gasification "tars": Their nature, formation and conversion. In: Report No. NREL/TP-570-25357. Golden, CO, USA: NREL; 1998

[3] Spath PL, Dayton DC. Preliminary screening-technical and economic assessment of synthesis gas to fuels and chemicals with emphasis on the potential for biomass-derived syngas. In: DTIC Document. 2003

[4] Bui T, Loof R, Bhattacharya S. Multi-stage reactor for thermal gasification of wood. Energy. 1994;**19**:397-404

[5] Mudge L, Baker E, Mitchell D, Brown M. Catalytic steam gasification of biomass for methanol and methane production. Journal of Solar Energy Engineering. 1985;**107**:88-92

[6] Kiel J, Van Paasen S, Neeft J, Devi L, Ptasinski K, Janssen F, et al. Primary measures to reduce tar formation in fluidised-bed biomass gasifiers. ECN, ECN-C-04-014. 2004

[7] Skoulou V, Kantarelis E, Arvelakis S, Yang W, Zabaniotou A. Effect of biomass leaching on H_2 production, ash and tar behavior during high temperature steam gasification (HTSG) process. International Journal of Hydrogen Energy. 2009;**34**:5666-5673

[8] Sarıoğlan A. Tar removal on dolomite and steam reforming catalyst: Benzene, toluene and xylene reforming. International Journal of Hydrogen Energy. 2012;**37**:8133-8142

[9] Dou B, Gao J, Sha X, Baek SW. Catalytic cracking of tar component from high-temperature fuel gas. Applied Thermal Engineering. 2003;**23**:2229-2239

[10] Rapagna S, Jand N, Foscolo P. Catalytic gasification of biomass to produce hydrogen rich gas. International Journal of Hydrogen Energy. 1998;**23**: 551-557

[11] Dayton D. A review of the literature on catalytic biomass tar destruction. US DOE NREL Report 510-32815. Golden, CO. 2002

[12] Abu El-Rub Z, Bramer E, Brem G. Review of catalysts for tar elimination in biomass gasification processes. Industrial & Engineering Chemistry Research. 2004;**43**:6911-6919

[13] Yu J, Tian F-J, McKenzie L, Li C-Z. Char-supported nano iron catalyst for water-gas-shift reaction: Hydrogen production from coal/biomass gasification. Process Safety and Environmental Protection. 2006;**84**:125-130

[14] Griffith DM, Mainhood J. Cracking of tar vapour and aromatic compounds on activated carbon. Fuel. 1967;**46**:167

[15] Chembukulam SK, Dandge AS, Rao NLK, Seshagiri K, Vaidyeswaran R. Smokeless fuel from carbonized sawdust. Industrial & Engineering Chemistry Product Research and Development. 1981;**20**:714-719

[16] Boroson ML, Howard JB, Longwell JP, Peters WA. Heterogeneous cracking of wood pyrolysis tars over fresh wood char surfaces. Energy Fuels. 1989;**3**:735-740

[17] Wang D, Yuan W, Ji W. Char and char-supported nickel catalysts for secondary syngas cleanup and

conditioning. Applied Energy. 2011;**88**:1656-1663

[18] Sueyasu T, Oike T, Mori A, Kudo S, Norinaga K, Hayashi J-i. Simultaneous steam reforming of tar and steam gasification of char from the pyrolysis of potassium-loaded woody biomass. Energy Fuels. 2011;**26**:199-208

[19] Hosokai S, Kumabe K, Ohshita M, Norinaga K, Li C-Z, Hayashi J-i. Mechanism of decomposition of aromatics over charcoal and necessary condition for maintaining its activity. Fuel. 2008;**87**:2914-2922

[20] Matsuhara T, Hosokai S, Norinaga K, Matsuoka K, Li C-Z, Hayashi J-i. In-situ reforming of tar from the rapid pyrolysis of a brown coal over char. Energy Fuels. 2009;**24**:76-83

[21] Feng D, Zhao Y, Zhang Y, Sun S, Meng S, Guo Y, et al. Effects of K and Ca on reforming of model tar compounds with pyrolysis biochars under H_2O or CO_2. Chemical Engineering Journal. 2016;**306**:422-432

[22] Yue B, Wang X, Ai X, Yang J, Li L, Lu X, et al. Catalytic reforming of model tar compounds from hot coke oven gas with low steam/carbon ratio over Ni/ MgO-Al_2O_3 catalysts. Fuel Processing Technology. 2010;**91**:1098-1104

[23] Devi L, Ptasinski KJ, Janssen FJ. Pretreated olivine as tar removal catalyst for biomass gasifiers: Investigation using naphthalene as model biomass tar. Fuel Processing Technology. 2005;**86**:707-730

[24] Min Z, Asadullah M, Yimsiri P, Zhang S, Wu H, Li C-Z. Catalytic reforming of tar during gasification. Part I. Steam reforming of biomass tar using ilmenite as a catalyst. Fuel. 2011;**90**:1847-1854

[25] Min Z, Yimsiri P, Asadullah M, Zhang S, Li C-Z. Catalytic reforming of tar during gasification. Part II. Char as a catalyst or as a catalyst support for tar reforming. Fuel. 2011;**90**:2545-2552

[26] Sathe C, Pang Y, Li C-Z. Effects of heating rate and ion-exchangeable cations on the pyrolysis yields from a Victorian brown coal. Energy Fuels. 1999;**13**:748-755

[27] Li C-Z, Sathe C, Kershaw J, Pang Y. Fates and roles of alkali and alkaline earth metals during the pyrolysis of a Victorian brown coal. Fuel. 2000;**79**: 427-438

[28] Morf P, Hasler P, Nussbaumer T. Mechanisms and kinetics of homogeneous secondary reactions of tar from continuous pyrolysis of wood chips. Fuel. 2002;**81**:843-853

[29] Stiles HN, Kandiyoti R. Secondary reactions of flash pyrolysis tars measured in a fluidized bed pyrolysis reactor with some novel design features. Fuel. 1989;**68**:275-282

[30] Jess A. Mechanisms and kinetics of thermal reactions of aromatic hydrocarbons from pyrolysis of solid fuels. Fuel. 1996;**75**:1441-1448

[31] Li C-Z, Nelson PF. Fate of aromatic ring systems during thermal cracking of tars in a fluidized-bed reactor. Energy Fuels. 1996;**10**:1083-1090

[32] Williams PT, Taylor DT. Aromatization of tyre pyrolysis oil to yield polycyclic aromatic hydrocarbons. Fuel. 1993;**72**:1469-1474

[33] Kong M, Yang Q, Fei J, Zheng X. Experimental study of Ni/MgO catalyst in carbon dioxide reforming of toluene, a model compound of tar from biomass gasification. International Journal of Hydrogen Energy. 2012;**37**:13355-13364

[34] Takahashi H, Iwatsuki M, Essaki K, Tsutsumi A, Chiba T. Rapid conversion

of tar and char from pyrolysis of a brown coal by reactions with steam in a drop-tube reactor. Fuel. 2000;**79**:439-447

[35] Gil J, Corella J, Aznar MAP, Caballero MA. Biomass gasification in atmospheric and bubbling fluidized bed: Effect of the type of gasifying agent on the product distribution. Biomass Bioenergy. 1999;**17**:389-403

[36] Jönsson O. Thermal cracking of tars and hydrocarbons by addition of steam and oxygen in the cracking zone. In: Fundamentals of Thermochemical Biomass Conversion. Springer; 1985. pp. 733-746

[37] Zhang Y, Kajitani S, Ashizawa M, Oki Y. Tar destruction and coke formation during rapid pyrolysis and gasification of biomass in a drop-tube furnace. Fuel. 2010;**89**:302-309

[38] Wang Y, Namioka T, Yoshikawa K. Effects of the reforming reagents and fuel species on tar reforming reaction. Bioresource Technology. 2009;**100**: 6610-6614

[39] Miura K, Kawase M, Nakagawa H, Ashida R, Nakai T, Ishikawa T. Conversion of tar in hot coke oven gas by pyrolysis and steam reforming. Journal of Chemical Engineering of Japan. 2003;**36**:735-741

[40] Dufour A, Girods P, Masson E, Rogaume Y, Zoulalian A. Synthesis gas production by biomass pyrolysis: Effect of reactor temperature on product distribution. International Journal of Hydrogen Energy. 2009;**34**:1726-1734

[41] Feng D, Zhang Y, Zhao Y, Sun S. Catalytic effects of ion-exchangeable K^+ and Ca^{2+} on rice husk pyrolysis behavior and its gas-liquid-solid product properties. Energy. 2018;**152**:166-177

[42] Keown DM, Li X, Hayashi J-i, Li C-Z. Characterization of the structural features of char from the pyrolysis of cane trash using Fourier transform-Raman spectroscopy. Energy Fuels. 2007;**21**:1816-1821

[43] Feng D, Zhao Y, Zhang Y, Sun S, Meng S, Guo Y, et al. Effects of K and Ca on reforming of model tar compounds with pyrolysis biochar under H_2O or CO_2. Chemical Engineering Journal. 2016;**306**:422-432

[44] Zhao Y, Feng D, Zhang Y, Tang W, Meng S, Guo Y, et al. Migration of alkali and alkaline earth metallic species and structure analysis of sawdust pyrolysis biochar. Korean Chemical Engineering Research. 2016;**54**:659-664

[45] Zhao Y, Feng D, Zhang Y, Huang Y, Sun S. Effect of pyrolysis temperature on char structure and chemical speciation of alkali and alkaline earth metallic species in biochar. Fuel Processing Technology. 2016;**141**:54-60

[46] Gilbert P, Ryu C, Sharifi V, Swithenbank J. Tar reduction in pyrolysis vapours from biomass over a hot char bed. Bioresource Technology. 2009;**100**:6045-6051

[47] Dufour A, Masson E, Girods P, Rogaume Y, Zoulalian A. Evolution of aromatic tar composition in relation to methane and ethylene from biomass pyrolysis-gasification. Energy Fuels. 2011;**25**:4182-4189

[48] Milne TA, Abatzoglou N, Evans RJ. Biomass Gasifier "Tars": Their Nature, Formation, and Conversion. Golden, CO: National Renewable Energy Laboratory; 1998

[49] Ledesma E, Kalish M, Nelson P, Wornat M, Mackie J. Formation and fate of PAH during the pyrolysis and fuel-rich combustion of coal primary tar. Fuel. 2000;**79**:1801-1814

[50] Dong G, Hüttinger K. Consideration of reaction mechanisms leading to

pyrolytic carbon of different textures. Carbon. 2002;**40**:2515-2528

[51] Guan Y, Luo S, Liu S, Xiao B, Cai L. Steam catalytic gasification of municipal solid waste for producing tar-free fuel gas. International Journal of Hydrogen Energy. 2009;**34**:9341-9346

[52] Feng D, Zhao Y, Zhang Y, Sun S. Effects of H_2O and CO_2 on the homogeneous conversion and heterogeneous reforming of biomass tar over biochar. International Journal of Hydrogen Energy. 2017;**42**:13070-13084

[53] Badger GM, Kimber RWL, Spotswood TM. Mode of formation of 3, 4-benzopyrene in human environment. Nature. 1960;**187**(4738):663-665

[54] Rybak W, Zembruski M, Smith I. Twenty-First Symposium (International) on Combustion. Munich, Germany: The Combustion Institute; 1986. p. 231

[55] Badger G. Pyrolysis of hydrocarbons. Progress in Physical Organic Chemistry. 1965;**3**:1

[56] Poutsma ML. A Review of Thermolysis Studies of Model Compounds Relevant to Processing of Coal. TN, USA: Oak Ridge National Lab; 1987

[57] Futamura S, Annadurai G. Plasma reforming of aliphatic hydrocarbons with CO_2. ITIA. 2005;**41**:1515-1521

[58] Futamura S, Kabashima H, Annadurai G. Roles of CO_2 and H_2O as oxidants in the plasma reforming of aliphatic hydrocarbons. Catalysis Today. 2006;**115**:211-216

[59] El-Rub ZA, Kamel Z. Biomass char as an in-situ catalyst for tar removal in gasification systems. Fuel. 1990;**69**(10):1219-1225

[60] Chen G-M, Zhang X-W, Mi Z-T. Effects of pressure on coke and formation of its precursors during catalytic cracking of toluene over USY catalyst. Journal of Fuel Chemistry and Technology. 2007;**35**:211-216

[61] Wang F-J, Zhang S, Chen Z-D, Liu C, Wang Y-G. Tar reforming using char as catalyst during pyrolysis and gasification of Shengli brown coal. Journal of Analytical and Applied Pyrolysis. 2014;**105**:269-275

[62] Wu H, Quyn DM, Li C-Z. Volatilisation and catalytic effects of alkali and alkaline earth metallic species during the pyrolysis and gasification of Victorian brown coal. Part III. The importance of the interactions between volatiles and char at high temperature. Fuel. 2002;**81**:1033-1039

[63] Klinghoffer NB, Castaldi MJ, Nzihou A. Catalyst properties and catalytic performance of char from biomass gasification. Industrial & Engineering Chemistry Research. 2012;**51**:13113-13122

[64] Kyotani T. Control of pore structure in carbon. Carbon. 2000;**38**:269-286

[65] Frackowiak E, Beguin F. Carbon materials for the electrochemical storage of energy in capacitors. Carbon. 2001;**39**:937-950

[66] Ariyadejwanich P, Tanthapanichakoon W, Nakagawa K, Mukai S, Tamon H. Preparation and characterization of mesoporous activated carbon from waste tires. Carbon. 2003;**41**:157-164

[67] Du C, Liu L, Qiu P. Importance of volatile AAEM species to char reactivity during volatile-char interactions. RSC Advances. 2017;**7**:10397-10406

[68] Franz M, Arafat HA, Pinto NG. Effect of chemical surface heterogeneity on the adsorption mechanism of dissolved aromatics on activated carbon. Carbon. 2000;**38**:1807-1819

[69] Hashimoto K, Miura K, Xu J-J, Watanabe A, Masukami H. Relation between the gasification rate of carbons supporting alkali metal salts and the amount of oxygen trapped by the metal. Fuel. 1986;**65**:489-494

[70] Wu W-g, Luo Y-h, Su Y, Zhang Y-l, Zhao S-h, Wang Y. Nascent biomass tar evolution properties under homogeneous/heterogeneous decomposition conditions in a two-stage reactor. Energy Fuels. 2011;**25**:5394-5406

[71] Anis S, Zainal ZA. Tar reduction in biomass producer gas via mechanical, catalytic and thermal methods: A review. Renewable and Sustainable Energy Reviews. 2011;**15**:2355-2377

[72] Song Y, Wang Y, Hu X, Xiang J, Hu S, Mourant D, et al. Effects of volatile-char interactions on in-situ destruction of nascent tar during the pyrolysis and gasification of biomass. Part II. Roles of steam. Fuel. 2015;**143**:555-562

[73] Howard J. Fundamentals of coal pyrolysis and hydropyrolysis. Chemistry of Coal Utilization. 1981;**2**:665-784

Chapter 5

Comparative Evaluation of Hydrochars and Pyrochars for Phosphate Adsorption from Wastewater

Aicha Slassi Sennou, Shuangning Xiu and Abolghasem Shahbazi

Abstract

Biochar represents the rich carbon residues that remains after thermally pyrolyzing or liquefying different biomass types in an oxygen-free environment. The availability of animal and agricultural wastes makes the biochar a low-cost product. It is, as a carbon-rich product, resistant to mineralization and decomposition. Biochar can be used as a multifunctional material in many applications in the environmental and agricultural sectors. Recently, a growing interest for the use of biochar in different fields is rising because of its use as a sorbent for organic and nonorganic contaminants from aqueous solutions. In this chapter, recent studies on pyrochar/hydrochar production, characterization, and phosphate adsorption are reviewed and summarized. The remediation technologies for phosphate removal from contaminated water using biochar are also discussed. The effects of reaction temperature and initial solution pH on phosphate adsorption onto biochar are compared. In addition, we highlighted the models that are used for adsorption kinetics and adsorption isotherms.

Keywords: biochar, hydrochar, pyrochar, pyrolysis, hydrothermal carbonization, phosphate, filtration

1. Introduction

Phosphorus (P) is an essential nutrient for the growth of plants; however, its excessive release into runoff water can impose a danger on environmental health [1]. Phosphor soluble in water is present under three forms, ortho-, poly-, and organic phosphate. Orthophosphate comprises HPO_4^{-}, $H_2PO_4^{2-}$, H_3PO_4, and PO_4^{3-} which encourages the evolution of aquatic microorganisms and macroorganisms leading to eutrophication. Eutrophication is caused by the presence of excessive phosphate amounts and has adverse impacts on the water ecosystem. In fact, even low amounts of phosphate 0.02 mg/L can cause algae to grow leading to reduced oxygen in water that could kill fish and damage wildlife [2]. The municipal sewage phosphate concentration is in the 4–15 mg-P/L range, while it exceeds 10 mg-P/L in industrial wastewaters. Consequently, phosphate concentration from wastewater needs to be reduced prior to the discharge in water bodies. On the other hand, phsophorus is a non-renewable source available in limited quantities in nature with Morocco owning approximately 75% of the market [3]. There is an uncertainty about the lifetime of the remaining reserves and their accessibility. Studies have shown that P reserves are limited and will deplete soon with the increasing use of phosphorus as a fertilizer for agronomic production. Thus, it is crucial to develop effective phosphate removal technologies from aqueous solution.

Many technologies have been studied for phosphate removal from industrial and municipal effluents; these have been categorized into chemical, physical, and biological wastewater treatments. Studies have proven the effectiveness of biological and chemical treatments. These treatment methods use plants such as algae or chemicals such as alum, lime, and ferric salts to remove phosphorus from water before discharge into water bodies. Studies have shown that 97% of the total phosphorus (TP) could be removed from aqueous solutions with the use of phosphate biological treatment methods which does not present the risk of adding chemical contaminants to the water. However, the phosphate removal efficiency may be low due to its sensitivity to operation conditions [1]. However, other treatment method costs remain expensive, and some of them necessitate continuous checking of the operating conditions in order to prevent the introduction of new contaminants into the water. Thus, developing a cost-effective technology to prevent eutrophication and manage phosphorus recovery will prevent the shortage of this important element as it can be recycled and reused if it is successfully recovered from wastewater.

Adsorption presents a low-cost and efficient method for phosphorus removal due to its cost-effectiveness, accessibility, and performance. However, the high cost of traditional adsorbents (e.g., active carbon and anion exchange resins) and the difficulty of disposal are problems that limit their use. Thus, researchers study the development possibility of low-cost adsorbents that could come from cheap and available products such as biomass. Biochar has been studied as a prospective adsorbent for its properties including low cost, availability, stability, high porosity, and non-costly preparation and upgrading [2]. However, powder carbon-based biochars were identified to be difficult to separate from aqueous solutions which requires the use of magnetic separation that requires the addition of magnetic iron oxide through chemical co-precipitation [4].

Biochar has many advantages over traditional adsorbents such as being environmentally friendly, recyclable, low cost, easy to prepare, and having a high porosity. Consequently, researchers have studied the potential application of biochar as a phosphate adsorbent from wastewater [5] and its use as a fertilizer for soil quality enhancement. However, non-modified biochar with a surface charged negatively has a limited adsorption capacity for anionic pollutants which requires impregnating some metals on the biochar surface (Fe, Mg, Al, etc.) [2]. The preparation process of magnetic adsorbents is considered complicated and costly due to the complicated steps involved starting by the precursor preparation, oxidation of activated carbon, and then iron grafting [6]. For this reason, biochar use as a phosphate adsorbent from wastewater needs to be examined. Factors affecting the effectiveness of the anionic pollutant adsorption such as the functional site amount, affinity, dispersion, and surface accessibility need to be considered [7]. Some researchers studied the phosphate adsorption capacity of modified biochar. Junk et al. prepared magnetic biochar by magnesium ferrite impregnation through co-precipitation of Mg and Fe and pyrolysis and found to have an adsorption capacity of 487.99 mg P/g where P is the symbol of phosphorus [8, 9]. Junk et al. adsorption rate is considered among the highest when compared to other biochars adsorption rates 41.16 mg P/g, 125.40 mg P/g, 135 mg P/g [13, 17, 40]; this is shown later in Section 4 [7].

Other techniques are also employed to improve biochar surface characteristics for better adsorption. These treatment methods can be divided into gas phase and liquid phase. Gas phase uses steam or carbon dioxide for achieving physical activation, whereas liquid phase uses chemicals [10]. Both liquid phase and gas phase treatment methods use activation temperatures higher than 450°C. Chemical activation has advantages over physical activation due to its less cost and time along with higher char porosity [11].

2. Phosphate in water bodies

P is a vital nutrient for the growth of aquatic organisms, plants, and animals [12, 13]. It also plays an essential role as a limiting nutrient, preventing the development of algae and aquatic plants in the water ecosystem [14]. TP amount needs to be less than 0.03 mg/L, while phosphate level between 0.005 and 0.05 mg/L as an excessive amount of phosphate in water bodies causes eutrophication. Eutrophication leads to the destruction of water ecosystem including rivers, lakes, and seas causing harm to human health and an increase to the water treatment cost [15]. Natural phosphate reserves are limited and will be depleted in 50–100 years [16]. Thus, phosphorus recovery from phosphate-rich wastewater can simultaneously solve the phosphorus depletion issue and improve water quality [17].

3. Biochar production through pyrolysis and hydrothermal carbonization (HTC)

3.1 Biochar production from hydrothermal carbonization

Biomass HTC is performed at temperatures ranging from 160–800°C. It is categorized into low-temperature reactions (lower than 300°C) and high-temperature reactions (300–800°C). During high-temperature HTC, the principal reaction is gasification, and the products are gases including hydrogen and methane [18] while carbonization takes place at low temperatures. Gasification favors the production of methane and hydrogen through

$$C_6H_{12}O_6 + 6H_2O \rightarrow 6CO_2 + 12H_2 \quad (1)$$

$$C_6H_{12}O_6 \rightarrow 3\,CH_4 + 3CO_2 \quad (2)$$

Hydrogen formation is endothermic and methane formation is exothermic. Therefore, at high temperatures hydrogen formation exceeds methane as illustrated in **Figure 1**.

Hydrothermal carbonization is used for high moisture content biomass contrarily to pyrolysis and gasification that deals with low moisture content biomass [19]. Many factors affect the hydrochar characteristics. These include the water-to-biomass ratio, reaction temperature, residence time, and pressure. Hydrochar is defined as the product of HTC reaction and has a 40–70 wt% yield. Hydrochars have lower yields but high higher heating value (HHV) than low-temperature pyrolysis (LTP) pyrochars [20]. The ultimate analysis confirmed these findings as it shows hydrochars having high carbon contents and low oxygen contents than pyrochars. Pyrochars have higher yields than hydrochars and thus higher-energy yields despite their lower HHVs. This indicates that biomass experienced a deep carbonization and decomposition in the LTP process. Oliveira et al. [21] showed that deeper carbonization of biochar can be achieved by combining different

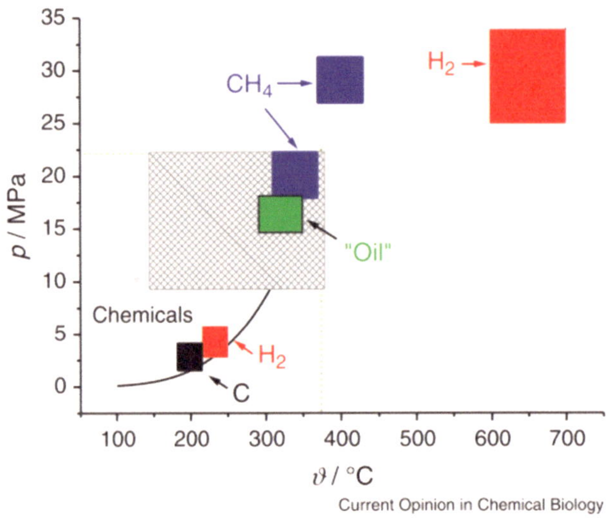

Figure 1.
Different biomass conversion processes overview and vapor-pressure water curve [18].

agricultural residues and different types of biomass. The use of lignocellulosic residues leads to hydrochars with high-energy yields.

In addition, hydrochars produced good dewatering and drying properties. The water involved in the process could be reused which will reduce the environmental impact and increase the energy efficiency. Different researchers have studied production conditions such as feedstock, temperature, and residence time on hydrochar. Sabio et al. [22] reported that the main factors affecting the hydrochar solid yield are residence time and temperature. Producing hydrochars in the ideal conditions can contribute to promote energy densification (increase in HHV).

3.2 Biochar production from biomass pyrolysis

Pyrolysis is used for the conversion of biomass into alternative products such as biochar, syngas, and bio-oil in an oxygen-free environment at temperatures ranging from 300 to 900°C. There are different types of reactors used to produce biochar including pyrolysis rotary drums, paddle pyrolysis kiln, and auger reactor [23].

The rotary drum reactor is considered a reliable system for biochar production. The reactor is composed of internal concentric steel tube and a rotary part made of insulated mantle. The mantle supports a sequence of radial steel fins and connects to the steel tube. Two fixed parts at the rotary end are responsible of the charge and the discharge of solid and gaseous products. The biomass moves through the carbonization kiln, passes inside the internal concentric tube, and is moved at increasing temperatures through various sections. The biomass temperature fumes enter the furnace and increase the temperature to 500°C. The carbonization process requires heat that is delivered by burning gases and pyrolysis vapors. The process is energy independent except the starting phase that uses external combustion. Fumes exit the reactor through the chimney while the produced biochar is gathered in a stocking silo [24] (**Figure 2**).

The auger pyrolysis reactor is normally nourished with wood waste by a hopper at the retort end. The wood waste is transported through an auger into other extremities for carbonization. The vapors and gases are sent to a condenser, and the biochar is collected in the retort through a single tube. The main structure at the lowest level includes a combustion chamber where carbonization occurs. The combustion chamber uses gases to provide heat [24] (**Figure 3**).

The paddle pyrolysis reactor is a low-temperature gasifier. Its main characteristics is to move and mix biomass increasing the heat transfer on the material surface. The system output is biochar and syngas; the reactor has a unique cleanup system to produce syngas for downstream applications [24].

Pyrolysis can be divided into two types, slow and fast pyrolysis, based on the reaction conditions (heating rate, residence time, pressure, and mainly temperature) [25]. Slow pyrolysis is conducted in temperatures ranging from 400 to 600°C and is used in applications that seek a high biochar yield and a low bio-oil and syngas products including CH_4, H_2, CO, CO_2, and C1–C2 hydrocarbons. The reaction is conducted at atmospheric pressure and residence time longer than 1 h at a heating rate in the range of 5–7°C/min [25, 26]. These conditions lead to more cracking reactions which decrease the liquid yield and consequently increase biochar yield [27]. Slow pyrolysis also called conventional carbonization process produces high biochar yields mainly for feedstocks with high ash content, lignin, and particle size [28]. Another factor that can contribute to high biochar yield is the increasing particle size of the sample.

This is a simple, cost-effective but powerfully built process which is mainly applicable to small-scale biochar production farms. On the other hand, fast pyrolysis has more opportunities toward increasing bio-oil yield up to 75% from original

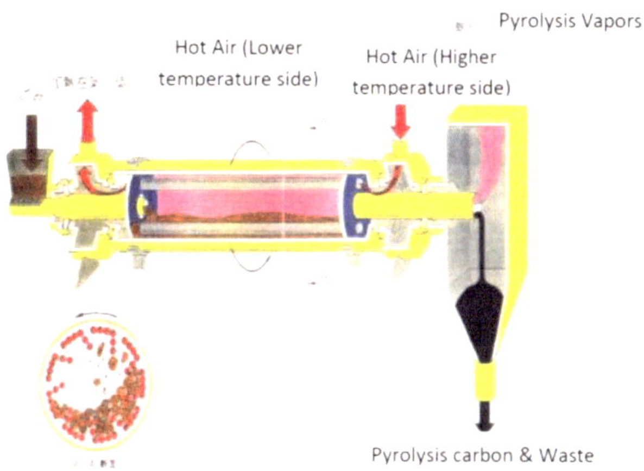

Figure 2.
Pyrolysis rotary drum [24].

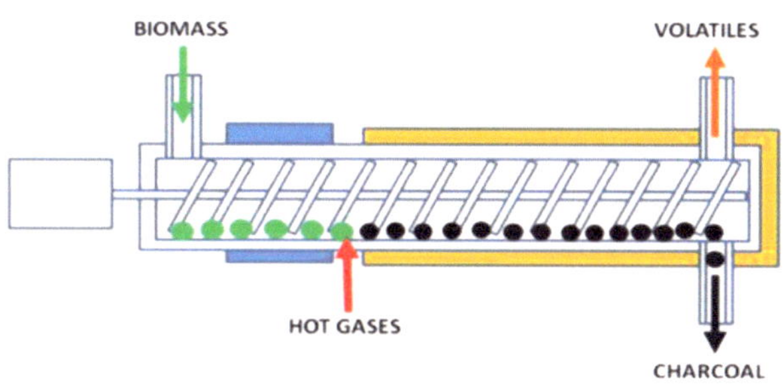

Figure 3.
Auger reactor [24].

raw biomass, and in contrast to slow pyrolysis the heating rate is higher than 200°C/min and residence time less than 10 s which favors the decrease in biochar yield [29].

4. Remediation technologies for phosphate removal in aqueous solutions using biochar

Many researchers have explored different paths for phosphate adsorption from wastewater. Methods have been studied including solvent extraction, chemical precipitation, gravity separation, ion exchange, reverse osmosis, solvent extraction, electrodialysis and electrocoagulation, adsorption, and flotation for phosphate removal from sewage system [30]. Adsorption is preferred for cost-effectiveness. However, waste sludge disposal resulting from adsorption of phosphate from wastewater is complicated and expensive which hinders the use of adsorption technique for phosphate removal in industrial scale.

The use of metal salts has been explored for phosphate removal. Most metals have a strong ability to adsorb contaminants using electrostatic attraction between the surface charged negatively and the metal ions charged positively [31]. Moreover, there is a limitation on transforming agricultural residues to green by-products since most farmers burn their crop stalk causing more greenhouse gas emissions and harm to the environment. Biochar could be used as a clean environmental by-product as it is produced in an oxygen-free environment and can be used for different applications [13]. The use of biochar and impregnation of metal oxides on its surface for phosphate adsorption was studied as a technique for contaminant removal.

Techniques for phosphorus recovery can be categorized as physical (electrodialysis, reverse osmosis, ion exchange) [1], chemical, and biological [32]. Some of the mostly used techniques are ion exchange, chemical precipitation, electrocoagulation, chemical precipitation, crystallization, and adsorption. Phosphorus chemical precipitation can be achieved by the deposition of some metal elements and metal oxides. Crystallization can be represented by attractive struvite (magnesium ammonium phosphate) which is a white crystal containing important elements for the growth of plants. Struvite is composed of Mg, P, and N with equal molar concentrations. Struvite can be crystallized and applied to the soil as a fertilizer or used for the recovery of phosphorus and nitrogen. However, the economic feasibility of struvite precipitation is influenced by the costs of reagents such as magnesium. Magnesium sources such as MgO are environmental materials; their advantage is the high alkalinity than other alkalis. Alkalinity plays an important role in acid neutralization as well as creation of adequate pH region for crystallization. The MgO available in the market is a solid mineral which releases low amounts of Mg due to slow dissolution [33]. Struvite crystallization can be used for sludge anaerobic digestion due to the high concentration of ammonia and phosphorus in reactors. Struvite formation formula is

$$Mg^{2+} + NH_4^{+} + PO_4^{3-} + 6H_2O \rightarrow MgNH_4PO_4 6(H_2O) \tag{3}$$

The struvite chemical precipitation has high adsorption of soluble phosphate 80–90% [34]. Struvite precipitation faces challenges in phosphorus recovery from wastewater when the phosphorus concentration is below 50 mg/L and the suspended solid concentration is higher than 2000 mg/L [35]. Mariana et al. [35] suggested the application of struvite precipitation in secondary streams with high phosphorus concentrations.

Adsorption is another attractive technology. It has been well-known for its simplicity, low cost, and high adsorption capacity. Different adsorbents have been tested for phosphate adsorption from wastewater including activated carbon, metal-based materials, and designed engineering particles. These adsorbents have shown some drawbacks such as high cost and inefficiency. Thus, increasing interest is going to agricultural by-products. During the past several years, research studies have been done on the removal of contaminants from aqueous solutions using biochar. Biochar is considered a potential great adsorbent because of its high ion exchange capacity, large surface area, and abundant functional groups. However, adsorption of biochar to anion pollutants such as $PO4^{3-}$ phosphate anion is limited due to the negatively charged surface of biochar and its low anion exchange capacity. Thus, studies suggested that the removal capacity can be improved by using chemical co-precipitation to deposit metal element cations (Ca^{2+}, Fe^{3+}, Al^{3+}, Mg^{2+}) or metal oxides (CaO, MgO, La_2O_3, Fe_2O_3, Al_2O_3). Metal oxides and elements react with oxyanions such as PO_4^{3-}, resulting in improved anion removal rate [36].

Ion exchange is carried to remove certain ions from an aqueous solution. It is used for purification of aqueous solutions with the use of polymeric ion exchange resin or other materials that have ion exchange properties. Ion exchangers can be anion exchangers (exchanging ions charged negatively) or cation exchangers (exchanging ions charged positively). Ion exchange is considered a reversible process where the ion exchanger could be loaded with appropriate ions through washing with these ions excess. Electrostatic attraction is a force that binds particles to non-conducting surfaces. Simple Coulombic attraction helps in the attraction of charged particles to oppositely charged surfaces. However, both surfaces do not need to be charged as particles can be attracted to neutral surfaces [37].

Biochar has been produced from different precursors at different conditions and tested in the laboratory to study its adsorption capacity. In recent years, it has been noticed a growing interest in converting anaerobic digestion residue to biochar [38]. Jiwei et al. [9] produced a MgO-modified biochar using chemical co-precipitation of Mg^{2+} and Fe^{3+} on biochar from anaerobic digestion residue. The modified biochar was pyrolyzed at different temperatures and tested for phosphate removal from aqueous solution. The modified biochar reached its maximum phosphate adsorption capacity of 149.25 mg/g. The influence of reaction parameters on adsorption was also investigated, including temperature, solution pH, phosphate concentration, and coexisting anions. Ci et al. [39] used corn biochar with impregnated magnesium for phosphate adsorption. The biochar was pyrolyzed at 300, 450, and 600°C and dipped in $MgCl_2$ and pyrolyzed again under same conditions. The maximum phosphate adsorption using the Langmuir-Freundlich model is 239 mg/g.

The biochar produced from anaerobic digested sugar beet tailing has been studied as a phosphate adsorbent from wastewater. Yao et al. [1] indicated that MgO particles present on the biochar surface allow better phosphate ion adsorption of 133.08 mg/g according to the Langmuir model. However, the anaerobic digestion of biochar makes it more costly as anaerobic digestion requires more time and control of the reactions.

Ningyuan et al. [13] prepared a wheat straw biochar impregnated with Bismuth at 400, 500, and 600°C for application in wastewater. The bismuth impregnation increased the specific surface area and micropores which played the role of adsorption sites for phosphorus. Bismuth biochars showed high phosphate sorption ability to phosphate with maximum P adsorption capacity of 125.40 mg/g. Hui et al. [17] prepared a hydrochar composite by hydrothermal carbonization of tobacco stalk. The hydrochar composite was prepared by adding the prepared feedstock to metal solution containing AlCl3 and MgCl2 and then putting the solution in autoclave at 180°C for 12 h. However, the highest P adsorption observed is 41.16 mg P/g at 45°C adsorption reaction time. Ming et al. [40] produced a biochar composite material which linked biochar with AlOOH nanoparticles. The biochar was produced through a slow pyrolysis of $AlCl_3^-$ pretreated biomass at 600°C. The characterization of the AlOOH biochar showed a uniform presence of AlOOH particles on biochar surface according to scanning electron microscope (SEM) studies. The Langmuir maximum capacity best describes phosphate adsorption isotherm data to be around 135 mg/g. This makes the biochar/AlOOH nanocomposite a very competitive and efficient adsorbent that can be used in the recovery of phosphate from polluted water.

Jung et al. [41] used a dried microalga as a feedstock to prepare a Mg-Al-assembled biochar. The biochar was prepared using an electro-assisted modification method by dipping the microalgae in $MgCl_2$ solution with $MgCl_2$ acting as an electrolyte. The solution pH was adjusted to 3 using NaOH solutions and 0.5 M H_2SO_4, and a current density was applied to the solid sample that was pyrolyzed at 600°C for 1 h at 5°C/min rate. This method reported the highest adsorption capacity of 887 mg/g according to the Langmuir-Freundlich model. Same authors [42] used the same electrochemical modification with changes in parameters such as using MgO nanocomposites instead of Mg-Al [41], resulting in a maximum adsorption capacity of 620 mg P/g.

5. Factors impacting phosphate adsorption

5.1 Influence of temperature

Liu et al. prepared a modified biochar using anaerobic digestion residue by chemical co-precipitation of Mg^{2+}/Fe^{3+}. Pyrolysis was performed at temperatures 500, 600, 700, and 800°C. **Figure 4** shows the effect of pyrolysis temperature on phosphate adsorption capacity. The maximum adsorption capacity on modified biochar is achieved at 600°C and adsorption decreases at 700 and 800°C. This could be explained by the disintegration of the carbon skeleton, the drop in functional groups, and reduction in surface area. It could also be explained by the pores' blockage due to their softening, carbonization, and melting during high pyrolysis temperatures [43].

On the other hand, there is no clear impact of the change of temperature on phosphate adsorption from wastewater. **Figure 5** shows no clear trend on phosphate adsorption with increasing hydrochar temperature for different tested feedstocks (W wood, D digested, and M *Miscanthus*).

5.2 Influence of pH

The influence of pH changes on P adsorption from aqueous solutions and wastewaters varies slightly between studies. Ci et al. [39] used magnesium-modified corn biochar for phosphorus removal from swine wastewater. Their study investigated the impact of initial solution pH on phosphate adsorption. **Figure 6(a)** shows that Mg-impregnated biochar adsorption increases as pH increases from 6 to 10. The adsorption reached its maximum at pH 9 and then dropped at 10. The non-modified biochar is not significantly impacted by the change in pH as it increases from 6 to 10, aside from a slight decrease in phosphate adsorption starting at pH 9. The non-modified biochar adsorption capability relied on the physical structure including the surface area, distribution and quantity of mesoporous structures, and organic functional groups. However,

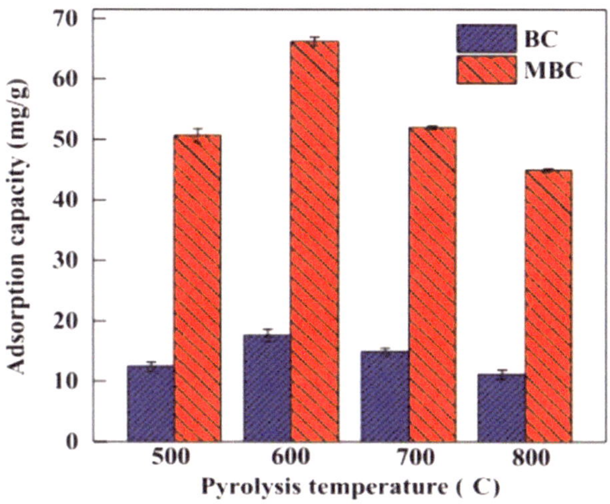

Figure 4.
The pyrolysis temperature influence on P adsorption on biochar and Mg/biochar [9].

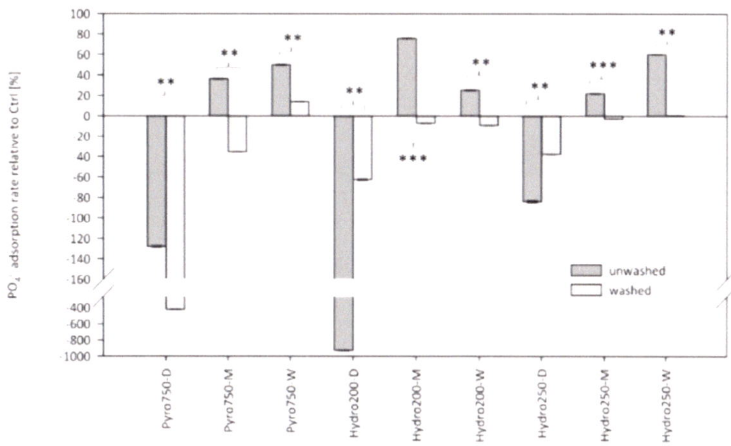

Figure 5.
The pyrolysis temperature influence on P adsorption on pyrochar and hydrochar adsorption [44].

Mg-modified biochar adsorption relies on both chemical and physical properties. P is considered as a ternary acid with the following ionization constants 2.15, 7.20, and 12.33. At an acidic solution environment (pH < 6), there is a low interaction between Mg biochar and P. $H_2PO_4^-$ is the superior in solution form at a pH above 6 and below 7.21, while HPO_4^{2-} is the superior in solution form at a pH above 7.21 and below 9. P adsorption increased with increasing chemical action, while it can decrease as adsorption sites are used. At pH between 9 and 10, P adsorption decreases as OH- competes with PO_4^{3-} on adsorption sites. The highest P adsorption amount by Mg/biochar is 239 mg/g [39].

On the other hand, Li et al. [2] found that phosphate adsorption continuously decreased when pH moved from 3.0 to 10.9 (**Figure 6(b)**). Other authors also reported that better adsorption is favored at lower pH for some adsorbents with metal oxides [11, 45]. The properties of the biochar and phosphate species distribution can explain the negative impact of the increase in pH on phosphate adsorption. At low pH levels, biochar has a more phosphate adsorption capacity.

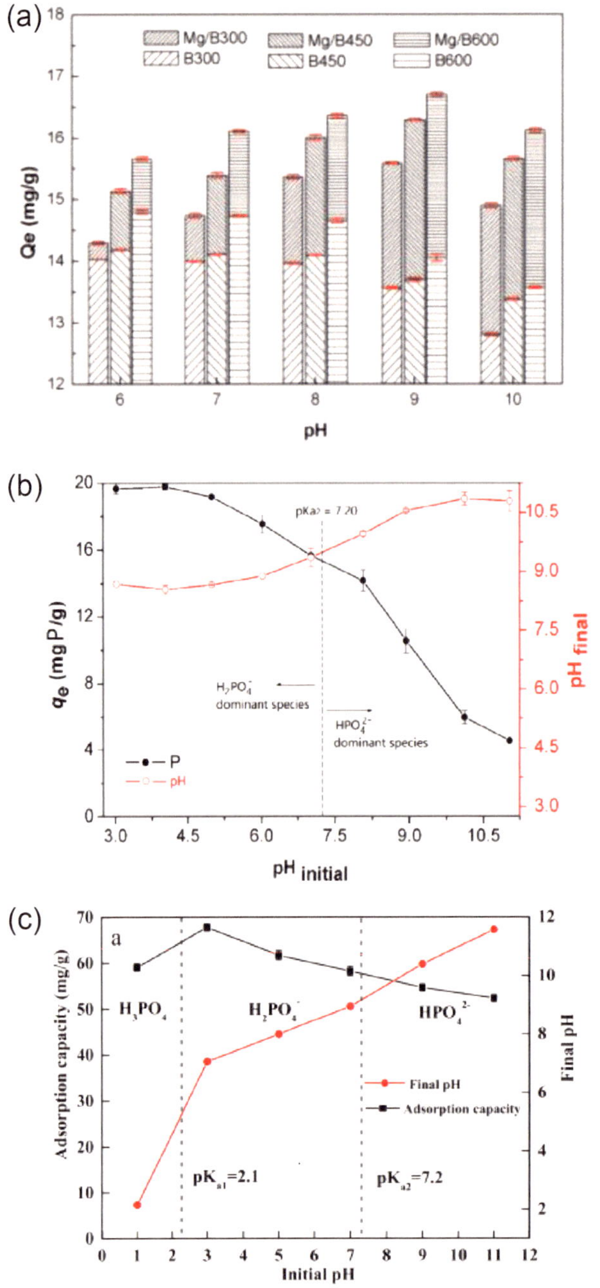

Figure 6.
The pH influence on P adsorption on biochar [2, 9, 39]. (a) effect of solution pH on the P adsorption of Mg modified corn biochar, (b) effect of solution pH on the P adsorption of Mg modified sugar cane harvest residue biochar, (c) effect of solution pH on the P adsorption of Mg modified anaerobic digestion residue biochar.

In a study by Li et al [2], phosphate existed in two forms, HPO_4^{2-} and $H_2PO_4^-$ over pH from 3 to 10.9. At lower initial pH, Mg and Fe oxides impregnated on biochar react with the solution to become $FeOH^+$ and $MgOH^+$ protons which can increase the pH of the solution. Those protons can interact with the anions HPO_4^{2-} and $H_2PO_4^-$ in an electrostatic interaction process resulting in better phosphate adsorption. The increasing pH would transform the surface to negatively charged which can cause an electrostatic repulsive interaction between phosphate anions and the surface [46]. A similar effect was observed by Liu et al. [9] who prepared a MgO-modified biochar and studied the pH effect on phosphate adsorption. **Figure 6(c)** shows that adsorption capacity increased from pH 1 to 3 and decreased from 3 to 11. The surface property is directly related to phosphate adsorption at different pH according to Li et al. [2]. As shown in **Figure 4(c)**, phosphate exists in different forms including H_3PO_4, $H_2PO_4^-$, HPO_4^{2-}, and $H_2PO_4^-$ attached better to a MgO-modified biochar in the pH ranges because it has lower free energy which resulting in higher adsorption in the pH range 2.15–7.21.

5.3 Characteristics of adsorbed surfaces

Biochar has different components: fixed carbon, labile carbon, moisture, volatiles, and ash content. The chemical environment of the carbon in the biochar is changed during the heating process allowing the production of aromatic structures that could resist microbial decomposition. Consequently, there is a stability in biochar C compounds for long periods of time that could reach thousands of years. The biochar skeletal structure consists of different pore size minerals and carbon. Micropores control high adsorption capacity and surface area, while mesopores control liquid-solid adsorption processes and macropores are responsible for the movement of roots, hydrology, aeration, and bulk soil structure. The biochar feedstock and pyrolysis temperature are directly responsible for the pattern and pore size. SEM is used to determine the biochar pore size distribution and morphology. Biochar porous structure is composed of aromatic compounds in addition to functional groups coming from lignin biomass production. This porous structure serves as channels for the flow of nutrients in solutions such as soil solutions [47].

During pyrolysis, O and H are lost to water followed by the formation of tar-rich vapors and hydrocarbons and gases (H_2, CO, and CO_2) [48]. During pyrolysis, some inorganic compounds volatilize while the major part does not as it takes part of the biochar surface. At low temperatures, N present in biomass, Cl, and k vaporize. At high temperatures, Mg, Si, and Ca are released while Mn, S, P, and Fe are retained in biochar. At pyrolysis temperature higher than 300°C, the biochar cross section appears as graphene sheets. The graphene is described as a polyaromatic, monolayer carbon atom structure produced at temperature 250–550°C, with high breakage resistance, stability index, and electrical conductivity [49]. Aromatic C-containing groups are dominant in biochars produced at temperatures 350°C and above; these are efficient adsorbents for hazardous molecules and heavy metals. P, S, H, N, and O related to the aromatic rings control the biochar electronegativity, which has a big influence on cation exchange capacity. The biochar surface charge contributes to the biochar interaction with its environment (soil, water, organic matter) [47].

At pyrolysis above 900°C, biochar surface is deformed as walls separating adjacent pores are destructed causing a widening in the micropores. Moreover, high pyrolysis temperature decreases the amount of volatile matter in the biochar and also its particle size. This results in a higher amount of graphene layers, which leads to an increase in the solid density. Overall, biochar properties depend on parameters such as heating rate, pyrolysis temperature, furnace residence time, and type of pyrolytic reactor of feedstock. Biochar derived from animal manure has more N than plant-derived biochar. On the other hand, plant-derived biochar has a more organized pore structure and was tested as a good-quality fertilizer and good heavy metal adsorbent [50]. The biochar efficiency is impacted when fungi, bacteria, or others enter the pores. The pores get clogged and the biochar adsorption capacity decreases leading to the deactivation of biochar [47].

6. Adsorption kinetic, isotherm, and thermodynamics

6.1 Adsorption kinetics

Researchers use adsorption kinetics to study phosphate adsorption over time, in terms of solute uptake rate, considered an important characteristic defining adsorption efficiency. Solute uptake by biochar can be calculated by the difference between the initial and final quantities of the solute (phosphate) concentration in the solution (mg/L) using

$$Q = V\,(C_0 - C_f)/M \tag{4}$$

where M is the mass of the biosorbent (biochar) in g, V is the solution volume (mg/L), and C_0-C_f represents the difference between the initial and equilibrium solute concentrations (mg/L) [40, 46, 51–53].

The behavior of biochar can be examined by studying phosphate adsorption kinetics. For that, experimental kinetics are calculated using mathematical models which are listed below; all these models were tested by [54]

$$\frac{dq_t}{dt} = k_1(q_e - q_t), \text{first - order} \tag{5}$$

$$\frac{dq_t}{dt} = k_2\,(q_e - q_t)^2, \text{second - order} \tag{6}$$

$$\frac{dq_t}{dt} = k_n\,(q_e - q_t)^N, \text{N\_th - order} \tag{7}$$

$$\frac{dq_t}{dt} = \alpha\,\exp(-\beta q_t), \text{Elevich} \tag{8}$$

where k_1 represents the first-order, k_2 is the second-order, and k_n is Nth-order apparent adsorption rate constants in (h^{-1}, kg/mg h, and $kg^N mg^{-N} h^{-1}$). For the Elevich model, α represents the initial adsorption rate (mg/kg), and β denotes the desorption constant (mg/kg). q_e characterizes the amount of phosphate adsorbed at equilibrium, and q_t is the phosphate adsorbed at time t, in (mg/kg). First-order, second-order, and Nth-order characterize the solid solution kinetics system based on mononuclear, binuclear, and N nuclear adsorption, respectively. The Elevich model is used if the researchers would like to consider desorption in their calculations.

Krishnan et al. [11] used pseudo-second order to study phosphate adsorption on modified coir pith at different initial phosphate solution concentrations over time. **Figure 7** shows that the initial phosphate adsorption rate increases with an increase in phosphate concentration. This can be explained by the increase in covalent interactions of the adsorbent with phosphate $H_2PO_4^-$. Similar conclusions are also drawn by [55]. Zhang et al. performed kinetic studies using all the described models above and found the best fit to be the first-order model as shown in **Figure 8** [40].

6.2 Adsorption isotherms

The following isotherms are used to simulate biochar phosphate adsorption [1, 40, 46, 52, 53]:

$$q_e = \frac{KQC_e}{1 + KC_e}, \text{Langmuir} \tag{9}$$

$$q_e = K_f C_e^n \text{ Freundlich} \tag{10}$$

$$q_e = \frac{KQC_e^n}{1 + KC_e^n} \text{ Langmuir - Freundlich} \tag{11}$$

where K represents the Langmuir bonding term for energy interactions (L/mg) and K_f represents the Freundlich affinity coefficient in $mg^{(1-n)}L^n kg^{-1}$. C_e symbolizes the equilibrium solution concentration of sorbate ($mg\ L^{-1}$). Q represents the Langmuir maximum capacity ($mg\ kg^{-1}$). The Langmuir model assumes homogenous surface and monolayer adsorption on its surface without molecule interactions, while Freundlich and Langmuir-Freundlich models are empirical equations which describe the adsorption on heterogeneous equations.

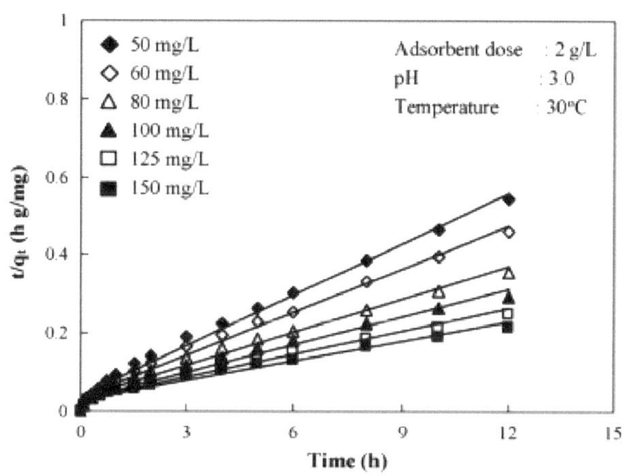

Figure 7.
Pseudo-second-order kinetic plots for phosphate adsorption on coir pith iron-modified biochar at different initial concentrations [11].

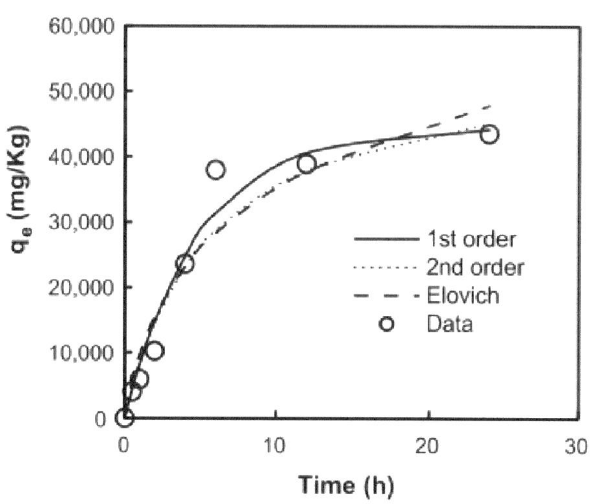

Figure 8.
Adsorption kinetic for phosphate on biochar/AlOOH nanocomposite [40].

Yao et al. [1] used these models to draw adsorption isotherm for phosphate on anaerobically digested sugar beet tailings. **Figure 9** shows that all models reproduced isotherm data correctly with correlation coefficients of 0.95. The highest adsorption capacity is presented by the Langmuir model at 133,085 mg/kg, while Freundlich and Langmuir-Freundlich models gave a better fit to the experimental data. It indicates that phosphate adsorption onto the biochar was determined by heterogeneous processes.

Zhang et al. [40] ran isotherm models of phosphate adsorption on biochar and found that both Freundlich model and Langmuir model described the isotherm data well, while the Freundlich model had a better fit for the data as shown in **Figure 10**. The maximum adsorption capacity was 135,000 mg/kg according to the Langmuir model.

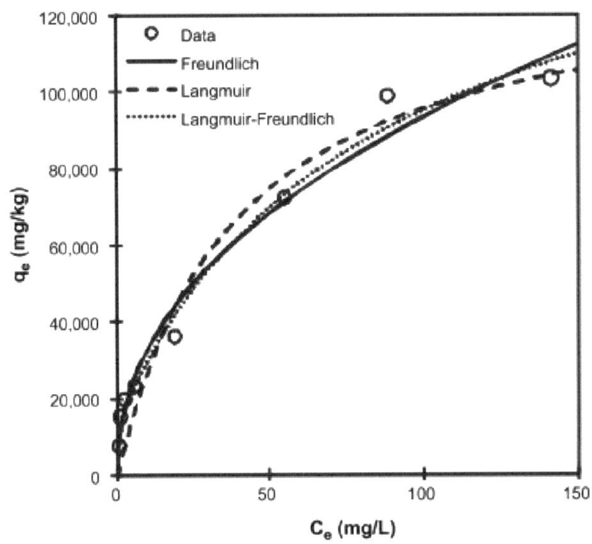

Figure 9.
Adsorption isotherm for phosphate on biochar [1].

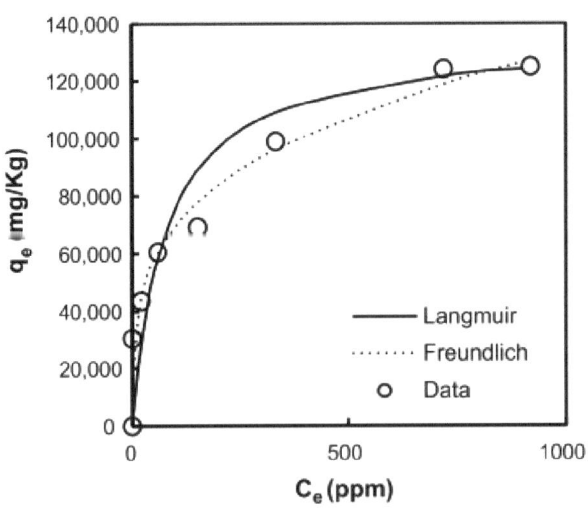

Figure 10.
Adsorption isotherm for phosphate adsorption onto biochar AlOOH nanocomposite [40].

7. Conclusion

Biomass conversion into pyrochar and hydrochar has seen a growing interest in the last years because of its use in different applications including phosphate adsorption from wastewater. Biochar has economic and sustainability benefits. In this chapter, an overview of hydrochar and pyrochar production techniques in addition to the application of biochar for phosphate adsorption from wastewater is discussed. Biochar needs to have adequate properties to be applied for phosphate adsorption from wastewater. Several factors influence the biochar properties including feedstock, pyrolysis temperature, solution pH, modification techniques, and treatment conditions. Studies have suggested that magnetic biochar has better adsorption properties than non-magnetic biochar. The biochar adsorption mechanisms are explained including ion exchange, electrostatic attraction, and chemical precipitation. Overall, biochar was proven to offer good phosphate adsorption rate along with environmental advantages such as low carbon emissions and renewability. However, further life cycle assessment studies of biochar with an evaluation of its economic benefits and environmental impacts are necessary for long-term applications.

Acknowledgements

The authors thank the USDA-CSREE-EVANS-ALLEN Project (NCX-272-5-13-130-1) for the financial support.

Author details

Aicha Slassi Sennou*, Shuangning Xiu and Abolghasem Shahbazi
Biological Engineering Program, Department of Natural Resource and Environmental Design, North Carolina A&T State University, Greensboro, North Carolina, USA

*Address all correspondence to: aslassisennou@aggies.ncat.edu

References

[1] Yao Y et al. Removal of phosphate from aqueous solution by biochar derived from anaerobically digested sugar beet tailings. Journal of Hazardous Materials. 2011;**190**(1):501-507

[2] Li R et al. Recovery of phosphate from aqueous solution by magnesium oxide decorated magnetic biochar and its potential as phosphate-based fertilizer substitute. Bioresource Technology. 2016;**215**:209-214

[3] Bensalah H et al. Removal of cationic and anionic textile dyes with Moroccan natural phosphate. Journal of Environmental Chemical Engineering. 2017;**5**(3):2189-2199

[4] Chen B, Chen Z, Lv S. A novel magnetic biochar efficiently sorbs organic pollutants and phosphate. Bioresource Technology. 2011;**102**(2):716-723

[5] Yao Y et al. Engineered carbon (biochar) prepared by direct pyrolysis of Mg-accumulated tomato tissues: Characterization and phosphate removal potential. Bioresource Technology. 2013;**138**:8-13

[6] Liu W-J et al. Mesoporous carbon stabilized MgO nanoparticles synthesized by pyrolysis of $MgCl_2$ preloaded waste biomass for highly efficient CO_2 capture. Environmental Science & Technology. 2013;**47**(16):9397-9403

[7] Arcibar-Orozco JA, Avalos-Borja M, Rangel-Mendez JR. Effect of phosphate on the particle size of ferric oxyhydroxides anchored onto activated carbon: As(V) removal from water. Environmental Science & Technology. 2012;**46**(17):9577-9583

[8] Jung K-W, Lee S, Lee YJ. Synthesis of novel magnesium ferrite ($MgFe_2O_4$)/biochar magnetic composites and its adsorption behavior for phosphate in aqueous solutions. Bioresource Technology. 2017;**245**:751-759

[9] Liu J et al. Removal of phosphate from aqueous solution using MgO-modified magnetic biochar derived from anaerobic digestion residue. Journal of Environmental Management. 2019;**250**:109438

[10] Takaya CA et al. Recovery of phosphate with chemically modified biochars. Journal of Environmental Chemical Engineering. 2016;**4**(1):1156-1165

[11] Krishnan KA, Haridas A. Removal of phosphate from aqueous solutions and sewage using natural and surface modified coir pith. Journal of Hazardous Materials. 2008;**152**(2):527-535

[12] Filippelli GM. The global phosphorus cycle. Reviews in Mineralogy and Geochemistry. 2002;**48**(1):391-425

[13] Zhu N et al. Adsorption of arsenic, phosphorus and chromium by bismuth impregnated biochar: Adsorption mechanism and depleted adsorbent utilization. Chemosphere. 2016;**164**:32-40

[14] Jiang J et al. Phosphorus removal mechanisms from domestic wastewater by membrane capacitive deionization and system optimization for enhanced phosphate removal. Process Safety and Environmental Protection. 2019;**126**:44-52

[15] Yin H et al. Phosphate removal from wastewaters by a naturally occurring, calcium-rich sepiolite. Journal of Hazardous Materials. 2011;**198**:362-369

[16] Zhang H et al. Roles of biochar in improving phosphorus availability in soils: A phosphate adsorbent and

a source of available phosphorus. Geoderma. 2016;**276**:1-6

[17] He H et al. Efficient phosphate removal from wastewater by MgAl-LDHs modified hydrochar derived from tobacco stalk. Bioresource Technology Reports. 2019;**8**:100348

[18] Kruse A, Funke A, Titirici M-M. Hydrothermal conversion of biomass to fuels and energetic materials. Current Opinion in Chemical Biology. 2013;**17**(3):515-521

[19] Wilk M, Magdziarz A. Hydrothermal carbonization, torrefaction and slow pyrolysis of Miscanthus giganteus. Energy. 2017;**140**:1292-1304

[20] Liu Z, Balasubramanian R. Upgrading of waste biomass by hydrothermal carbonization (HTC) and low temperature pyrolysis (LTP): A comparative evaluation. Applied Energy. 2014;**114**:857-864

[21] Oliveira I, Blöhse D, Ramke H-G. Hydrothermal carbonization of agricultural residues. Bioresource Technology. 2013;**142**:138-146

[22] Sabio E et al. Conversion of tomato-peel waste into solid fuel by hydrothermal carbonization: Influence of the processing variables. Waste Management. 2016;**47**:122-132

[23] Zhang Z et al. Insights into biochar and hydrochar production and applications: A review. Energy. 2019;**171**:581-598

[24] Fuch M. Literature Review of Pyrolysis Reactors. United States: Washington State Department of Ecology; 2011

[25] Liu W-J, Jiang H, Yu H-Q. Development of biochar-based functional materials: Toward a sustainable platform carbon material. Chemical Reviews. 2015;**115**(22):12251-12285

[26] Al Arni S. Comparison of slow and fast pyrolysis for converting biomass into fuel. Renewable Energy. 2018;**124**:197-201

[27] Duku MH, Gu S, Hagan EB. Biochar production potential in Ghana—A review. Renewable and Sustainable Energy Reviews. 2011;**15**(8):3539-3551

[28] Demirbas A. Effects of temperature and particle size on bio-char yield from pyrolysis of agricultural residues. Journal of Analytical and Applied Pyrolysis. 2004;**72**(2):243-248

[29] Qian K et al. Recent advances in utilization of biochar. Renewable and Sustainable Energy Reviews. 2015;**42**:1055-1064

[30] Hua M et al. Heavy metal removal from water/wastewater by nanosized metal oxides: A review. Journal of Hazardous Materials. 2012;**211-212**:317-331

[31] Loganathan P, Vigneswaran S, Kandasamy J. Enhanced removal of nitrate from water using surface modification of adsorbents – A review. Journal of Environmental Management. 2013;**131**:363-374

[32] Yin Z et al. Activated magnetic biochar by one-step synthesis: Enhanced adsorption and coadsorption for 17β-estradiol and copper. Science of the Total Environment. 2018;**639**:1530-1542

[33] Xia P et al. Struvite crystallization combined adsorption of phosphate and ammonium from aqueous solutions by mesoporous MgO-loaded diatomite. Colloids and Surfaces A: Physicochemical and Engineering Aspects. 2016;**506**:220-227

[34] Le Corre KS et al. Phosphorus recovery from wastewater by struvite

crystallization: A review. Critical Reviews in Environmental Science and Technology. 2009;**39**(6):433-477

[35] Chrispim MC, Scholz M, Nolasco MA. Phosphorus recovery from municipal wastewater treatment: Critical review of challenges and opportunities for developing countries. Journal of Environmental Management. 2019;**248**:109268

[36] Ahmed MB et al. Progress in the preparation and application of modified biochar for improved contaminant removal from water and wastewater. Bioresource Technology. 2016;**214**:836-851

[37] Welker RW. Developments in Surface Contamination and Cleaning: Detection, Characterization, and Analysis of Contaminants. Elsevier; 2012. ISBN: 978-1-4377-7883-0

[38] Bogusz A et al. Synthesis of biochar from residues after biogas production with respect to cadmium and nickel removal from wastewater. Journal of Environmental Management. 2017;**201**:268-276

[39] Fang C et al. Application of magnesium modified corn biochar for phosphorus removal and recovery from swine wastewater. International Journal of Environmental Research and Public Health. 2014;**11**(9):9217-9237

[40] Zhang M, Gao B. Removal of arsenic, methylene blue, and phosphate by biochar/AlOOH nanocomposite. Chemical Engineering Journal. 2013;**226**:286-292

[41] Jung K-W et al. Phosphate adsorption ability of biochar/Mg–Al assembled nanocomposites prepared by aluminum-electrode based electro-assisted modification method with $MgCl_2$ as electrolyte. Bioresource Technology. 2015;**198**:603-610

[42] Jung K-W, Ahn K-H. Fabrication of porosity-enhanced MgO/biochar for removal of phosphate from aqueous solution: Application of a novel combined electrochemical modification method. Bioresource Technology. 2016;**200**:1029-1032

[43] Jung K-W et al. Influence of pyrolysis temperature on characteristics and phosphate adsorption capability of biochar derived from waste-marine macroalgae (Undaria pinnatifida roots). Bioresource Technology. 2016;**200**:1024-1028

[44] Gronwald M et al. Effects of fresh and aged chars from pyrolysis and hydrothermal carbonization on nutrient sorption in agricultural soils. The Soil. 2015;**1**(1):475-489

[45] Zhang L et al. Removal of phosphate from water by activated carbon fiber loaded with lanthanum oxide. Journal of Hazardous Materials. 2011;**190**(1):848-855

[46] Ren J et al. Granulation and ferric oxides loading enable biochar derived from cotton stalk to remove phosphate from water. Bioresource Technology. 2015;**178**:119-125

[47] Qambrani NA et al. Biochar properties and eco-friendly applications for climate change mitigation, waste management, and wastewater treatment: A review. Renewable and Sustainable Energy Reviews. 2017;**79**:255-273

[48] Antal MJ, Grønli M. The art, science, and Technology of Charcoal Production. Industrial & Engineering Chemistry Research. 2003;**42**(8):1619-1640

[49] Geim AK, Novoselov KS. The rise of graphene. Nature Materials. 2007;**6**(3):183-191

[50] Uchimiya M et al. Influence of pyrolysis temperature on biochar

property and function as a heavy metal sorbent in soil. Journal of Agricultural and Food Chemistry. 2011;**59**(6):2501-2510

[51] Tafakori V et al. Equilibrium isotherm, kinetic modeling, optimization, and characterization studies of cadmium adsorption by surface-engineered *Escherichia coli*. Iranian Biomedical Journal. 2017;**21**(6):380-391

[52] Özacar M. Adsorption of phosphate from aqueous solution onto alunite. Chemosphere. 2003;**51**(4):321-327

[53] Zhang M et al. Phosphate removal ability of biochar/MgAl-LDH ultra-fine composites prepared by liquid-phase deposition. Chemosphere. 2013;**92**(8):1042-1047

[54] Gerente C et al. Application of chitosan for the removal of metals from wastewaters by adsorption-mechanisms and models review. Critical Reviews in Environmental Science and Technology. 2007;**37**(1):41-127

[55] Namasivayam C, Prathap K. Recycling Fe(III)/Cr(III) hydroxide, an industrial solid waste for the removal of phosphate from water. Journal of Hazardous Materials. 2005;**123**(1):127-134

Section 3

Applications of Biochar for Soil Fertility Improvement

Chapter 6

Biochar Application for Improved Resource Use and Environmental Quality

Stephen Yeboah, Patricia Oteng-Darko, Joseph Adomako and Abdul Rauf Alhassan Malimanga

Abstract

Agroecosystems have become very important not only for their role in achieving food security but also in mitigation of greenhouse gas emissions. This agro-ecological function has become very important since society expects agriculture to be more sustainable, by decreasing fertilizer inputs while reducing greenhouse gas emission. Mitigation measures to reduce net GHG emissions include increasing soil C sequestration by reduced tillage, biochar and straw application, and increased crop-use efficiency of fertilizer-N. An adequate management of soils and crops could result in a reduction of GHG emissions through complex interactive factors. However, which factors are mainly responsible for the differences in emissions across soil and environment type remain unclear and the mechanism underlying GHG emissions are complex. It is therefore imperative to determine how biochar could mitigate greenhouse gas emissions without adverse effect on crop yield. This chapter will predominantly review biochar application for improved resource use and reduce greenhouse gas emission in sub-Saharan Africa, except in some cases where specific mechanisms have been elucidated in other regions. We consider future perspectives on whether biochar application offers economic opportunities for smallholder farmers in developing countries, with a particular focus on Ghana. These issues provided the motivation for this chapter.

Keywords: biochar, greenhouse gas, resource use, crop productivity, soil quality

1. Introduction

Global demand for food will increase during the coming decades, yet agricultural systems are already strained across the globe. The agricultural systems are either too extensive or intensive, which is a threat to sustainable food production. Agriculture is also both a major contributor to greenhouse gas emission (GHG) and very susceptible to climate change. This increases the urgency for making agriculture climate smart, both through adaptation and mitigation by reducing GHG emissions. At the same time, arable land and other production resources are limited, and area expansion for food production is not desirable. Increasing the overall production of agricultural productivity without further increase of the area used for food production or its environmental footprint in a climate smart way is therefore essential. This means that food production and the type of food produced has to change to conform to good practices.

Sustainably producing sufficient, safe, and nutritious food implies that we should focus on increasing the efficiency along the production chain and across multiple resources (including land, water, nutrients, energy, labor) and recapturing waste into useful resource such as fertilizer and pesticides. This requires a radical change from the traditional linear "take-make-use-waste-recycle production model toward a sustainable production system with optimal use of resources and full reuse of wastes as shown in **Figure 1**.

Biochar is a carbon-rich residue that is important for an optimal use of resources with a focus on the lowest footprint per unit of quality food. Biochar is a recalcitrant source of C, which when applied to the soil slows down the turnover of native SOC, enhances the use efficiency of applied fertilizer-N, and therefore, reduces fertilizer-induced GHG emissions [1]. The soil incorporation of crop residues, particularly with high C/N ratio, improves soil organic C levels, enhances biological activity, and increases nutrient availability [2]. Recalcitrant C-rich biochar is a suitable means to mitigate climate change and improve soil fertility [3] and crop productivity [4]. These functions of biochar are collaborated by Yeboah et al., who reported improved soil organic carbon and moisture when biochar was applied in semi-arid Loess plateau of China. However, the effects have been shown to vary depending upon the type of biochar used and the environmental and soil conditions under which the material is applied.

These responses have limited widespread use of such management practices on cropping lands. Varied results have been obtained depending on soil and environmental conditions under which the technology is applied. The research results achieved are very diverse, and it is credible that the application of sustainable soil management technologies such as biochar, residue, and farmyard manure could imply higher yields of crops and lower greenhouse gas emissions compared to conventional agricultural practices. In addition, application of these technologies as a GHG mitigation strategy requires the understanding of the mechanism underlying the production of the greenhouse gas emission and developing the necessary

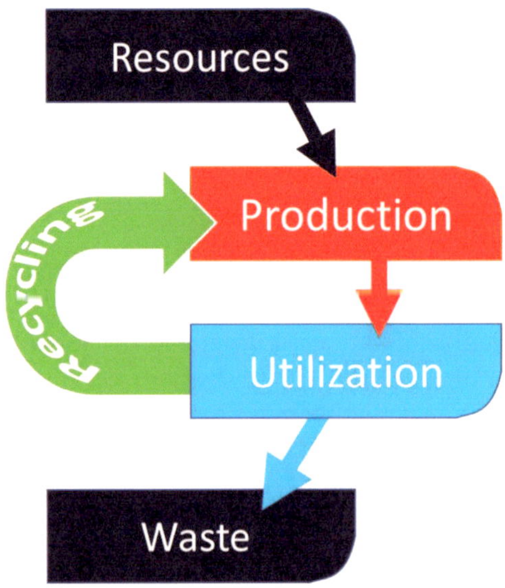

Figure 1.
Production system with recycling. Authors' personal communication.

component technologies to reduce the emissions without confounding effect on the agricultural yields. We consider future perspectives on whether biochar application offers economic opportunities for smallholder farmers in developing countries by considering the case study of Ghana. These issues provided motivation for this chapter.

2. Biochar for C-use efficiency

It is well known that the global atmospheric concentrations of the major greenhouse gases (GHGs) have been increasing [5], the largest coming from agriculture and land-use changes like urbanization and industrialization. This is an important issue in agriculture, both because of the impacts on agricultural production and agriculture being a major contributor to buildup of greenhouse gases in the atmosphere [4]. But net GHG emissions from farming-related activities can be reduced by increasing carbon (C) sequestration in soil and crop biomass. Current increases in atmospheric GHG levels require that novel approaches are undertaken to mitigate impacts of climate change, such as management practices conducive to improved soil C sequestration [6]. Recently, different means have been proposed to increase soil C in soil and thus decrease CO_2 emission. One such mitigation strategy is to sequester atmospheric CO_2 captured through photosynthesis in biomass and convert into a more stable form of carbon called biochar.

The sequestration of C and N in soils could be achieved through the adoption of crop residue retention. In drylands, the application of crop residues, among other measures, is recommended for the management of soil organic matter [7]. The application of biochar plays a dual role of sequestrating organic C and enhancing soil productivity [8], mainly because biochar contains high C content and could protect organic C from utilization. A similar result was found by Yeboah et al. [4] as shown below (**Figure 2**).

The significance of retaining crop residues was emphasized in the study by the difference of organic C between the organic amended soils. The authors attributed the increased C content in soil to its high C content and the fact that biochar could slow down organic C utilization by microbes. The higher organic C produced by the biochar-treated soils could be related to its ability to stabilize the native carbon

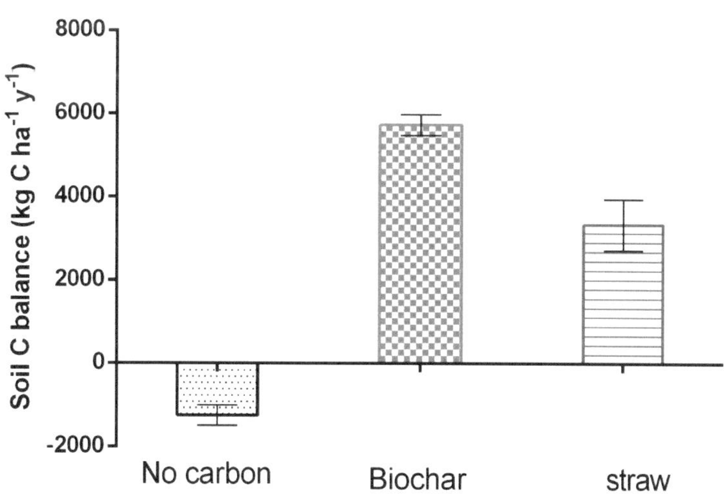

Figure 2.
Soil C balance under different treatments. Data replotted from [4] thesis, unpublished.

and recalcitrant to microbial decomposition. The resistance of biochar to microbial decomposition is dependent on its chemical composition resulting from the heat treatment and properties of the initial biomass [8]. There is consensus that biochars produced at higher temperatures contain more aromatic structures, which confer intrinsic recalcitrance. Due to its high surface area and porosity, biochar may also stabilize native SOC by reducing microbial activity as reported by several other researchers [7, 8].

Biochar amendment could exert high carbon recalcitrance against microbial decay, which in turn may reduce emission of GHGs. However, the effect of biochar on carbon emissions in soils are very complex and changes in emissions can be a response of diverse mechanisms. It is believed that biochar amendment affects CO_2 emissions by changing the characteristics of the soil and of the microbial diversity [8]. The increase in soil microbial biomass could be due to the increased C-use efficiency following the accumulation of soil organic C and microbes on the biochar surface. This is possible since crop residues serve as a precursor of the soil organic C pool, and returning more crop residues to the soil in the form of biochar is associated with increases in organic C concentration [2]. This function is particularly important in the stressful environment like water- and nutrient-limited conditions where the factors that limit yields of global agriculture production are many.

3. Biochar increases crop productivity

The challenge of meeting the demand for food has received great attention worldwide. The current increases in food production in the last four decades may be due to increased N fertilization and area of cultivation. However, the increased utilization of agricultural lands including the indiscriminate use of N fertilizer has resulted in negative effects on agriculture, socio-economic and environmental quality such as global warming [9]. Agricultural and environmental sustainability issues have stimulated attempts to increased crop yields while decreasing N fertilization. The potential to increase C inputs to soils is associated with high yield agriculture. It is within this framework that the ability to develop and implement innovative soil management practices becomes a key to improving the productive capacity of soils. This will enhance the resilience of the agroecosystem which is a key priority to crop production.

Aside the carbon sequestration potential of biochar amendment to agricultural soils, the production of biochar and its application to soil will deliver immediate benefits through increased crop production [10]. Biochar additions to agricultural fields are expected to increase yields [11] and reduce loss of nutrients [10]. A reduced number of studies have examined application of different carbon sources and patterns on crop productivity of loess soils.

In a study by Yeboah et al. [4], the greatest grain yield of spring wheat was recorded on biochar-treated soils and the lowest on soils without carbon amendment (**Figure 3**).

Yield increases with biochar application have been documented in controlled environments as well as in the field [7, 11], and several underlying mechanism have been attributed to this phenomenon. Bruun et al. [12] noted that improvements to the habitat for beneficial soil microbes are the most likely causes of productivity improvements associated with the application of biochar. However, other authors [9, 13] have reported that when biochar and inorganic fertilizers are applied together, an increased nutrient supply to plants may be the most important factor in increasing crop yields. The mechanism may be complex but the effect of biochar on soil quality could be prominent in influencing yield. It is therefore opined that

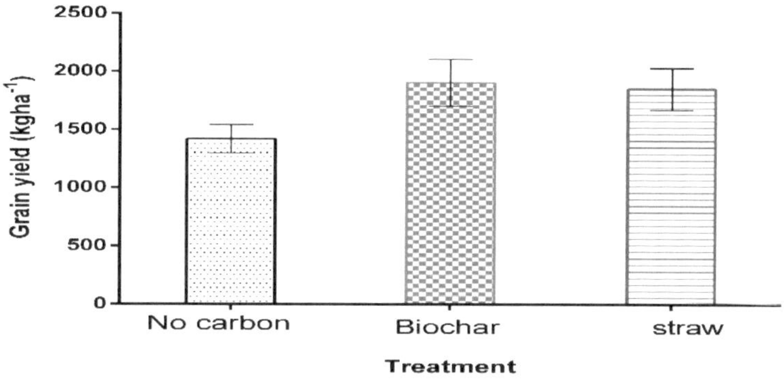

Figure 3.
Effect of different carbon sources on grain yield of spring wheat. Data replotted from [4].

improved crop yields in biochar amended study could be attributed to increased nutrient availability through enhanced soil quality. Our study evidenced a positive effect of biochar amendment on soil quality and spring wheat yield consistent over 3 years.

Such sustainable increasing effect could also be supported by other field experiments on dry croplands. Steiner et al. [14] reported cumulative yield increases of rice and sorghum in Brazil after four cropping seasons when 11 t ha^{-1} biochar made from rice straw was applied. Steiner et al. [14] found increased maize yield after three repeated maize stubble biochar applications of 7 t ha^{-1} of over 2 years. However, Asai et al. [15] reported a decreased grain yield following the application of biochar amendment without N fertilization in a soil that had poor N availability.

4. Biochar for mitigating greenhouse gas emissions

In an effort to reduce the concentrations of greenhouse gases (GHGs) in the atmosphere in order to reduce the potential effects, considerable attention has been paid to soil management practices. According to Snyder et al. [16], improving cropland management practices such as reduced tillage and residue retention has the potential to reduce agricultural greenhouse gas emission irrespective of type of cultivation. Recent increases in atmospheric GHG levels require that novel approaches are undertaken to mitigate impacts of climate change, such as management practices conducive to improved soil C sequestration [17]. Nitrogen fertilization and crop residue retention play a major role in GHG emission. Soil carbon sequestration through the application of recalcitrant C-rich biochar is mentioned as a suitable means to mitigate climate change.

A number of mechanisms have been proposed in the literature to explain the effect of biochar amendment on soil N_2O emissions, with limited amounts of evidence to support them. Since biochar has significant impact on soil environment and affects many soil parameters, such as pH or the availability of soil minerals, it is very likely that biochar will have significant effects on the production of N_2O. Diverse studies confirm this—most of them reporting reduced N_2O emissions from soil following biochar application [18], similar effect in the field [19], and no suppression of soil N_2O emissions [20]. Biochar amendment has been observed to modify soil physical properties such as reduced soil bulk density or increased water holding capacity, therefore increasing soil aeration. This may lead to lower soil N_2O emissions as soil aeration influenced both nitrifier and denitrifier activity. By changing the physical properties of the soil, biochar may suppress N_2O production from denitrification by increasing the air content of the soil or by absorbing water from the soil, thus improving aeration of the soil [18]. Biochar amendment may increase soil pH (**Figure 4**) when applied to soil [11]. Changes in soil pH may result in changes in nitrifier or denitrifier enzymatic activity and therefore soil N_2O emissions.

There is limited evidence, mostly from studies in rice paddies to suggest that biochar amendment affects soil CH_4 emissions [20]. The greater uptake of methane may be attributed to the protected environment created for the CH_4 oxidizers and improved soil porosity. The aerobic, well-drained soils can be a sink for CH_4 due to the possible high rate of CH_4 diffusion and ensuing oxidation by methanotrophs. Improved soil physical properties such as low bulk density and the associated increase in total porosity, mostly due to the relative increase in macroporosity [21], may significantly decrease CH_4 emissions. Increased availability of labile C substrates following biochar addition stimulates the activities of methanogenic bacteria that may account for increased CH_4 emissions [22]. However, this could be a short-term effect since labile carbon fraction in biochar could be mineralized rapidly. Karhu et al. [23] observed increased soil CH_4 consumption in arable soil due to increased soil aeration following biochar application. Biochar addition to soil has been assumed to increase soil temperature and soil pH. However, the effect of biochar on soil temperature and soil pH has not been suggested as mechanisms to explain differences in overall soil CH_4 [10].

Carbon dioxide is produced mainly from the decomposition of plant residues and organic matter by soil microbes and respiration from microbes and roots. Carbon dioxide can be divided into autotrophic and heterotrophic respiration based on different biological sources [10]. The effect of biochar on carbon emissions in soil are very complex and changes in CO_2 emissions can be a response of diverse mechanisms. Biochar amendment affects CO_2 emissions by changing the physical and chemical characteristics of the soil and of the microbial diversity [10]. CO_2 emission could be reduced through the effect of biochar application on C-mineralizing enzymes [25]. Some studies have shown that biochar addition could stimulate the mineralization of soil organic carbon (SOC) [11, 12] and correspondingly increase emissions of CO_2. However, conversely, the suppression of SOC mineralization has also been reported [25], thereby causing a decrease in CO_2 emissions. Biochar application could also stimulate CO_2 emission by enhancing soil properties [10]. As indicated in Refs. [16, 22], transforming carbon in plant residues into stable

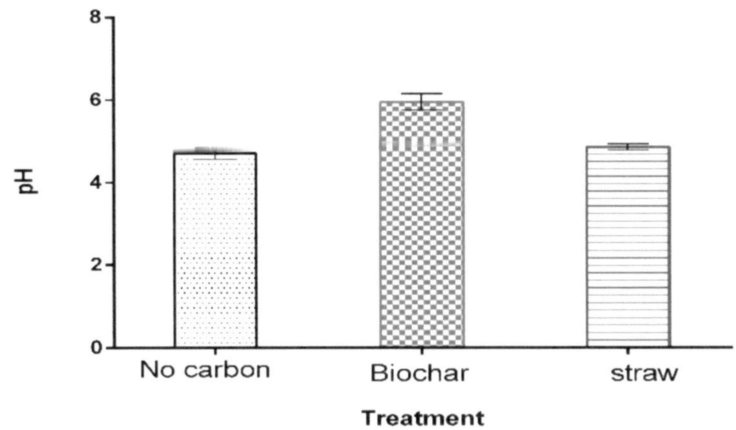

Figure 4.
Soil pH of different treatments. Data replotted from [24].

C form is the main role for decreasing CO_2 emission compared to the natural plant residues. Therefore, the mechanism underlying biochar effect on CO_2 emission is unclear, because it could have occurred because of several interactive factors.

5. Biochar for soil fertility

The decline of soil fertility is a major problem confronting crop production and environmental sustainability. The functions of soil depend on three main properties, physical, chemical, and biological, which influence global cycles of organic C and N [26]. The adoption of sound soil management strategies such as appropriate tillage methods, crop residues practices, biochar application, and efficient N fertilization has been suggested to improve soil properties [27]. These strategies can be achieved by increased input of crop residues while minimizing C loses by erosion, decomposition, and carbon emission. While conservation agriculture systems have been noted to improve soil organic C [27], conventional plow-based farming systems could accelerate carbon mineralization and thus reduce soil C content, which are attributed to soil aggregates disruption and increased oxidization through soil disturbances [28]. The incorporation of biochar into soil varies soil structure, porosity, and bulk density. According to Oguntunde et al. [29], this may in turn have consequences for important soil functions such as soil aeration and plant growth. In Ref. [30], it is postulated that biochar application results in an increase in soil C.

The expectation of increased soil fertility attributed to biochar application emanated from the studies of the terra preta that contains high proportions of black carbon [29]. The high soil organic matter content of the terra preta provides the evidence of the enhancement due to biochar application. In contrast to mainstream chemical fertilizer, biochar also contains bioavailable elements such as selenium that has potential to assist in enhancing crop growth. It is not clear concerning the potential effects of biochar on microbial activity in soil. Assuming that plant inputs and hence microbial substrate remain unchanged, enhanced microbial activity alone would diminish soil organic matter. However, this is contrary to the observation in terra preta, where soil organic matter is generally higher than in similar surrounding soil [26]. However, a change in the balance of microbial activity between different functional groups could benefit crop nutrition, specifically enhancement of mycorrhizal fungi [11], and this could feedback into higher net primary productivity and carbon input.

There are several reasons why biochar might be expected to decrease the potential for nutrient leaching in soils, and thus enhance nutrient cycling and also protect against leaching loss. In field studies where positive yield response to biochar application has been observed [7, 13], enhanced nutrient dynamics could be the reason for the observed results.

6. Biochar improves nitrogen use efficiency

Biochar is a recalcitrant source of C, which when applied to the soil slows down the turnover of native SOC, enhances the use efficiency of applied fertilizer-N, and, therefore, reduces fertilizer-induced GHG emissions [1]. Biochar improves N use efficiency through indirect processes including the improvement of soil conditions to maximize nitrogen uptake. This means through the application of biochar, nitrogen in the soil is conserved or the nitrogen that is applied through fertilization (added nitrogen) is conserved. Calys-Tagoe et al. [24] found higher N conserved

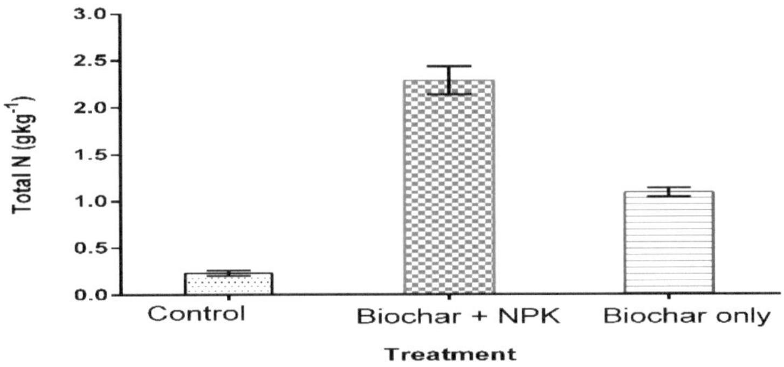

Figure 5.
Total N content of different treatments. Data replotted from [24].

in the soil when biochar was applied in combination with nitrogen fertilizer (**Figure 5**). Biochar amendment is said to conserve N in the soil and improve soil biological and physicochemical properties, which increases the ability of the soil to utilize nitrogen and other essential nutrients in the soil.

There are several reasons to expect that biochar might improve nitrogen use efficiency. It is believed that biochar improves nitrogen uptake by plants through the following ways: (1) retaining N in soil through adsorption of ammonia (NH_3) and ammonium (NH_4^+), (2) reducing nitrous oxide (N_2O) emissions, (3) reducing nitrate (NO_3^-) leaching, and (4) enhancing biological fixation of N in the soil [19]. These functions of biochar in soils have shown to increase nitrogen uptake by plants [37] and improve nitrogen use efficiency [19], either through the action of biochar as fertilizer and/or the improvement of soil condition for enhanced nitrogen uptake. Biochar application resulted in increased N uptake by plants, which was attributed to the ability of biochar to supply more N.

The improved N use efficiency function of biochar is very significant as it can impact the overall carbon balance of agricultural activities. A low ratio of fertilizer nitrogen application to crop nitrogen uptake can, therefore, influence the overall C balance. Higher fertilizer use efficiency may lead to a lower fertilizer requirement per unit yield and usually lower nitrous oxide emission resulting in improved environmental quality.

7. Biochar for pests and disease suppression

Crop pests and diseases constitute important threat to food security and reduce income levels along the value chain of both producers and consumers worldwide. With the increasing global population, the need to maximize food production to feed the world is of utmost importance. To achieve this, there is the need to minimize losses caused by pests and diseases to mitigate the possibility of global food insecurity. Controlling biotic stresses to maximize crop production must, however, be sustainable and environmentally friendly. In view of this, sustainable crop production through integrated pest and disease management (IP&DM) strategies has gained worldwide recognition and the use of organic soil amendments (OSA) such as biochar is an integral tool for the success of this strategy. Apart from managing pest and disease pathogens, OSA have the potential of improving soil tilth, nutrient availability, water-holding capacity, soil microbial diversity, and population dynamics, and reduces nutrient-leaching loss [31].

The desire to effectively control plant diseases without compromising environmental quality and safety has increased the demand for research into finding sustainable ways of managing pests and diseases. In the context of sustainability, research into the use of organic substances such as animal manure, green manure, organic agro wastes, and compost management have gained tremendous interest among researchers and various stakeholders. OSA do not only have the advantage of improving soil structure and quality but also found to increase the suppressive potential of treated soils thereby inhibiting diseases caused by plant pathogens [32]. OSA have successfully been used to reduce the activities of several plant pathogens such as plant parasitic nematodes, *Fusarium* spp., *Phytophthora* spp., *Pythium* spp., *Rhizoctonia solani*, *Sclerotinia* spp., *Sclerotium* spp., and *Verticillium dahlia* [32]. The use of untreated OSAs have, however, been found to exacerbate activities of disease pathogens and can increase their incidence and severity. Also, the release of certain phytotoxic compounds such as xanthatin and 4-epiisoxanthanol during the decomposition process can damage the roots of plants [32] and make them vulnerable to pathogen attack [33]. The benefits of OSA however, outweigh the constraints and the search for numerous agents for use as OSA is continuous with reports emanating from several research works globally. One such OSA agent that has received global attention and interest is the biochar. This is obtained from the slow pyrolysis of biomass in the absence of air. It is a by-product from the biofuel industry and a high-carbon material [34]. As an OSA, there has been an increased interest in its importance as soil health and disease management agent. Current studies have proven its role in carbon sequestration leading to the removal of desirable carbon from the atmosphere. In addition to this, current studies have pointed to the fact that apart from increasing the cation exchange capacity in organic matter deficient soils, improving the pH status of acidic soils, and increasing nutrition and water holding capacities of soils, it also enriching microorganisms in the soil that improve soil suppression potential against pathogens. Suppressive soils are able to inhibit disease development by stimulating biota activity, increasing and favoring populations of biocontrol agents, and reduce the inoculum potential of both foliar and soilborne pathogens [35]. With regard to soilborne pathogens, biochar has successfully been used to suppress the activities and infection capabilities of pathogens. Data available shows that the application of biochar reduced the infection potential of *Meloidogyne graminicola* in rice [36] and other plant-parasitic nematodes, while it increased the population of free-living nematodes that may be beneficial to the improvement of soil health. Bonanomi et al. [37] reported that the application of biochar was found to be effective in suppressing foliar pathogens such as *Rhizoctonia solani*, *Fusarium*, and *Phytophthora* species. Similarly, the application of biochar was reported to reduce the severity of gray mold disease on both *Lycopersicon esculentum* and *Capsicum annuum* [38]. Also, *Fusarium* root rot disease incidence in asparagus reduced following the application of biochar inoculated with mycorrhizal fungi [39]. Nerome et al. [40] reported that the application of biochar obtained from municipal organic waste inhibited the infection capacity of *Ralstonia solanacearum* to cause disease, increasing the advocacy to use biochar and its amended composts to control fungi and bacterial disease in crop production. The disease suppression potential of biochar is influenced by several mechanisms. According to Rawat et al. [41], the effectiveness of biochar to inhibit disease is linked to the presence of calcium compounds as well as improving the physical, chemical, and biological characteristics of the soil. Similarly, Noble and Coventry [42] hypothesized the induction of systemic resistance in the host plants, enhanced abundance and activities of beneficial microbes, modification of soil quality in terms of nutrient availability, and abiotic conditions; direct fungitoxic effect and sorption of allelopathic and phytotoxic compounds as mechanisms by which biochar suppresses plant diseases.

Different pesticides are used in crop production to reduce the impact of targeted agent. In addition to its disease suppressive potential, biochar also reduces the impact of pesticides on the environment through the absorption and adsorption of different pesticides [43] to reduce the bioavailability of pesticides due to its large surface area and high porosity. The sorption capacity of biochar amendment is however dependent on age as Martin et al. [44] reported that the herbicide, atrazine adsorption in biochar amended soil decreased with the age of biochar an indication that aging has influence on its usefulness in pesticide sorption. Biochar can again protect roots of plants from phytotoxic compounds in the soil released by roots of other plants, through the decomposition of plant residues and soil amendment as well as agro-waste products and immature compost [45]. The high sorption rate of biochar, on the other hand, can negatively affect the efficacy of agrochemicals and increase the application rates of pesticides.

8. Biochar application: a case study in Ghana

In recent years, agricultural growth in Ghana has seen accelerated growth, but most of this growth is driven by the expansion of the cultivated area rather than by increased yield per unit area. It has been suggested that yield should be increased by at least 20% annually across the staple crops to meet the food needs of the people [46]. This is envisaged to be difficult with the current impact of climate change (evident by the rising temperatures and increased mid-season drought), resource scarcity, and environmental degradation. In sub-Saharan African (SSA), soil fertility decline, mainly through continuous cropping and rapid organic matter mineralization, is the main cause of food insecurity and poverty. Smallholder agricultural production systems in most SSA countries including Ghana are characterized by low productivity due to low and erratic rainfall patterns, outdated agricultural practices, and low application of nutrient inputs. To intensify agriculture, chemical fertilizer is highly utilized.

To pursue the fastest and most practical route to sustainable food production, substantial improvement in crop and soil management practices, which are currently suboptimal, is required. The deployment of soil management technologies such as biochar application is a surest means to reversing the rapid decline in soil fertility in sub-Saharan Africa, particularly Ghana. The large availability of biomass resources in Ghana gives a great potential for biochar production and utilization in the country. This is very important in the tropics since turnover rates of organic matter are much faster. Waste management as a social problem has spared neither the developed nor developing nations as statistics have proven that some developed nations are seriously grappling with this bane. Ghana produces 1.7 billion tons of waste annually (source: ghananewsagency.org).

Biochar research is a recent development in Ghana and so there is a paucity of information regarding its effects on soil properties, crop growth, and yield in Ghanaian soils. The effect of biochar and/or compost applications on the soil pH of the Aiyinase and Cape Coast soils in Ghana after the 14-day incubation period is shown in **Figure 6**. The results revealed a significant increase in the soil pH, following sole and combined applications of compost and biochar in both soils. Also, the application of biochar and compost, alone or in combination, increased soil total organic carbon (TOC) contents in both the Aiyinase and Cape Coast soils (**Figure 6**). The application of biochar significantly improved soil chemical properties with reference to the control.

A case study in Ghana illustrates the significance of biochar application in augmenting water retention in a dryland crop (**Figure 7**). A farmer reports 100%

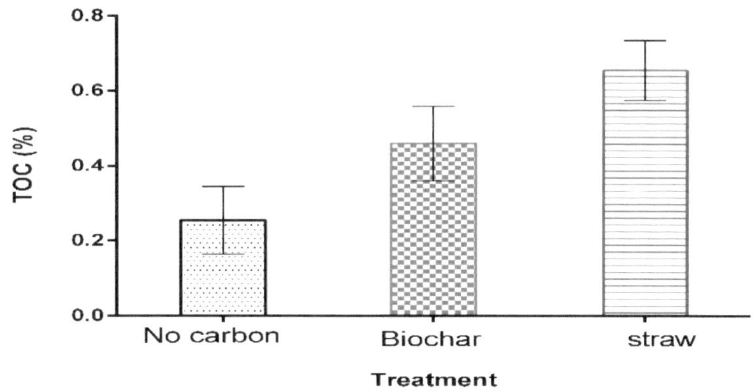

Figure 6.
Total organic carbon as affected by different treatments. Data replotted from [24].

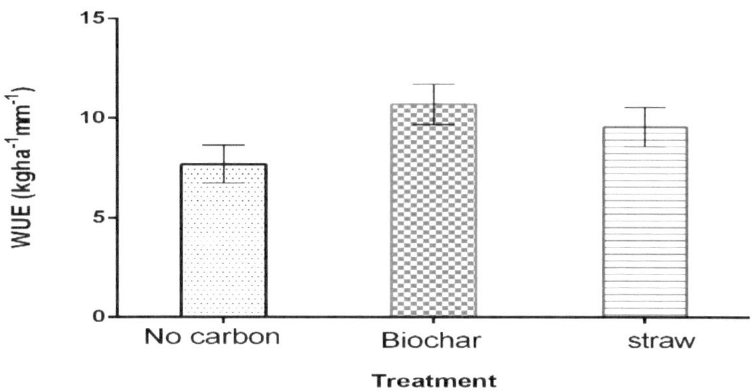

Figure 7.
Total organic carbon as affected by different treatments. Data replotted from [4].

increase in yields. "Her perception is that the underlying mechanism for the effects she sees is entirely physical, citing two factors: enhanced rainwater infiltration and enhanced soil moisture retention" [10]. Given the fact that drought-susceptible sandy soils are prevalent in Ghana, crop performance is much influenced by rainfall (timing and intensity).

The results of studies in Ghana indicate the potential exist for farmers to produce their own biochar on-farm, although various factors must be considered in deciding whether this methodology is appropriate in a particular context. Factors such as capacity of farmers determine the pyrolysis conditions and activation methods.

9. Conclusion

Rainfed agricultural ecosystem in Ghana is extremely fragile, improving soil fertility and crop productivity, and reducing greenhouse gas emission (GHG) is a key factor for developing sustainable agriculture. The review provides insights into the potential of biochar in improving the agroecological system with reference to Ghana. Given the fact that yield gaps are greater in many developing countries, there is considerable need for better soil management technologies to ensure higher yields for improved food security. Biochar application offers a great potential in improving tropical soils and crop productivity.

Conflict of interest

Authors declare no competing interest.

Author details

Stephen Yeboah[1*], Patricia Oteng-Darko[1], Joseph Adomako[1] and Abdul Rauf Alhassan Malimanga[2]

1 CSIR - Crops Research Institute, Kumasi, Ghana

2 Department of Agriculture, Tamale, Ghana

*Address all correspondence to: proyeboah@yahoo.co.uk

References

[1] Lehmann J. A handful of carbon. Nature. 2007;**447**:143-144. DOI: 10.1038/447143a

[2] Russell AE, Cambardella CA, Laird DA, Jaynes DB, Meet DW. Nitrogen fertilizer effects on soil carbon balances in Midwestern U.S. agricultural systems. Ecological Applications. 2009;**19**(5):1102-1113. DOI: 10.1890/07-1919.1

[3] Laird D, Fleming P, Wang B, Horton R, Karlen D. Biochar impact on nutrient leaching from a Midwestern agricultural soil. Geoderma. 2010;**158**:436-444. DOI: 10.1016%2Fj.geoderma.2010.05.012

[4] Yeboah S, Zhang R, Cai L, Li L, Xie J, Luo Z, et al. Soil water content and photosynthetic capacity of spring wheat as affected by soil application of nitrogen-enriched biochar in a semiarid environment. Photosynthetica. 2017;**55**(3):532-542. DOI: 10.1007/s11099-016-0672-1

[5] IPCC. Climate Change: Mitigation of Climate Change. Intergovernmental Panel on Climate Change [Internet]. 2014. Available from: http://www.ipcc.ch/report/ar5/wg3/

[6] Woolf D, Amonette JE, Street-Perrott FA, Lehmann J, Joseph S. Sustainable biochar to mitigate global climate change. National Communication. 2010;**1**:1-9. DOI: 10.1038/ncomms1053

[7] Yeboah S, Zhang RZ, Cai LQ, Jun W. Different carbon sources enhance system productivity and reduce greenhouse gas intensity. Plant, Soil and Environment. 2018;**64**(10):463-469. DOI: 10.17221/83/2018-pse

[8] Zimmerman AR, Gao B, Ahn MY. Positive and negative carbon mineralization priming effects among a variety of biochar amended soils. Soil Biology and Biochemistry. 2011;**43**(1):1169-1179. DOI: 10.1016/j.soilbio.2011.02.005

[9] Rasouli S, Whalen JK, Madramootoo CA. Review: Reducing residual soil nitrogen losses from agroecosystems for surface water protection in Quebec and Ontario, Canada: Best management practices, policies and perspectives. Canadian Journal of Soil Science. 2014;**94**:109-127. DOI: 10.4141/cjss2013-015

[10] Lehmann CJ, Joseph S. Biochar systems. In: Lehmann CJ, Joseph S, editors. Biochar for Environmental Management: Science and Technology. London: Earthscan; 2009. p. 448

[11] Blackwell P, Evelyn K, Greg B, Allan H, Zakaria S. Effect of banded biochar on dryland wheat production and fertiliser use in south–western Australia: An agronomic and economic perspective. Australian Journal of Soil Research. 2010;**48**(7):531-545

[12] Bruun EW, Muller-Stover D, Ambus P, Hauggaard-Nielsen H. Application of biochar to soil and N_2O emissions: Potential effects of blending fast–pyrolysis biochar with anaerobically digested slurry. European Journal of Soil Science. 2011;**62**:581-589. DOI: 10.1111/j.1365-2389.2011.01377x

[13] Borchard N, Wolf A, Laabs V, Aeckersberg R, Scherer HW, Moeller A, et al. Physical activation of biochar and its meaning for soil fertility and nutrient leaching—A greenhouse experiment. Soil Use and Management. 2012;**28**:177-184. DOI: 10.1111/j.1475-2743.2012.00407.x

[14] Steiner C, Teixeira WG, Lehmann J, Nehls T, de Macêdo JLV, Blum WEH, et al. Long term effects of manure, charcoal and mineral fertilization on

crop production and fertility on a highly weathered central Amazonian upland soil. Plant and Soil. 2007;**291**(1-2):275-290. DOI: 10.1007/s11104-007-9193-9

[15] Asai H, Samson BK, Stephan HM, Songyikhangsuthor K, Homma K, Kiyono Y, et al. Biochar amendment techniques for upland rice production in northern Laos. Field Crops Research. 2009;**111**:81-84. DOI: 10.1016/j.fcr.2008.10.008

[16] Snyder CS, Bruulsema TW, Jensen TL, Fixen PE. Review of greenhouse gas emissions from crop production systems and fertilizer management effects. Agriculture, Ecosystems and Environment. 2009;**133**:247-266. DOI: 10.1016/j.agee.2009.04.021

[17] Powlson DS, Gregory PJ, Whalley WR, Quinton JN, Hopkins DW, Whitmore AP, et al. Soil management in relation to sustainable agriculture and ecosystem services. Food Policy. 2011;**36**:S72-S87. DOI: 10.1016/j.foodpol.2010.11.025

[18] Yanai Y, Toyota K, Okazaki M. Effect of charcoal addition on N_2O emissions from soil resulting from rewetting air–dried soil in short–term laboratory experiments. Soil Science and Plant Nutrition. 2007;**53**:181-188. DOI: 10.1111/j.1747-0765.2007.00123.x

[19] Zhang A, Cui L, Pan G, Li L, Hussain Q, Zhang X, et al. Effect of biochar amendment on yield and methane and nitrous oxide emissions from a rice paddy from Tai Lake plain, China. Agriculture, Ecosystems and Environment. 2010;**139**:469-475. DOI: 10.1016/j.agee.2010.09.003

[20] Clough TJ, Bertram JE, Ray JL, Condron LM, O'Callaghan M, Sherlock RR, et al. Unweathered wood biochar impact on nitrous–oxide emissions from a bovine-urine-amended pasture soil. Soil Science Society of America Journal. 2010;**74**:852-860. DOI: 10.1111/j.1747-0765.2007.00123.x

[21] Yeboah S, Zhang R, Cai L, Song M, Li L, Xie J, et al. Greenhouse gas emissions in a spring wheat–field pea sequence under different tillage practices in semi-arid Northwest China. Nutrient Cycling in Agroecosystems. 2016;**106**:77-91. DOI: 10.1007/s10705-016-9790-1

[22] Wang L, Tian H, Song C. Xu X, Chen G, Ren W, Lu C. Net exchanges of CO_2, CH_4 and N_2O between marshland and the atmosphere in Northeast China as influenced by multiple global environmental changes. Atmospheric Environment, 2012, 63:77-85. doi:10.1016/j.atmosenv.2012.08.069

[23] Karhu K, Mattila T, Bergstrom I, Regina K. Biochar addition to agricultural soil increased CH_4 uptake and water holding capacity—Results from a short–term pilot field study. Agriculture, Ecosystems and Environment. 2011;**140**:309-313. DOI: 10.1016/j.agee.2010.12.005

[24] Calys-Tagoe E, Sadick A, Yeboah E, Amoah B. Biochar effect on maize yield in selected farmers fields in the northern and upper east regions of Ghana. Journal of Experimental Agriculture International. 2019;**30**(6): 1-9. DOI: 10.9734/jeai/2019/44168

[25] Jin H. Characterization of Microbial Life Colonizing Biochar and Biochar–Amended Soils. 2010. Available from: http://search.proquest.com.ezprozy.webfeat.lib.ed.ac.uk/docview/741717654

[26] Lal R. Soil carbon sequestration to mitigate climate change. Geoderma. 2004;**121**:1-22. DOI: 10.1016/j.geoderma.2004.01.032

[27] Andruschkewitsch R, Geisseler D, Koch HJ, Ludwig B. Effects of tillage on

contents of organic carbon, nitrogen, water–stable aggregates and light fraction for four different long–term trials. Geoderma. 2013;**192**:368-377. DOI: 10.1016/j.geoderma.2012.07.005

[28] Ussiri D, Lal R, Jarecki MK. Nitrous oxide and methane emissions from long–term tillage under a continuous corn cropping system in Ohio. Soil and Tillage Research. 2009;**104**:247-253

[29] Oguntunde PG, Abiodun BJ, Ajayi AE, van de Giesen N. Effects of charcoal production on soil physical properties in Ghana. Journal of Plant Nutrition and Soil Science. 2008;**171**(4):591-596. DOI: 10.1002/jpln.200625185

[30] Keith A, Singh B, Singh BP. Interactive priming of biochar and labile organic matter mineralization in a smectite-rich soil. Environmental Science & Technology. 2011;**45**:9611-9618. DOI: 10.1021/es202186j

[31] Eyles A, Bound SA, Oliver G, Corkrey R, Hardie M, Green S, et al. Impact of biochar amendment on the growth, physiology and fruit of a young commercial apple orchard. Trees. 2015;**29**(6):1817-1826. DOI: 10.1007/s00468-015-1263-7

[32] Bonanomi G, Sicurezza MG, Caporaso S, Assunta E, Mazzoleni S. Phytotoxicity dynamics of decaying plant materials. New Phytologist. 2006;**169**:571-578. DOI: 10.1111/j.1469-8137.2005.01611.x

[33] Bonanomi G, Antignani V, Barile E, Lanzotti V, Scala F. Decomposition of *Medicago sativa* residues affects phytotoxicity, fungal growth and soil-borne pathogen diseases. Journal of Plant Pathology. 2011;**93**:57-69. DOI: 10.4454/jpp.v93i1.274

[34] Chan KY, Van Zwieten L, Meszaros I, Downie A, Joseph S. Agronomic values of greenwaste biochar as a soil amendment. Soil Research. 2007;**45**:629-634. DOI: 10.1071/SR07109

[35] Graber ER, Frenkel O, Jaiswal AK, Elad Y. How may biochar influence severity of diseases caused by soilborne pathogens? Carbon Management. 2014;**5**:169-183. DOI: 10.1080/17583004.2014.913360

[36] Huang WK, Ji HL, Gheysen G, Debode J, Kyndt T. Biochar-amended potting medium reduces the susceptibility of rice to root-knot nematode infections. BMC Plant Biology. 2015;**15**(1):267. DOI: 10.1186/s12870-015-0654-7

[37] Bonanomi G, Antignani V, Pane C, Scala F. Suppression of soilborne fungal diseases with organic amendments. Journal of Plant Pathology. 2007;**89**:311-324

[38] Mehari ZH, Elad Y, Rav-David D, Graber ER, Harel YM. Induced systemic resistance in tomato (*Solanum lycopersicum*) against *Botrytis cinerea* by biochar amendment involves jasmonic acid signaling. Plant and Soil. 2015;**395**(1-2):31-44. DOI: 10.1007/s11104-015-2445-1

[39] Thies JE, Rillig M. Characteristics of biochar: Biological properties. In: Lehmann M, Joseph S, editors. Biochar for Environmental Management Science and Technology. London: Earthscan; 2009. pp. 85-105. DOI: 10.4324/9781849770552-13

[40] Nerome M, Toyota K, Islam TM, Nishimima T, Matsuoka T, Sato K, et al. Suppression of bacterial wilt of tomato by incorporation of municipal biowaste charcoal into soil. Soil Microorganisms. 2005;**59**(1):9-14

[41] Rawat J, Saxena J, Sanwal P. Biochar: A sustainable approach for improving plant growth and soil properties. In:

Abrol V, Sharma P, editors. Biochar—An Imperative Amendment for Soil and the Environment. London: IntechOpen; 2019. DOI: 10.5772/intechopen.82151

[42] Noble R, Coventry E. Suppression of soil-borne plant diseases with composts: A review. Biocontrol Science and Technology. 2005;**15**:3-20. DOI: 10.1080/09583150400015904

[43] Kookana RS, Sarmah AK, Van Zwieten L, Krull E, Singh B. Biochar application to soil: Agronomic and environmental benefits and unintended consequences. In: Donald LS, editor. Advances in Agronomy 112. New York: Academic Press; 2011. pp. 103-143. DOI: 10.1016/B978-0-12-385538-1.00003-2

[44] Martin SM, Kookana RS, Van Zwieten L, Krull E. Marked changes in herbicide sorption-desorption upon ageing of biochars in soil. Journal of Hazardous Materials. 2012;**231-232**:70-78. DOI: 10.1016/j.jhazmat.2012.06.040

[45] Tiquia SM. Reduction of compost phytotoxicity during the process of decomposition. Chemosphere. 2010;**79**:506-512. DOI: 10.1016/j.chemosphere.2010.02.040

[46] International Food Policy Research Institute (IFPRI). New GSSP policy note. In: Arhin BG, editor. Ghana Strategy Support Program, GSSP. Ghana Agricultural News Digest; 2014

Chapter 7

Importance of Biochar in Agriculture and Its Consequence

Hanuman Singh Jatav, Satish Kumar Singh, Surendra Singh Jatav, Vishnu D. Rajput, Manoj Parihar, Sonu Kumar Mahawer, Rajesh Kumar Singhal and Sukirtee

Abstract

Climate change is affecting all four dimensions of food security: food availability, food accessibility, food utilization, and food systems stability. It is also affecting human health, livelihood assets, food production, and distribution channels, as well as changing purchasing power and market flows. Keeping in view, the present chapter is focusing mostly on biochar. Biochar is usually produced by pyrolysis of biomass at around temperature range of 300–600°C. It is under investigation as an approach to carbon sequestration to produce negative carbon emissions. Present agriculture is leading mining of nutrients and reduction in soil organic matter levels through repetitive harvesting of crops. The most widespread solution to this depletion is the application of soil amendments in the form of fertilizers containing the three major nutrients. The nitrogen is considered the most limiting nutrient for plant growth useful for protein builds, structures, hormones, chlorophyll, vitamins, and enzymes. Biochar may be added to soils to improve soil health, improve soil fertility, and sequester carbon. However, the variable application rates, uncertain feedstock effects, and initial soil state provide a wide range of cost for marginally improved yield from biochar additions, which is often economically impracticable. There is a need for further research on optimizing biochar application to improve crop yields.

Keywords: soil health, soil fertility, nutrients, carbon, biochar, soil properties

1. Introduction

Natural organic biomass burning creates black carbon which forms a considerable proportion of the soil's organic carbon. Due to black carbon's aromatic structure, it is recalcitrant and has the potential for long-term carbon sequestration in soil. Soils within the Amazon-basin contain numerous sites where the "dark earth of the Indians" (Terra preta deIndio, or Amazonian Dark Earths [ADE]) exist and are composed of variable quantities of highly stable organic black carbon waste ("biochar") [1]. The intensification of agricultural production on a global scale is necessary to secure the food supply for an increasing world population. As a result, fallow periods are often reduced in shifting cultivation in the humid tropics leading to irreversible soil degradation and increased

IntechOpen

destruction of remaining natural forests due to the cultivation of new areas after slash and burn [2].

The incorporation in soils influences soil structure, texture, porosity, particle size distribution and density. The molecular structure of biochar shows a high degree of chemical and microbial stability [3]. A key physical feature of most biochar is its highly porous structure and large surface area [4]. This structure can provide refugia for beneficial soil micro-organisms such as mycorrhizae and bacteria and influences the binding of important nutritive cations and anions. This binding can enhance the availability of macro-nutrients such as N and P. Other changes in soil by biochar applications include alkalization of soil pH and increases in electrical conductivity (EC) and cation exchange capacity (CEC) [5–7]. Ammonium leaching is reduced, along with N_2O soil emissions. There may also be reductions in soil mechanical impedance. Terra preta soils contain a higher number of "operational taxonomic units" and have highly distinctive microbial communities relative to neighboring soils [8]. The apparent high agronomic fertility of these sites, relative to tropical soils in general, has attracted interest. Biochar can be produced by "burning" organic matter under low oxygen (pyrolysis). Principally biochar is produced through various thermo-chemical conversion methods such as low pyrolysis, fast pyrolysis, gasification, and torrefaction, under different process parameters [9]. The quantities of key mineral elements within this biochar can be directly related to the levels of these components in the feedstock before burning [10]. The potential importance of biochar soil incorporation on mycorrhizal fungi has also been noted with biochar providing a physical niche devoid of fungal grazers. Improvements in soil field capacity have been recorded upon biochar additions [4].

2. Applications of biochar and their effect on soil properties

Evidence shows that bioavailability and plant uptake of key nutrients increases in response to biochar application, particularly when in the presence of added nutrients. A systematic representation of the potential of biochar in soil and plant system is presented in **Figure 1**.

Depending on the quantity of biochar added to soil significant improvements in plant productivity have been achieved, but these reports derive predominantly from studies in the tropics [11, 12]. As yet there is limited critical analysis of possible agricultural impacts of biochar application in temperate regions, nor on the likelihood of utilizing such soils as long-term sites for carbon. On the other hand, soil application of biochar can permanently appropriate C in the soil and reduce net emissions of carbon dioxide gas improve crop productivity through enhanced physio-chemical and biological properties, nutrient release pattern, reduce denitrification and soil pollutants [10]. Biochar application can be a means of not only sequestering carbon in the soil but also returning essential organic matters lost with biomass removal from agro- and/or forestry systems for energy production. Thus, biochar can potentially provide two simultaneous economic benefits. One, it may improve the agronomic and environmental sustainability of biomass production systems. Two, it may improve the economic sustainability of bioenergy enterprises by offsetting feedstock purchases with revenue from biochar sales [9]. Biochar has the capacity to produce revenue and boost the sustainability of agriculture and environment. The agricultural and bioenergy industries will be reluctant to pay for biochar until its precise effects on soil properties and crop production are shown. Complete development of biochar as a commercial product

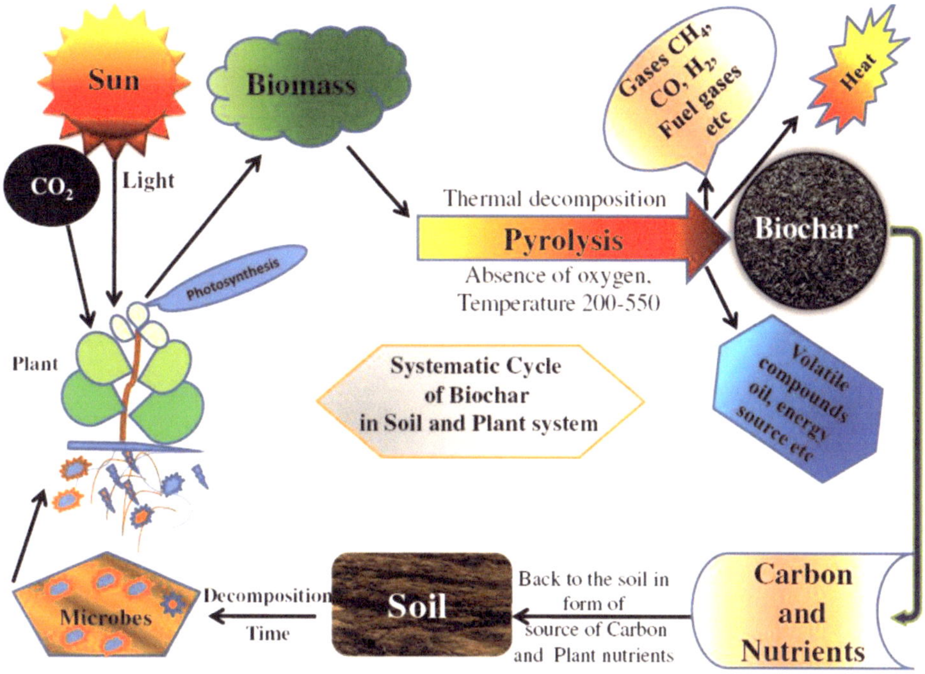

Figure 1.
Systemic potential mechanism of biochar in soil and plant system.

must establish concrete benefits of the product to soil properties and crop production and link all these benefits to biochar properties and its appropriate use and economic value. One of the most important factors to make this a reality is the understanding of how this product is made and how the production process affects its performance. Its benefits on crop production, environment, and soil will be a moot point if it is not reproducible and consistent. Biochar and its beneficial component are presented in **Table 1**.

2.1 Physical properties

Biochar itself is a porous material thus it can be adsorbed and retain a huge amount of water. Dugan et al. also reported that the maize stover biochar and sawdust biochar increased the water holding capacity (WHC) of loamy sand in Ghana when it was applied at the rate 5, 10, and 15 ton ha^{-1}. The WHC increased because small pores in biochar retain moisture and there are largely absent in coarse texture soils. The increased moisture retention depends on the higher porosity of biochar. Soils amendment with biochar is more ineffective improving WHC in sandy soils than in loamy and clay soils by improved water holding capacity [18, 19]. Pietikäinen et al. reported that two biochars, one prepared from humus and one from wood, had a similar water-holding capacity (WHC) (2.9 mL g^{-1} dry matter) than activated carbon (1.5 mL g^{-1} dry matter). Smaller pores will attract and retain capillary soil water much longer than larger pores (larger than 10–20 μm) in both the biochar and the soil. During thermal conversion, the mineral and carbon skeleton formed retains the rudimentary porosity and structure of the original material. Microscopy analysis proves the presence of aligned honeycomb-like groups of pores on the order of 10 μm in diameter, most likely the carbonaceous skeleton from the biological capillary structure of the raw material [20]. Brunauer-Emmett-Teller

Properties	Effect of biochar application on various factors	References
Soil fertility	Biochar can improve soil fertility, stimulating plant growth, which then consumes more CO_2 in a positive feedback effect.	[13]
Reduced fertilizer inputs	Biochar can reduce the need for chemical fertilizers, resulting in reduced emissions of greenhouse gases from fertilizer manufacture.	[14]
Reduced N_2O and CH_4 emissions	Biochar can reduce emissions of nitrous oxide (N_2O) and methane (CH_4), two potent greenhouse gases from agricultural soils.	[15]
Enhanced soil microbial life	Biochar can increase soil microbial life, resulting in more carbon storage in soil.	[16]
Reduced emissions from feed stocks	Converting agricultural and forestry waste into biochar can avoid CO_2 and CH_4 emissions otherwise generated by the natural decomposition or burning of the waste.	[15]
Energy generation	The heat energy and also the bio oils and synthesis gases generated during biochar production can be used to displace carbon positive energy from fossil fuels.	[17]

Table 1.
Biochar and its beneficial component in the environment.

(BET) surface areas of olive kernel biochars increased with increasing mass loss (burn-off) regardless of the activation temperature [21]. Micropores (<2 nm in diameter) are responsible for adsorption and high surface area the total pore volume of the biochar will be divided as microspores (pores of internal diameter less than 2 nm), mesopores (pores of internal width between 2 and 50 nm) and macropores (pores of internal width greater than 50 nm) [22].

2.2 Chemical properties

Soil application of biochar resulted in a significant increase in soil pH. Van et al. suggested that biochar derived from poultry litter facilitates liming in soil resulting in the rise of pH of acidic or neutral soils. Hoshi et al. in his experiment suggested that the 20% increase in height and 40% increase in the volume of tea trees were partly due to the ability of the biochar to maintain the neutral pH of the soil. Such ability is related to the liming value of the biochar. Van et al. reported a nearly 30–40% increase in wheat height when biochar produced from paper mill sludge was applied at a rate of 10 t ha^{-1} to an acidic soil but not to neutral soil. The increase in soil organic carbon with the application of biochar might have resulted from the recalcitrant nature of carbon found in biochar which is largely resistant to decomposition [1, 23–25] also reported that soil carbon increased significantly over control. Available N, P and K applying biochar to forest soils along with natural or synthetic fertilizers have been found to increase the bioavailability and plant uptake of P, alkaline metals and some trace metals [2, 19, 25] but the mechanisms for these increases are still a matter of speculation. Lehmann et al. demonstrated the ability of biochar to retain applied fertilizer against leaching with resulting increase in fertilizer use efficiency. In the manufacture of the N-enriched biochar Day et al. suggested that biochar produced at a lower temperature of 400–500°C is more effective in adsorbing ammonia than that produced at higher temperatures (700–1000°C). Total N content depends on pyrolysis temperature initially at low pyrolysis temperature N content increases which further decreases with higher temperature due to volatilization of N whereas C/N ratio of biochar (63–80) varies less with pyrolysis temperature but varies significantly with the type of feedstock material [26]. Biochar influence the dynamics of different nutrients indirectly by its high surface area and high cation exchange capacity. Changes in the dynamics of N with the application of biochar are not fully understood [27]. Weathering of biochar in soil fastens immobilization of nitrogen on its surface, studies have shown that high application rate of biochar (10% or 20%, w/w) significantly decreased NH_4 volatilization due to its high cation exchange capacity [25] but in case of NO_3^- the leaching increased especially if the initial N content of biochar is high [28]. Biochar itself is a very good source of several essential plant nutrients. Chemical, physical properties, and nutrient content status of biochar are shown in **Table 2**.

2.3 Biological activity

Ameloot et al. showed that the type of biochar alone has a significant effect on soil enzymatic activity. The quoted authors proved that poultry litter biochar produced at 400°C and amended to soil 20 t ha^{-1} caused a significant increase in the activity of dehydrogenases. Biochar has a positive effect on mycorrhizal

Parameter	Value		References
	Minimum	Maximum	
pH	4.5	12.9	[29]
Electrical conductivity(mS cm^{-1})	20	10,260	[30]
Cation exchange capacity (cmolþ kg^{-1})	3.8	272	[31]
Surface area (m2 g^{-1})	0.1	410	[32]
Bulk density (g cm^{-3})	0.05	0.7	[33]
Volatile matter (%)	0.6	85.7	[30]
N (g kg^{-1})	0.1	6.4	[17]
K (g kg^{-1})	0.3	74.0	[34]
P (g kg^{-1})	0.005	59	[34]
Ca (g kg^{-1})	0.04	92	[34]
Mg (g kg^{-1})	0.009	37	[34]
Carbon (%)	17.7	92.7	[35]
Hydrogen (%)	0.05	5.30	[35]
Oxygen (%)	0.01	39.2	[36]
H/C	<0.01	1.14	[30]
O/C	0.02	1.11	[36]

Table 2.
Various physico biochemical properties of biochar.

association when applied to soil [37–39] evaluated the increase in microbial biomass when biochar is applied to the soil and its efficacy as a measure of CO_2 released per.

Microbial biomass carbon in the soil increase in basal respiration due to addition of the carbon in soil. Biochar does not contribute directly to the microbial population in the soil. Hence higher porosity of biochar creates a favorable environment for microbes to make a habitat in soil [40] researchers have suggested that biochar benefits microbial communities by providing suitable habitats for microorganisms that protect them from predation [41–43]. Microbial cells typically range in size from 0.5 to 5 μm and consist predominantly of bacteria, fungi, actinomycetes, lichens, and algae species are from 2 to 20 μm [44]. The microscopic studies indicate that biochar in soil serve as habitat for microorganisms [3]. The loss of volatile and condensable compounds from biochars and the concomitant relative increase in the organized phase formed by graphite-like crystallites leads to the increase in solid density (or true density) of the round 1.5–1.7 g cm^{-3}. Increasing anthropogenic activities have mainly resulted into buildup of non-essential heavy metals in agricultural soils. Recently chromium (Cr) contamination in water and soil is a serious concern [44].

3. Application of biochar in decontamination/removal of organic pollutants from soil and water

In this era of high population and modernization, the contamination of soil and water resources due to organic contaminants is a major concern. Biochar from different sources has a porous carbons, the pore network of biochar is typically composed of micropores <2 nm, mesopores 2–50 nm, and macropores >50 nm. But, micropores and small mesopores (2–20 nm) are suggested to contribute the majority to the surface area and excellent adsorption capacity of biochar. Because of such excellent properties, it can be a good tool to remove organic pollutants from contaminated soil and water resources. There are several evidences available in the literature about the use of biochar for the removal of organic pollutants from contaminated soil and water in **Table 3** [54].

Organic contaminants	Biochar type	Mechanisms	References
Carbamazepine	Loblolly pine chips	Hydrophobic adsorption	[45]
Ethinylestradiol	poultry litter	Pore-filling	[46]
2,4-Dichlorophenoxyacetic acid	Wood chips	Surface adsorption	[47]
Diazinon	Rice straw	Hydrogen bonding with polar groups	[48]
Atrazine	Dairy manure	Partitioning	[49]
Nitrobenzene	Pine needles	Pore-filling	[50]
Humic acid	Grass	Hydrophobic interactions	[51]
Perfluorooctane sulfonate	Maize	Hydrophobic adsorption	[52]
p-Coumaric acid	Hardwood litter	Hydrogen bonding	[53]

Table 3.
Organic contaminants sorbet by different biochars and their abstraction mechanisms.

4. Application of biochar for soil carbon sequestration and mitigate GHGs emission

The current availability of biomass in India (2010–2011) is estimated at 500 Mtpa. Annual bio-manure production (in tons) is 32,582. A potential 61.1 MMT of fuel crop residue and 241.7 MMT of fodder crop residue are being consumed by farmers themselves. In India total biomass power generation capacity is 17,500 MW. At present power being generated is 2665 MW which include 1666 MW by cogeneration. Studies sponsored by the Ministry of New and Renewable Energy, Govt. of India have estimated surplus biomass availability at about 120–150 Mtpa. Of this, about 93 Mt. of crop residues are burnt each year. The generation of crop residues is highest in Uttar Pradesh (60 Mt) followed by Punjab (50 Mt). Efficient utilization of this biomass by converting it as a valuable source of soil amendment is one approach to manage soil quality, fertility, mitigate GHGs emissions and increase carbon sequestration [55]. Biochar has a condensed aromatic structure that makes it a stable solid rich in carbon content which is known to be highly resistant to microbial decomposition, thus it can be used to lock carbon in the soil. Biochar application has received a growing interest as a sustainable technology to improve highly weathered or degraded tropical soils [10]. Biochar can reduce N_2O emission from the soil which might be due to inhibition of either stage of nitrification and/or inhibition of denitrification, or encouragement of the decrease of N_2O and these impacts could occur simultaneously in a soil. Several workers have reported that applications of biochar to soils have shown positive responses for the yield of several crops. Similarly, biochar has also been found to have significant positive interaction with plant growth-promoting rhizobacteria for improving total dry matter yield of rice. Biochar from different sources has several other important roles other than the above mentioned depending on its source such as the role in plant growth enhancement, quality and quantity improvement of several crop species, improvement of water holding capacity, soil porosity, etc.

5. Biochar prospects and essential research

The global potential of biochar reaches far beyond *slash and char*. Inspired by the recreation of *Terra Preta*, most biochar research was restricted to the humid tropics. More information is needed on the agronomic potential of biochar, the potential to use alternative biomass sources (crop residues) and the production of by-products to evaluate the opportunities for adopting a biochar system on a global scale. Biochar as soil amendment needs to be studied in different climate and soil types. Today, crop residue biomass represents a considerable problem as well as new challenges and opportunities. A system converting biomass into energy (hydrogen-rich gas) and producing biochar as a by-product might offer an opportunity to address these problems. Biochar can be produced by incomplete combustion from any biomass, and it is a by-product of the pyrolysis technology used for biofuel and ammonia production [56]. The acknowledgment of biochar as a carbon sink would facilitate C-trading mechanisms. Although most scientists agree that the half-life of biochar is in the range of centuries or millennia, a better knowledge of the biochar's durability in different ecosystems is important to achieve this goal. The systematic recycling of biochar in the environment has been depicted in **Figure 2**.

Access to the C trade market holds out the prospect to reduce or eliminate the deforestation of the primary forest because using intact primary forest would reduce the farmer's C credits. It is estimated the above-ground biomass of unlogged

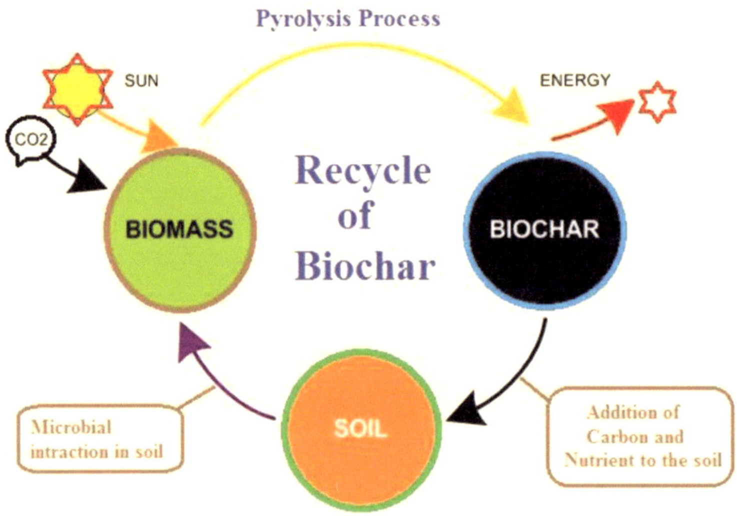

Figure 2.
Systematic recycle of biochar in the environment.

forests to be 434 Mg ha^{-1}, about half of which is C. This C is lost if burned in the slash-and-burn scenario and lost at a high percentage if used for biochar production. The Trade could provide an to cease further deforestation; instead, reforestation and recuperation of degraded land for fuel and food crops would gain magnitude. As tropical forests account for between 20 and 25% of the world terrestrial C reservoir [57] this consequently reduces emissions from tropical forest conversion, which is estimated to contribute globally as much as 25% of the net CO_2 emissions [58]. Today most biomass gasification systems tend to suppress the creation of residuals, like total organic carbon (TOC) and ashes. The C-emission trading options and a better knowledge of biochar as soil additive would add value to these residues. Further, this would facilitate the use of alternative biomass, those which are currently avoided due to their higher TOC residuals. The tarry vapors constitute a significant loss of carbon during carbonization [59] although representing another valuable product. Japanese researchers attempt to produce biochar with a specific pore size distribution to favor desired microorganisms. Pore structure, surface area, and adsorption properties are strongly influenced by the peak temperature during biochar production [59]. Increasing porosity is achieved with increasing temperature but the functional groups are gradually lost. In this context, it is also important to discern the mechanisms of nutrient retention (mainly N) due to biochar applications. The biochar's low biodegradability [60], low nutrient content [59], and high porosity and specific surface area [61] make biochar a rather exceptional SOM constituent. *Terra Preta's* research has shown that oxidation on the edges of the aromatic backbone and adsorption of other organic matter to biochar is responsible for the increased CEC, though the relative importance of these two processes remains unclear [21].

6. Conclusion

Energy from crop residues could lower fossil energy consumption and CO_2-emissions, and become a completely new income source for farmers and rural regions. A global analysis by revealed that up to 12% of the total anthropogenic C emissions by land-use change (0.21 Pg C) can be off-set annually in the soil if slash and burn are replaced by slash and char. Agricultural and forestry wastes add a conservatively estimated 0.16 Pg C $yr.^{-1}$. If the demand for renewable fuels by the year 2100 was met through pyrolysis, biochar sequestration could exceed current emissions from fossil fuels (5.4 Pg C $yr.^{-1}$). The described mixture of driving forces and technologies has the potential to use residual waste carbon-rich residues to reshape agriculture, balance carbon and address nutrient depletion.

Author details

Hanuman Singh Jatav[1*], Satish Kumar Singh[2], Surendra Singh Jatav[2], Vishnu D. Rajput[3], Manoj Parihar[4], Sonu Kumar Mahawer[5], Rajesh Kumar Singhal[5] and Sukirtee[6]

1 S.K.N. Agriculture University, Jobner, Rajasthan, India

2 Department of Soil Science and Agricultural Chemistry, Institute of Agricultural Sciences, Banaras Hindu University, Varanasi, India

3 Academy of Biology and Biotechnology, Southern Federal University, Rostov-on-Don, Russia

4 ICAR-Vivekananda Parvatiya Krishi Anusandhan Sansthan, Almora, India

5 ICAR-Indian Grassland and Fodder Research Institute, Jhansi, India

6 Department of Soil Science and Agricultural Chemistry, Chaudhary Charan Singh Haryana Agricultural University, Hissar, India

*Address all correspondence to: hanumaniasbhu@gmail.com; hsjatav.soils@sknau.ac.in

References

[1] Masulili A, Utomo WH, Syechfani MS. Rice husk biochar for rice based cropping system in acid soil 1. The characteristics of rice husk biochar and its influence on the properties of acid sulfate soils and rice growth in West Kalimantan, Indonesia. Journal of Agricultural Science. 2010;**2**(1):39

[2] Hauck FW. Shifting cultivation and soil conservation in Africa. In: FAO/SIDA/ARCN Regional Seminar Held at Ibadan, Nigeria, 2-21 July, 1973. FAO; 1974

[3] Gorovtsov AV, Minkina TM, Mandzhieva SS, Perelomov LV, Soja G, Zamulina IV, et al. The mechanisms of biochar interactions with microorganisms in soil. Environmental Geochemistry and Health. 2019;**14**:1-24

[4] Atkinson CJ, Fitzgerald JD, Hipps NA. Potential mechanisms for achieving agricultural benefits from biochar application to temperate soils: A review. Plant and Soil. 2010;**337**(1-2):1-8

[5] Jatav HS, Singh SK, Singh YV, Paul A, Kumar V, Singh P, et al. Effect of biochar on yield and heavy metals uptake in rice grown on soil amended with sewage sludge. Journal of Pure and Applied Microbiology. 2016;**10**(2):1367-1378

[6] Jatav HS, Singh SK, Singh Y, Kumar O. Biochar and sewage sludge application increases yield and micronutrient uptake in rice (*Oryza sativa* L.). Communications in Soil Science and Plant Analysis. 2018;**49**(13):1617-1628

[7] Jatav HS, Singh SK. Effect of biochar application in soil amended with sewage sludge on growth, yield and uptake of primary nutrients in Rice (*Oryza sativa* L.). Journal of the Indian Society of Soil Science. 2019;**67**(1):115-119

[8] Van Zwieten L, Kimber S, Downie A, Chan KY, Cowie A, Wainberg R, et al. Papermill char: Benefits to soil health and plant production. In: Proceedings of the Conference of the International Agrichar Initiative 2007 Apr 30. Vol. 30. 2007

[9] Wang D, Jiang P, Zhang H, Yuan W. Biochar production and applications in agro and forestry systems: A review. Science of The Total Environment. 2020;**10**:137775

[10] Lehmann J, Gaunt J, Rondon M. Bio-char sequestration in terrestrial ecosystems–a review. Mitigation and Adaptation Strategies for Global Change. 2006;**11**(2):403-427

[11] Rajput VD, Gorovtsov AV, Fedorenko GM, Minkina TM, Fedorenko AG, Lysenko VS, et al. The influence of application of biochar and metal-tolerant bacteria in polluted soil on morpho-physiological and anatomical parameters of spring barley. Environmental Geochemistry and Health. 2020;**27**:1-3

[12] Rawat J, Saxena J, Sanwal P. Biochar: A sustainable approach for improving plant growth and soil properties. In: Biochar-An Imperative Amendment for Soil and the Environment. IntechOpen; 2019

[13] Schulz H, Dunst G, Glaser B. Positive effects of composted biochar on plant growth and soil fertility. Agronomy for Sustainable Development. 2013;**33**(4):817-827

[14] Qian L, Chen L, Joseph S, Pan G, Li L, Zheng J, et al. Biochar compound fertilizer as an option to reach high productivity but low carbon intensity in rice agriculture of China. Carbon Management. 2014;**5**(2):145-154

[15] Huang Y, Wang C, Lin C, Zhang Y, Chen X, Tang L, et al. Methane and nitrous oxide flux after biochar

application in subtropical acidic paddy soils under tobacco-rice rotation. Scientific Reports. 2019;**9**(1):1

[16] Krishnakumar S, Rajalakshmi AG, Balaganesh B, Manikandan P, Vinoth C, Rajendran V. Impact of biochar on soil health. International Journal of Advanced Research. 2014;**2**(4):933-950

[17] Available from: https://biochar-international.org/sustainability-climate-change/

[18] Dugan E, Verhoef A, Robinson S, Sohi S. Bio-char from sawdust, maize stover and charcoal: Impact on water holding capacities (WHC) of three soils from Ghana. In: 19th World Congress of Soil Science, Symposium. Vol. 4, No. 2. 2010. pp. 9-12

[19] Glaser B, Lehmann J, Zech W. Ameliorating physical and chemical properties of highly weathered soils in the tropics with charcoal–a review. Biology and Fertility of Soils. 2002;**35**(4):219-230

[20] Liang B, Lehmann J, Solomon D, Kinyangi J, Grossman J, O'Neill B, et al. Black carbon increases cation exchange capacity in soils. Soil Science Society of America Journal. 2006;**70**(5):1719-1730

[21] Zabaniotou A, Stavropoulos G, Skoulou V. Activated carbon from olive kernels in a two-stage process: Industrial improvement. Bioresource Technology. 2008;**99**(2):320-326

[22] Rouquerol F, Rouquerol J, Sing K. Chapter 6–Assessment of surface area. In: Adsorption by Powders and Porous Solids: Principles, Methodology and Applications. London, UK: Academic Press; 1999. p. 165

[23] Van Zwieten L, Kimber S, Morris S, Chan KY, Downie A, Rust J, et al. Effects of biochar from slow pyrolysis of papermill waste on agronomic performance and soil fertility. Plant and Soil. 2010 Feb 1;**327**(1-2):235-246

[24] Hoshi T. Growth promotion of tea trees by putting bamboo charcoal in soil. In: Proceedings of 2001 International Conference on O-cha (Tea) Culture and Science 2001 Oct. 2001. pp. 147-150

[25] Lehmann J, da Silva JP, Steiner C, Nehls T, Zech W, Glaser B. Nutrient availability and leaching in an archaeological Anthrosol and a Ferralsol of the Central Amazon basin: Fertilizer, manure and charcoal amendments. Plant and Soil. 2003 Feb 1;**249**(2):343-357

[26] Kloss S, Zehetner F, Dellantonio A, Hamid R, Ottner F, Liedtke V, et al. Characterization of slow pyrolysis biochars: Effects of feedstocks and pyrolysis temperature on biochar properties. Journal of Environmental Quality. 2012;**41**(4):990-1000

[27] Clough TJ, Bertram JE, Ray JL, Condron LM, O'Callaghan M, Sherlock RR, et al. Unweathered wood biochar impact on nitrous oxide emissions from a bovine-urine-amended pasture soil. Soil Science Society of America Journal. 2010;**74**(3):852-860

[28] Singh BP, Hatton BJ, Singh B, Cowie AL, Kathuria A. Influence of biochars on nitrous oxide emission and nitrogen leaching from two contrasting soils. Journal of Environmental Quality. 2010;**39**(4):1224-1235

[29] Zhang J, Liu J, Liu R. Effects of pyrolysis temperature and heating time on biochar obtained from the pyrolysis of straw and lignosulfonate. Bioresource Technology. 2015;**176**:288-291

[30] Ramola S, Mishra T, Rana G, Srivastava RK. Characterization and pollutant removal efficiency of biochar derived from baggase, bamboo and Tyre. Environmental Monitoring and Assessment. 2014;**186**(12):9023-9039

[31] Frišták V, Pipíška M, Lesný J, Soja G, Friesl-Hanl W, Packová A. Utilization of biochar sorbents for Cd 2+, Zn

2+, and Cu 2+ ions separation from aqueous solutions: Comparative study. Environmental Monitoring and Assessment. 2015;**187**(1):4093

[32] Song XD, Xue XY, Chen DZ, He PJ, Dai XH. Application of biochar from sewage sludge to plant cultivation: Influence of pyrolysis temperature and biochar-to-soil ratio on yield and heavy metal accumulation. Chemosphere. 2014;**109**:213-220

[33] Alburquerque JA, Salazar P, Barrón V, Torrent J, del Campillo MD, Gallardo A, et al. Enhanced wheat yield by biochar addition under different mineral fertilization levels. Agronomy for Sustainable Development. 2013;**33**(3):475-484

[34] Vaughn SF, Kenar JA, Thompson AR, Peterson SC. Comparison of biochars derived from wood pellets and pelletized wheat straw as replacements for peat in potting substrates. Industrial Crops and Products. 2013;**51**:437-443

[35] Ehsan M, Barakat MA, Husein DZ, Ismail SM. Immobilization of Ni and Cd in soil by biochar derived from unfertilized dates. Water, Air, & Soil Pollution. 2014;**225**(11):2123

[36] Novak JM, Lima I, Xing B, Gaskin JW, Steiner C, Das KC, et al. Characterization of designer biochar produced at different temperatures and their effects on a loamy sand. Annals of Environmental Science. 2009;**3**:195-206

[37] Ameloot N, Graber ER, Verheijen FG, De Neve S. Interactions between biochar stability and soil organisms: Review and research needs. European Journal of Soil Science. 2013;**64**(4):379-390

[38] Warnock DD, Lehmann J, Kuyper TW, Rillig MC. Mycorrhizal responses to biochar in soil–concepts and mechanisms. Plant and Soil. 2007;**300**(1-2):9-20

[39] Steiner C, Das KC, Garcia M, Förster B, Zech W. Charcoal and smoke extract stimulate the soil microbial community in a highly weathered xanthic Ferralsol. Pedobiologia. 2008;**51**(5-6):359-366

[40] Thies JE, Rillig MC. Characteristics of biochar: Biological properties. In: Biochar for Environmental Management. Routledge; 2012. pp. 117-138

[41] Pietikäinen J, Kiikkilä O, Fritze H. Charcoal as a habitat for microbes and its effect on the microbial community of the underlying humus. Oikos. 2000;**89**(2):231-242

[42] Andrey G, Rajput V, Tatiana M, Saglara M, Svetlana S, Igor K, et al. The role of biochar-microbe interaction in alleviating heavy metal toxicity in *Hordeum vulgare* L. grown in highly polluted soils. Applied Geochemistry. 2019;**104**:93-101

[43] Lal R, editor. Encyclopedia of Soil Science. CRC Press; 2006

[44] Kumar V, Sharma PK, Jatav HS, Singh SK, Rai A, Kant S, et al. Organic amendments application increases yield and nutrient uptake of mustard (*Brassica juncea*) grown in chromium-contaminated soils. Communications in Soil Science and Plant Analysis. 2020;**51**(1):149-159

[45] Jung C, Park J, Lim KH, Park S, Heo J, Her N, et al. Adsorption of selected endocrine disrupting compounds and pharmaceuticals on activated biochars. Journal of Hazardous Materials. 2013;**263**:702-710

[46] Sun K, Ro K, Guo M, Novak J, Mashayekhi H, Xing B. Sorption of bisphenol a, 17α-ethinyl estradiol and phenanthrene on thermally and hydrothermally produced biochars. Bioresource Technology. 2011;**102**(10):5757-5763

[47] Kearns JP, Wellborn LS, Summers RS, Knappe DR. 2, 4-D adsorption to biochars: Effect of preparation conditions on equilibrium adsorption capacity and comparison with commercial activated carbon literature data. Water Research. 2014;**62**:20-28

[48] Taha SM, Amer ME, Elmarsafy AE, Elkady MY. Adsorption of 15 different pesticides on untreated and phosphoric acid treated biochar and charcoal from water. Journal of Environmental Chemical Engineering. 2014;**2**(4):2013-2025

[49] Cao X, Ma L, Gao B, Harris W. Dairy-manure derived biochar effectively sorbs lead and atrazine. Environmental Science & Technology. 2009;**43**(9):3285-3291

[50] Chen B, Zhou D, Zhu L. Transitional adsorption and partition of nonpolar and polar aromatic contaminants by biochars of pine needles with different pyrolytic temperatures. Environmental Science & Technology. 2008;**42**(14):5137-5143

[51] Kasozi GN, Zimmerman AR, Nkedi-Kizza P, Gao B. Catechol and humic acid sorption onto a range of laboratory-produced black carbons (biochars). Environmental Science & Technology. 2010;**44**(16):6189-6195

[52] Chen X, Xia X, Wang X, Qiao J, Chen H. A comparative study on sorption of perfluorooctane sulfonate (PFOS) by chars, ash and carbon nanotubes. Chemosphere. 2011;**83**(10):1313-1319

[53] Ni J, Pignatello JJ, Xing B. Adsorption of aromatic carboxylate ions to black carbon (biochar) is accompanied by proton exchange with water. Environmental Science & Technology. 2011;**45**(21):9240-9248

[54] Inyang M, Dickenson E. The potential role of biochar in the removal of organic and microbial contaminants from potable and reuse water: A review. Chemosphere. 2015;**134**:232-240

[55] Srinivasarao C, Venkateswarlu B, Lal R, Singh AK, Kundu S. Sustainable management of soils of dryland ecosystems of India for enhancing agronomic productivity and sequestering carbon. In: Advances in Agronomy. Vol. 121. Academic Press; 2013. pp. 253-329

[56] Day D, Evans RJ, Lee JW, Reicosky D. Valuable and stable carbon co-product from fossil fuel exhaust scrubbing. Preprint Papers - American Chemical Society, Division of Fuel Chemistry. 2004;**49**(1):352

[57] Bernoux M, Graça PM, Cerri CC, Fearnside PM, Feigl BJ, Piccolo MC. Carbon storage in biomass and soils. In: McClain ME, Victoria RL, Richey JE editors. The Biogeochemistry of the Amazon Basin. New York: Oxford University Press; 2001. pp. 165-184

[58] Palm C. Color texture classification by integrative co-occurrence matrices. Pattern Recognition. 2004;**37**(5):965-976

[59] Antal MJ Jr, Allen SG, Dai X, Shimizu B, Tam MS, Grønli M. Attainment of the theoretical yield of carbon from biomass. Industrial and Engineering Chemistry Research. 2000;**39**(11):4024-4031

[60] Kuhlbusch TA, Crutzen PJ. Toward a global estimate of black carbon in residues of vegetation fires representing a sink of atmospheric CO2 and a source of O2. Global Biogeochemical Cycles. 1995;**9**(4):491-501

[61] Braida WJ, Pignatello JJ, Lu Y, Ravikovitch PI, Neimark AV, Xing B. Sorption hysteresis of benzene in charcoal particles. Environmental Science & Technology. 2003;**37**(2): 409-417

Chapter 8

Biochar Potential in Improving Agricultural Production in East Africa

Godfrey Omulo

Abstract

Biochar is among the environmentally friendly bio-products possible of enhancing agricultural productivity due to inherent properties. Despite the increased biochar research output, the sustainability of biochar production and its applicability in developing countries is mostly uncertain. This chapter underscores the biochar production process, its environmental usefulness, and the prediction of its potential impact on agricultural productivity in East African countries. Currently, pyrolysis technology is the most effective means of biochar production. Predominantly, biochar is useful in carbon sequestration, soil amendment, and as a solid fuel source. In-depth analysis of crop residues production in East African countries vis-à-vis the potential for biochar production and the total planted areas strongly indicate that biochar could be sustainably produced and applied in agriculture without compromising the forests and the environment. This knowledge is vital in guarantying the feasibility of biochar technology among policymakers as a sustainable alternative to the exorbitant mineral fertilizers.

Keywords: biochar, agricultural productivity, East Africa, carbon sequestration, fertilizer, sustainable, climate-friendly

1. Introduction

One of the most significant bottlenecks to increased agricultural productivity in developing economies is continuous soil degradation due to land use change and erosion [1, 2]. Human activities have primarily destabilized the distribution of carbon in the universe that has released too much carbon to the atmosphere than what plants can utilize via the photosynthesis process (**Figure 1**). As a consequence, climate-change-related risks such as erratic rainfalls, floods, and fluctuating temperatures have ensued, causing soils to lose their nutrients through erosion and leaching. This has further led to the depletion of soil productivity, increased soil acidity, and a heightened need for mineral fertilizer application [4]. Acidity in soils is caused by factors ranging from nature of the soil, agroecological condition, and fertilization systems. For instance, the non-calcareous parent materials that are intrinsically acidic naturally undergo bleaching, especially in humid climates like East Africa and in high rainfall conditions. Further, reclaimed swampy soils (peats) and soils that have been highly treated with nitrogenous fertilizers tend to be acidic over time [5, 6].

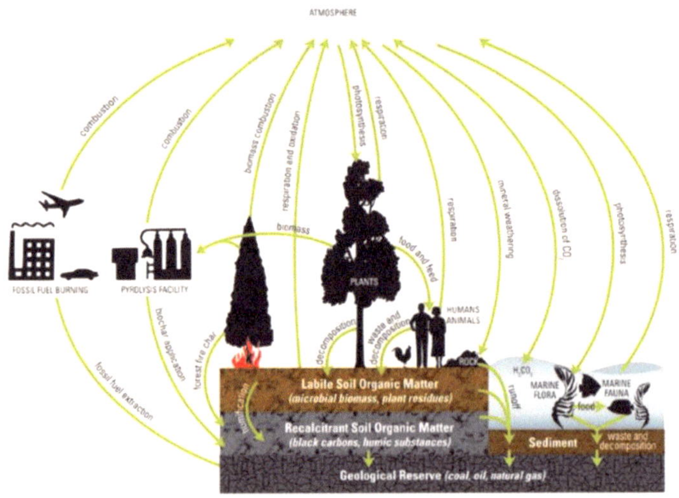

Figure 1.
Carbon cycle representing the global natural and anthropogenic contributions. Source: Brewer [3].

As a consequence, the production of biochar from various biomass sources has attracted immense attention among scientists and agricultural practitioners. This is because of its potential in revitalizing the fertility of degraded soils by sequestering carbon and reducing greenhouse gas emissions, thereby mitigating climate change [7].

Biochar is a black carbonaceous solid product that results from thermochemical decompositions of various biomass feedstocks at elevated temperatures under oxygen-deficient conditions [3, 8, 9]. Pyrolysis is the most common thermochemical process that yields biochar depending on the type of feedstock used and the variations in temperature regimes and the rate of heat application. Biochar from various types of biomass has been used predominantly as a soil amendment, carbon sequestrating tool, an agent for nutrient recycling, waste management tool [10–12], and as a solid fuel source [13]. Charcoal, one of the ancient products used for cooking, has also been investigated on its influence on agricultural productivity. In general, unlike biochar, charcoal has not been effective in fixing carbon into the soil except when it is mixed with mineral fertilizers or other organic manures. This is because mixing charcoal with organic fertilizers has the potential of enhancing nutrient accumulations at the crops root zone. Further, this mixture can minimize nutrient leaching in the vastly weathered tropical soils besides boosting crop productivity in acidic soils [14, 15].

According to Obi [16], more than 998 million tons of agricultural wastes result from crops, livestock, and aquaculture productions annually across entire Africa. Most of these wastes are reused as fuel sources, and others are left to decompose on the farms as organic manures or as feedstock for anaerobic digestions like biogas generation. From crop production alone, enormous amounts of plant-based biomasses are generated annually across East African countries. However, a clear focus toward their use as precursors of value-added products as biochar is mostly missing. The primary reason is the lack of appropriate technology to employ and limited informed strategies to spur biochar production in the region. Prudent implementation of sustainable biochar production is a potent stimulant to spanning agricultural productivity, better economic growth while minimizing negative environmental impacts in the area. Notably, the steady population increase in the region is exerting much pressure on the exponentially shrinking arable land besides other implications as climate change and land use change. Thus, ardent efforts to restore the degraded soils through the use of biochar are a potential remedy. This must be done while ensuring that the present and future regional agricultural production standards, food security, and renewable energy sources are uncompromised. This is a fundamental element in the biochar bio-economy discourse, which is aimed at revolutionizing agronomic operations in East Africa while underscoring the spectrum of its usefulness and viability [11].

2. Biochar production process

In this chapter, we highlight the overview of the biochar production process, its usefulness, and potentials in improving agricultural productivity in East Africa.

2.1 Biochar sources

In principle, biochar can be produced from a range of carbonaceous feedstocks subjected to various thermochemical processes. The feedstocks can include agricultural wastes, municipal solid wastes, residues from forests, used building materials, and hydrocarbon substances like used tires, among others. Important to note is that the suitability of various feedstocks principally depends on their availability, biosafety regulations, and the targeted market conditions. Depending on the desired end use, biochar production for agricultural production should take into considerations the environmental aspects and an understanding of soil condition as well as its properties. Discrete processes employed in biochar production are outlined in **Table 1**. The methods span from slow pyrolysis, fast pyrolysis, flash pyrolysis, intermediate pyrolysis, vacuum pyrolysis, hydropyrolysis, torrefaction, and gasification, receiving varied treatments according to the quality and quantity of the desired final product [17]. Generally, temperature, pressure, heating rate, residence time, reactive or inactive environment, type of the purifying gas, and its flow rate are engineered to yield the targeted products. In all the pyrolysis processes, three main products are generated: solid biochar or ash, bio-oil or tar liquid, and non-condensable gases or syngas [1, 3].

The principle behind pyrolysis and volatilization is combustion reaction of biomass in an inert atmosphere. The inert atmosphere is ensured by flushing through the reactors with argon or nitrogen gases [9]. The application of heat to biomass feedstock causes disintegration of chemical bonds leading to smaller molecules vaporizing into gas oxidation state [18]. Due to the oxygen-deficient condition, the products formed are water, methane, carbon monoxide, and carbon dioxide, otherwise, in the presence of excess oxygen, heat and light results. Thus, the lack of oxygen causes the volatiles to form into dense gases or liquid tar and soot. Consequently, once all the volatile components are eliminated or oxidized, the remaining slow-burning residue undergoes the final stage of combustion called solid-phase oxidation to yield radiant coal. Therefore, each thermochemical decomposition process is dependent on the heat energy applied, pressure, the quantity of oxygen supplied, type of precursor, and the residence time [9].

2.2 Slow pyrolysis

Slow pyrolysis takes place at low heating temperatures of 400°C and a long solids residence time, causing secondary cracking of the primary products. In a slow pyrolysis process, biochar yields are higher (up to 45%) compared to bio-oil (30%).

Thermochemical process	Temperature range (°C)	Heating rate	Pressure	Residence time	Primary product
Slow pyrolysis	350–800	Slow (<10°C/min)	Atmospheric	Hours–days	Biochar
Fast pyrolysis	500–1250	Very fast (10–200°C/sec)	Vacuum–atmospheric	10–20 s	Bio-oil
Flash pyrolysis	900–1200	Fast	Elevated	0.1–1 s	Biocarbon/char
Intermediate pyrolysis	500–650	0.1–10°C/sec	Low (0.1 MPa)	5–35 min	Bio-oil/biochar
Vacuum pyrolysis	450–600	Slow	Low (0.05–0.20 MPa)	Hours	Biochar
Hydropyrolysis	500–1000	Very fast in hydrogen	High (5–20 MPa)	10–20 s	Bio-oil
Torrefaction	200–300	Slow (<10°C/min)	Atmospheric	Minutes–hours	Stabilized, friable biomass
Gasification	700–1500	Moderate–very fast	Atmospheric-elevated	Seconds–minutes	Syngas/producer gas

Source: Omulo [17].

Table 1.
Thermochemical processes, their representative reaction conditions, particle residence times, and primary products.

The lower heating rates and longer retention time enable vapor formed from complete secondary reactions to be eliminated, thus forming the solid carbonaceous biochar [10, 11, 19–21].

2.3 Slow pyrolysis versus traditional charcoal making

Charcoal has been used as a perennial fuel for domestic heating. In practice, charcoal is made by slow burning of wood in the absence of oxygen at mild to high temperatures [22]. Even though the charcoal making process can be referred as slow pyrolysis, the initial heat required to ignite the reaction is generated by burning part of the wood or the feedstock making it hard to achieve inert environment. For a typical slow pyrolysis process, the heat needed to decompose the feedstock thermally is supplied externally via an indirect heating medium. In contrast to the charcoal making process, the feedstock remains in airtight vessels or reactors [8]. Thus, the goal of slow pyrolysis is to yield a biochar product with high energy and carbon content. This is besides other by-products like pyroligneous acid or wood tar and non-combustible or syngas.

3. Environmental usefulness of biochar

3.1 Biochar from biomass as a fertilizer and a soil conditioner

The use of biochar from plant biomass as a soil fertilizer or conditioner has received significant attention in the recent past [12]. Formerly, extractions from the fermentation of bioethanol and flavonoids as well as recoveries from chemicals have also been applied as organic soil fertilizers [22]. However, the residues from fermentation processes have short lifespans. They are uneconomical to use as fertilizer because of their high moisture content. Thus, the need to develop alternative products from sustainable thermal conversion ways has put the use of biochar into perspective [23]. This is key since the conventional method of leaving raw biomass wastes on the soil to degrade naturally has remarkable risks primarily due to high bulk density, high moisture content, and the hygroscopic nature. Further, the biomasses contribute to air and water pollution and greenhouse effects via smoke resulting from burning. On the other hand, the use of biochar has been cited as a viable way of stabilizing soil organic carbon while minimizing greenhouse gas (GHG) emissions [2].

Biochar has been identified as a carbon-neutral bioenergy resource capable of enhancing soil conditions for better agriculture. It can also aid in curbing greenhouse emission effects and global warming [24]. Biochar as a carbon sequester can significantly contribute to agricultural productivity through the improvement of soil fertility and controlled pollution of rivers and groundwater, which are threatened by continued unsustainable agrarian practices [25]. Besides influencing carbon content in the soil, fresh biochar is instrumental in immobilization of nitrogen, improvement of soil pH and soil structure [1]. Further, soils affected by continuous leaching due to herbicides application, research has shown that biochar can curb the leaching process and assist in reigniting microbial activity in the soil [24, 26]. The following biomass feedstocks have been used for biochar production to utilize it as a fertilizer: microalgae [24], eucalyptus crop residues, castor meal, coconut pericarp, sugarcane bagasse [27], water hyacinth [28], and banana wastes [29]. **Table 2** illustrates the biochar nutrient contents of various biomass feedstocks.

3.2 The potential of biochar in carbon sequestration

The potential of biochar as a viable tool to carbon sequestration has recently been centered on the common discourse of climate change. Biochar has been pointed out to enhance carbon sinks, especially in dry regions [32]. However, the degree with which biochar achieves carbon sequestration depends on various factors. Most importantly, it depends on the desired soil carbon content and the rate of carbon dioxide removal from the atmosphere [3]. There are considerable large sizes of arable lands (estimated at 6% of the earth's surface); thus, they require relatively high amounts of biochar to be incorporated therein. Ideally, up to 90 tons of biochar per hectare should be incorporated into the farms compared to the current recommendations of 50 tons of biochar in a hectare [1] to help in reducing the level of carbon dioxide in the atmosphere. Since it takes long to sequester carbon dioxide

Residue	N	P	K	C	Ca	Mg
Wheat straw		0.21	2.90	18.29	7.70	4.30
Maize cob	10.8	0.45	9.40	429	0.18	1.70
Maize stalk	8.1	2.10	0.03	427	4.70	5.90
Forest residue	1.6	0.29	0.11	39	130	19.0
Peanut	15.0	2.4	—	429	—	—
Soybean	23.8	0.9	—	441	—	—
Rice husk	0.3	0.16	0.48	36	1.63	—

Table 2.
Biochar nutrients proportion from various biomass feedstocks (in g kg^{-1}) [27, 30, 31].

from the atmosphere, the predisposition asserted by industrial activities makes the process even longer. This means that even though biochar has the potential to sequester carbon, sustainable land use change and pollution control are indispensable. With improved and cheaper innovations like pyrolysis techniques, biochar productions potentially depend on biomass availability [3, 17, 27]. **Table 3** highlights the average agricultural wastes across East African countries generated from the major food crops [33].

3.3 The extent of biochar use in agriculture

The continued awareness of the benefits of biochar as a carbon sequester and soil conditioner has propelled its demand and use in the agricultural sector worldwide [34]. Research institutions and organizations have championed evidence-based research as incentives to upscale biochar acceptability and salability to farmers. One such organization is the International Biochar Initiative (IBI), which is a non-profit organization founded in 2006. Even though it is the biggest biochar promoter, several other establishments exist in different countries and regions of the world [25, 35, 36]. These biochar promoter organizations have been at the forefront in organizing scientific conferences to share insights on the latest research on biochar. Most importantly, they have been instrumental in proposing policies regarding biochar legislation. One such milestone is the Post-Kyoto Climate Agreements under the UN Framework Convention on Climate Change (UNFCCC), where biochar was unilaterally accepted a viable mitigation strategy [30, 32]. The Kyoto protocol was further intended to aid small economy countries to achieve sustainable development goals and to secure compliance with GHG emission minimization targets [37].

3.4 How does biochar improve soil properties?

The historical background about the use of biochar as soil amendment tool can be traced back to as earlier as 1929 when John Morley working with the US National

Crop	Burundi (×1000MT)	Kenya (×1000MT)	Rwanda (×1000MT)	S. Sudan (×1000MT)	Tanzania (×1000MT)	Uganda (×1000MT)
Beans, dry	3.83	13.12	6.42	0.011	14.41	9.96
Maize	2.55	32.59	3.93	1.06	61.57	27.47
Millet	0.12	0.69	0.069	0.059	3.83	2.50
Potatoes	0.60	5.62	2.85	—	6.27	1.00
Rice, paddy	1.59	1.20	1.45	—	44.59	3.89
Sorghum	0.37	3.02	2.53	9.27	13.60	6.51
Soybeans	0.072	0.044	0.84	—	0.116	0.80
Wheat	0.12	2.27	0.17	—	1.59	0.34
Barley	—	0.93	—	—	0.18	—
Oats	—	0.059	—	—	—	—
Total	**9.26**	**59.54**	**18.26**	**10.40**	**146.15**	**52.48**

Source: FAOSTAT [33].

Table 3.
Comparison of crop residues among East African countries (×1000MT) nitrogen content.

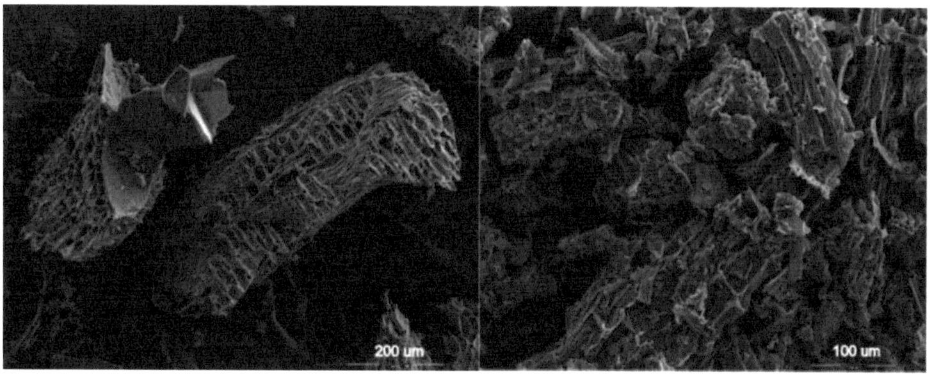

Figure 2.
Scanning electron micrographs of biochar particles showing porosity. Source: Brewer [3].

Greenkeeper realized that addition of traces of charcoal enhanced soil porosity [3]. Different kinds of biochar porosity exist depending on pore size. Pores can be categorized into micropores (diameter < 2 nm), mesopores (diameter 2–50 nm), and macropores (diameter > 200 nm) [3]. Mostly, macropores are susceptible to water, plant roots, and fungal hyphae penetration. Thus, the large pores influence the soil's hydrology and microbial ecosystem. It is easy to see the biochar pore size distributions using the scanning electron micrographs depending on the parent plant structure, see **Figure 2**. Therefore, it is the high porosity property of biochar that makes it contributes to the susceptibility of soil to water infiltration and increased micropore network in the soil [1, 30]. Thus, water retention in both sandy and silty soils can be significantly improved with the incorporation of biochar [1, 30].

Biochar's larger surface area to volume ratio also plays a significant role in cation exchange capacity (CEC) and the extent to which biochar can be integrated into the soil. The bigger the biochars' surface area, the greater the chemical exchanges; it can accommodate per unit gram [38]. Thus, it potentially curbs any form of nutrient leaching while boosting nutrients uptake [7, 31]. Biochar's bulk density is relatively low compared to soil bulk density; this encourages ease of nutrient release to plants and also lowering the effects of soil compatibility [3, 4].

Biochar is alkaline; this may influence the type of soil upon which it can be applied [39]. Depending on the type of feedstock pyrolyzed, biochar contains both primary and trace mineral elements useful for plants development [4]. Nonetheless, it has been noted that the presence of various functional hydrocarbon groups in biochar limits its release of water to plant roots especially in water stress conditions [36].

4. Potential of biochar use to boost East African agricultural productivity

East Africa is among the countries with the highest nutrient loss across sub-Saharan Africa with annual nutrient depletion rate of 41 kg N, 4 kg P and 31 kg K per hectare [40]. Even though soil fertility is quite dynamic, its inherent chemical, biological, physical, and anthropogenic characteristics play a significant role too [40]. Most soils in East Africa are acidic without enough nutrients to support sustainable crop production. This is because a bigger portion of the soils are extremely weathered, making them nutrient-deficient, especially with a limited stock of phosphorus, potassium, calcium, magnesium, and sulfur [41]. Similarly, soil acidity is influenced by the robust soluble aluminum, which is poisonous to most crops. Therefore, to ensure sustained crop productivity, improved soil fertility management is inevitable. This calls for sustainable and cheaper soil fertilization ventures like the use of biochar in these resource-constrained countries [40].

However, despite the known benefits of biochar to the scientific world, other stakeholders have concerns that are yet to be addressed. One of the significant issues is the uncertainty regarding the sustainable supply of feedstock for biochar production. Policymakers argue that mass production of biochar would need vast land for the feedstock required [1]. On the other hand, farmers still seem not to acknowledge that the crop residues within their farms can serve as biochar feedstock. The lack of reliable evidence in the literature regarding the sufficiency of crop residues as feedstock for biochar production in the context of East Africa is a gap. This section reiterates the fact that every field, farm, or region has the potential of generating enough biomass feedstock for biochar production. The analyses were carried out with a focus on the East African region. A few recent field and pot trials on the effectiveness of biochar on soil fertility enhancement have yielded very positive results [41]. Thus, the empirical evidence illustrated in this chapter reinforces the feasibility and sufficiency of biochar technology to impact the agricultural performance of smallholder farmers in East Africa amidst the effects of climate change.

According to the United Nations (UN) regional boundary delineation, East African region spans from the Red Sea coast in Eritrea, through to the Horn of Africa (Somalia) transcending the Indian Ocean coast line up to Mozambique. It further stretches inwards to Zimbabwe on the south, Zambia on the southwest and along the western rift valley encompassing Burundi, Rwanda, Uganda, South Sudan, and Sudan [4]. It also includes Indo-Oceanian Islands like Madagascar, Seychelles, Mauritius, and Comoros [42]. However, based on this chapter, a close focus will be given to the six countries forming the East African Community block. These countries include Kenya, Tanzania, Uganda, Rwanda, Burundi, and South-Sudan (although scarce information is available) (**Figure 3**). Agriculture still contributes substantially to the economic growth of East African, and it offers job opportunities to more than 70% of the region's population [44]. **Figure 4** illustrates the percentage of contributions of agriculture to the GDP of the countries in comparison with other sectors.

Despite this immense reliance on agriculture, agricultural production in East African countries is predominantly under subsistence basis. The bulging population, perennial low productivity recorded, alarming food insecurity and high demand for rich nutrient grains compound the desperate state of the agricultural sector in the region. Notwithstanding, the region is still estimated to harbor excellent agricultural potential. However, to harness this potential, sustainable and increased agricultural extensification via appropriate mechanization, improved land use management, and climate-adapted farming methods are inevitable [45].

The soils in East Africa have been affected by unsustainable continuous land use, non-conservation tillage methods employed, and native volcanic soils that are prone to degradation [4]. Consequently, the antidote to this menace is embracing sustainable and climate-friendly farming methods capable of increasing and preserving soil fertility. Biochar technology is one such viable means to achieve this noble vision [3].

4.1 How abundant are agricultural residues in East Africa?

The scalability of pyrolysis technologies in East African countries is yet at a very dismal stage. The primary reason is the low level of industrialization and little

Figure 3.
Map of East African countries. Source: United Nations [43].

appreciation of such technologies. This, however, does not disannul pyrolysis' potential to foster sustainable agricultural production through biochar production and use [7]. To underscore this hypothetical biochar usefulness in the region, critical scenario-based analysis seeking to disentangle the uncertainty on the availability and sustainability of biomass feedstock is paramount. The revelation will further contribute to changing the perception of both policymakers and farmers about biochar potential in the region [1]. As earlier stated, the feedstock can be generated from a range of biomass residues; the author, however, explore the potential in crops specifically on maize [46].

Maize is the most popular crop grown in East Africa because it is the staple food. Despite this, its productivity per hectare is below the estimated regional average (2.5 tons per hectare). Maize is a very high nutrient feeder crop; thus, its production requires extensive fertilizer use [47]. Other crops grown in the region include cash

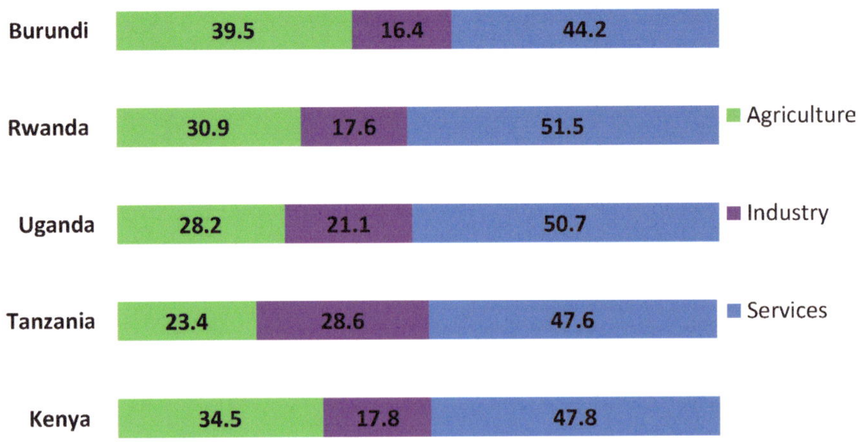

Figure 4.
Percentage contribution of agriculture to countries GDP compared to other sectors based on 2017 estimates. Source: The Centre of Intelligence, CIA [44].

crops such as tea, coffee, sugarcane, pyrethrum, and cotton; and food crops such as rice, wheat, beans, groundnuts, millet, sorghum, bananas, and potatoes, among others [48]. Taking maize as the most common crop, an illustration of its residues adequacy for biochar production is highlighted. **Table 4** outlines the estimates of maize production in the year 2019 versus the area planted across the selected East African countries.

The greatest challenge has been the inability to estimate the quantity of residues resulting from maize farming accurately. Even though proper records and monitoring of maize grains and other crops exists, very little has been done to quantify crop residues [49]. Nevertheless, in this section, the residues-to-product ratios (RPRs) estimation method is used. It is one of the most reliable ways to compute residue mass of any crop [49, 50]. Generally, maize crop generates 12% husks, 27% leaves, 49% stem, and 12% cob residues of the total plant mass [50]. Thus, the RPRs for maize residue are as follows: 2 for the stalk, 0.273 for cob, and 0.2 for husks. These factors are systematically used to quantify the possible maize residue that can be generated based on the current maize production rates in the East African countries

Country	Production (×1000 MT)	Planted area (×1000 Ha)	Yield (MT/Ha)
Kenya	3400	2000	1.7
Tanzania	6200	4200	1.5
Uganda	2800	1150	2.4
Rwanda	411	250	1.6
Burundi	260	180	1.4
South Sudan*	400	330	1.2
Total	**13,471**	**8110**	**1.7**

Source: The US Department of Agriculture, USDA [27].
**South Sudan data are based on 2016 statistics.*
Source: McKee [26].

Table 4.
Maize production, area planted and yield estimates in East Africa as of 2019.

(**Table 4**). Since the recorded crop masses are weighed in N kg, the respective residue masses are as follows: 76% representing stalk residue (leaves 27% and stem 49%) is 2.0 N kg at 15% moisture content, 12% representing cob residue is 0.273 N kg at 7.53% moisture content, and 12% representing husk residue is 0.2 N kg at 11.11% moisture content. When summed up, the expected total residue mass from the maize crop is approximately 2.47 N kg. Thus, expected total residue mass is 2.47 N kg [4, 50].

Consequently, a total of 13.5 million tons of maize is produced in East Africa under the total planted area of 8.1 million hectares. This implies that 33.3 million tons (13,471,000 × 2.47) of residues are generated annually. These residues are potential feedstock for pyrolysis for biochar production. To further estimate the possible amount of biochar that can be generated from these residues, pyrolysis parameters like residence time and heating temperature are paramount. Omulo et al. [29], Cantrell et al. [39] and Djurić et al. [51] noted that subjecting residues to temperature regimes of 300–650°C can lead to a biochar yield of between 40 and 28%. Thus, supposing that maize residues are pyrolyzed at a low temperature of 300°C, it is possible to generate up to 40% biochar as by-products. Thus, the rate of conversion of residues to biochar is taken as 0.4. This means that from the total mass of residues generated, about 13.3 million tons of biochar can be produced.

However, what percentage of the planted area in the region can be sustained by the produced biochar? In principle, biochar application can be made in two ways: a one-time application where biochar is applied at the required rate or an intermittent application where biochar is progressively applied until the acceptable threshold is achieved. Assuming that one-time biochar application is employed, Major [35] recommends the rate of 5 tons of biochar per hectare. Thus, based on the probable biochar yield in East Africa, a total of 2.7 million hectares (approximately 30% of the total planted area) can be adequately fertilized by biochar every season. Moreover, noting that biochar has a long decay life, even a one-time application is estimated to have long time effects on the soil [3].

4.2 Economic savings and biochar adoption potential for agricultural use

A plethora of evidence has shown that biochar use has the potential to improve agricultural productivity, especially among the acidic weathered soils [46]. Further proofs indicate that biochar application can double maize yield by application of only 4 tons per hectare [52]. Because soil degradation in East Africa has not escalated to irredeemable limits, proper biochar application is projected to have more profound impacts on crops productivity. Consequently, utilizing biochar to fertilize the staple maize farms can minimize the cost of fertilizers in the region while the saved revenue is used to extensify farming operations. Based on the estimations by Berazneva [41, 53], the mean shadow value of maize residues for farm soil fertility amendment is 0.07 USD/kg.

In comparison, an estimated cost of 0.04 USD/kg of fertilizer is conserved when the same residues are left on the field as mulch [53]. These values may differ slightly when pyrolysis costs are factored. Nevertheless, biochar production and utilization will potentially maximize the residue used to improve soil quality.

Hypothetically, considering the current low maize yield potential of the East African region, 1.7 tons per hectare, every hectare of maize would generate a total of 4199 kg of residues (**Table 5**). If these residues were to be utilized in biochar generation via pyrolysis process, then approximately $67.18 cost of fertilizer can be saved in 1 hectare of land. This implies that with the current price of $29.75 per 50 kg bag of diammonium phosphate (DAP), $24.05 per 50 kg bag of urea, and $29.15

Country	Production (×1000 MT)	Planted area (×1000 Ha)	Residue (×1000T)	Biochar (×1000T)	Area applied (×1000Ha)	Covered area as percentage of planted area
Kenya	3400	2000	8398	3359.2	671.84	33.59
Tanzania	6200	4200	15,314	6125.6	1225.12	29.17
Uganda	2800	1150	6916	2766.4	553.28	48.11
Rwanda	411	250	1015.17	406.068	81.2136	32.49
Burundi	260	180	642.2	256.88	51.376	28.54
South Sudan*	400	330	988	395.2	79.04	23.95
Total	**13,471**	**8110**	**33273.37**	**13309.348**	**2661.8696**	32.82

Source: USDA [48].
**Estimates are based on 2017 statistics except for South Sudan it is 2016.*

Table 5.
Possible area under biochar application as a percentage of the planted area at country level.

per 50 kg bag of NPK fertilizers [54], the saved cost can enable farmers to buy two more bags of fertilizers respectively. Ideally, this will reduce the amount of chemical fertilizers applied to the farm by two bags but with the same prospect of crop yield.

The basis of biochar production and use adoption in East Africa hangs on the current practices and use of crop residues in the region, especially the principal food crop, maize. According to Berazneva [41], most of the residues are burnt in situ on the farms to clean the land for the subsequent season and also to sterilize traces of pests and diseases. On the other hand, most farmers leave the residues on the farms for soil amendment even though their farms are susceptible to open grazing. Still, other farmers use the residues as a fuel source besides feeding their animals. Therefore, adoption of biochar technology to sustainably manage the generation of residues for fertilization can empower an Integrated Soil Fertility Management (ISFM) system.

This model of quantifying maize residues for biochar production underscores the potential of every country to achieve biochar application rate of 5 tons per hectare. The percentage areas that can be sustained by biochar in comparison to the planted areas depict that potential crop yield increment of the region (**Table 5**). Even the smallholder farmers have the opportunity to boost their productivity due to biochar application [55].

5. Possible barriers to biochar production in East Africa

Promotion and adoption of biochar technology in East Africa and across entire sub-Saharan Africa (SSA) is hampered by several obstacles ranging from policy and legal frameworks, institutional, socio-economic, fiscal, ecological, health, and technical issues [1]. This is majorly due to the lack of workable local policies and legal frameworks highlighting the rationale, the terms and conditions of biochar production as well as the associated technological aspects [41]. As a consequence, necessary measures to fast-track these impediments are deemed paramount and thus demands urgent actions as illustrated in **Figure 5**.

Borrowing a leaf from promotion and adoption of other renewable energy technologies, it is paramount that biochar technology is adapted to the local

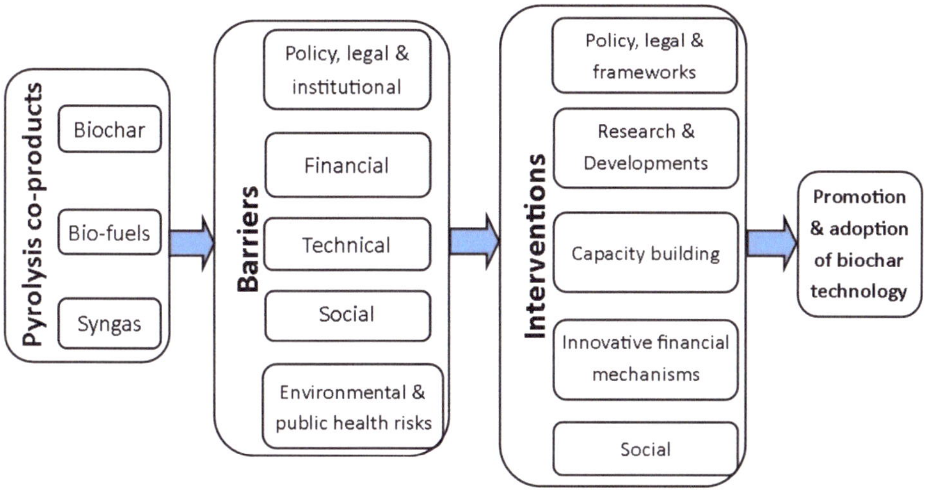

Figure 5.
Barriers to biochar and co-products production and use in SSA and viable interventions actions. Source: Gwenzi [1].

conditions and realities and should be affordable as possible. Deliberate capacity building on biochar technology through research and innovations channeled through various social strata can also break these constraints. Thus, for biochar technology to be feasible in East Africa, smallholder farmers should be able to understand the technology and afford the production and investment costs involved [1].

Another bottleneck to upscaling of biochar technology is the negative attitudes and perception surrounding it, especially by the majority risk-averse smallholder farmers. Discourses compound these cynicisms on nature conservation, competing interests, and deprivation of the scarce animal feeds [53]. The general belief that biochar production leads to deforestation besides being a complicated technology is quite difficult to disentangle. Nevertheless, proper knowledge sharing, supported by evidenced-based research, can serve as the most persuasive argument against such antagonistic ideologies. Further, the real potential and benefits of sustainable production and use of biochar for crop production can be underscored based on the perceived usefulness. Improved crop production implies better food security, poverty reduction, and reduced mortality rates. These challenges are faced by a majority of resource-constrained smallholder families across entire sub-Saharan Africa.

6. Future of biochar production and use

Biochar technology continues to offer numerous opportunities for developing economies. Apart from its suitability in agricultural production, biochar is highly an efficient and safe source of heat energy for small households compared to the current conventional use of charcoal [3]. Charcoal fires are usually operated openly, exposing then to inefficient heat transfer and air pollution, which can cause health complications too. Therefore, harnessing biochar production from crop residues offers excellent prospects for both energy and income generation even among smallholder families across the region. Government-led compensation schemes based on carbon credits on the amount of carbon sequestered and participation in climate change mitigation would highly incentivize biochar use.

With proper organization, biochar producers can significantly benefit from the improved market where they can sell their products and even via cooperatives to get better bargaining power. Access to the improved energy source for domestic use like biochar can create time for women to be involved in other more income-generating activities. Consequently, increased income will lead to improved quality of life among their households [3]. Biochar use can be diversified without interfering with the necessary amounts needed for soil amendments and increased crop productivity [4].

Critics have pointed out that sufficient biochar production may potentially lead to deforestation and that the use of inappropriate production methods may also result in air pollution [1]. Nevertheless, as earlier stated, sustainable biochar generation might not be dependent on a single biomass source. Instead, sourcing biomass residues from a wide range as forest products, crop wastes, animal wastes, biodegradable landfills, urban, and construction bio-wastes are more viable [17, 29]. This means that in future, pyrolysis techniques employing efficient bio-reactors for biochar production will potentially minimize any heat loss and pollution [3] while maximizing the yields. A more informed decision among users and action-oriented policy frameworks, as well as research development, can reinforce the desired production and use of biochar in the region.

7. Conclusion

The problems of land degradation and climate change effects are spread uniformly across East African countries. Farmers desire amicable solutions to these challenges. Biochar technology has proven to be such a feasible solution that doubles as a soil conditioner and a climate-friendly product. Based on the current maize production trends in the region, it is possible to generate enough biomass residues for biochar production. This has the potential of reducing fertilizer use by farmers in the region by up to 30% cushioning them from the exorbitant mineral fertilizers while still getting the desired yield. Thus, with proper uptake and implementation, driven by sound policies and good governance, biochar production has a great potential to boost farmers production and improve their quality of life. Nevertheless, it is paramount that the technology is adapted to the local conditions, be backed up with current research evidence, proper capacity building, and concerted efforts among the stakeholders.

Acknowledgements

The author would like to appreciate the auditors for their time and input throughout the review process of this chapter.

Conflict of interest

No conflict of interest to declare.

Author details

Godfrey Omulo
Institute of Agricultural Sciences in the Tropics (Hans-Ruthenberg-Institute), Universität Hohenheim (490), Stuttgart, Germany

*Address all correspondence to: omuloh@gmail.com

References

[1] Gwenzi W, Chaukura N, Mukome FND, Machado S, Nyamasoka B. Biochar production and applications in sub-Saharan Africa: Opportunities, constraints, risks and uncertainties. Journal of Environmental Management. 2015;**150**:250-261

[2] Duku MH, Gu S, Ben Hagan E. Biochar production potential in Ghana—A review. Renewable and Sustainable Energy Reviews. 2011;**15**(8):3539-3551

[3] Brewer CE. Biochar Characterization and Engineering. United States: Iowa State University; 2012

[4] Ndhlovu M, Banadda N. Determination of sufficiency of crop residue for biochar application. African Journal of Agricultural Research. 2017;**4**(1):254-262

[5] Kanyanjua SM, Ireri L, Wambua S, Nandwa SM. Acidic Soils in Kenya: Constraints and Remedial Options. Nairobi, Kenya: Kenya Agricultural Research Institute; 2002. p. 11

[6] Abate E, Hussein S, Laing M, Mengistu F. Soil acidity under multiple land-uses: Assessment of perceived causes and indicators, and nutrient dynamics in smallholders' mixed-farming system of Northwest Ethiopia. Acta Agriculturae Scandinavica Section B Soil and Plant Science. 2017;**67**(2):134-147

[7] Kong SH, Loh SK, Bachmann RT, Rahim SA, Salimon J. Biochar from oil palm biomass: A review of its potential and challenges. Renewable and Sustainable Energy Reviews. 2014;**39**:729-739

[8] Nartey OD, Zhao B. Biochar preparation, characterization, and adsorptive capacity and its effect on bioavailability of contaminants: An overview. Advances in Materials Science and Engineering. 2014;**2014**:1-12

[9] Zaman CZ et al. Pyrolysis: A sustainable way to generate energy from waste. In: Pyrolysis. London: Intech Open Science; 2017. p. 35

[10] Brewer CE, Schmidt-Rohr K, Satrio JA, Brown RC. Characterization of biochar from fast pyrolysis and gasification systems. Environmental Progress & Sustainable Energy. 2009;**28**(3):386-396

[11] Spokas KA et al. Biochar: A synthesis of its agronomic impact beyond carbon sequestration. Journal of Environmental Quality. 2012;**41**(4):973-989

[12] Klinar D. Universal model of slow pyrolysis technology producing biochar and heat from standard biomass needed for the techno-economic assessment. Bioresource Technology. 2016;**206**:112-120

[13] Sadaka S, Sharara M, Ashworth A, Keyser P, Allen F, Wright A. Characterization of biochar from Switchgrass carbonization. Energies. 2014;**7**(2):548-567

[14] Steiner C et al. Long term effects of manure, charcoal and mineral fertilization on crop production and fertility on a highly weathered central Amazonian upland soil. Plant and Soil. 2007;**291**(1-2):275-290

[15] Steinera C, Das KC, Garciac M, Försterd B, Zech W. Charcoal and smoke extract stimulate the soil microbial community in a highly weathered xanthic Ferralsol. Pedobiologica. 2008;**51**:359-366

[16] Obi F, Ugwuishiwu B, Nwakaire J. Agricultural Waste

Concept, Generation, Utilization and Management. Nigerian Journal of Technology. 2016;**35**(4):957

[17] Omulo G. Optimizing Slow Pyrolysis of Banana Wastes Using Response Surface Methodology to Enhance Bio-Infrastructure Products Yield. Kampala, Uganda: Makerere University; 2017

[18] Bridgwater V. Principles and practice of biomass fast pyrolysis processes for liquids. Journal of Analytical and Applied Pyrolysis. 1999;**51**(1):3-22

[19] Brennan L, Owende P. Biofuels from microalgae — A review of technologies for production, processing, and extractions of biofuels and co-products. Renewable and Sustainable Energy Reviews. 2010;**14**:557-577

[20] Demirbas A. Biofuels sources, biofuel policy, biofuel economy and global biofuel projections. Energy Conversion and Management. 2008;**49**:2106-2116

[21] Tripathi M, Sahu JN, Ganesan P. Effect of process parameters on production of biochar from biomass waste through pyrolysis: A review. Renewable and Sustainable Energy Reviews. 2016;**55**:467-481

[22] Chew KW, Chia SR, Yen HW, Nomanbhay S, Ho YC, Show PL. Transformation of biomass waste into sustainable organic fertilizers. Sustainability. 2019;**11**:1-19

[23] Volpe M, Panno D, Volpe R, Messineo A. Upgrade of citrus waste as a biofuel via slow pyrolysis. Journal of Analytical and Applied Pyrolysis. 2015;**115**:66-76

[24] Chang YM, Tsai WT, Li MH. Chemical characterization of char derived from slow pyrolysis of microalgal residue. Journal of Analytical and Applied Pyrolysis. 2015;**111**:88-93

[25] Barrow CJ. Biochar: Potential for countering land degradation and for improving agriculture. Applied Geography. 2012;**34**:21-28

[26] Hagner M, Penttinen O, Tiilikkala K, Setälä H. The effects of biochar, wood vinegar and plants on glyphosate leaching and degradation. European Journal of Soil Biology. 2013;**58**:1-7

[27] Abdelhafez AA, Abbas MHH, Li J. Biochar: The black diamond for soil sustainability, contamination control and agricultural production. In: Engineering Applications of Biochar. London: Intech Open Science; 2017. p. 23

[28] Doumer ME et al. Slow pyrolysis of different Brazilian waste biomasses as sources of soil conditioners and energy, and for environmental protection. Journal of Analytical and Applied Pyrolysis. 2015;**113**:434-443

[29] Omulo G, Banadda N, Kabenge I, Seay J. Optimizing slow pyrolysis of banana peels wastes using response surface methodology. Environmental Engineering Research. 2019;**24**(2):354-361

[30] Verheijen F, Jeffery S, Bastos AC, Van Der Velde M, Diafas I. Biochar Application to Soils: A Critical Scientific Review of Effects on Soil Properties, Processes and Functions. Vol. 8. No. 4. Luxembourg: Functions. EUR 24099 EN, Office for the Official Publications of the European Communities; 2009

[31] Yuan JH, Xu RK, Zhang H. The forms of alkalis in the biochar produced from crop residues at different temperatures. Bioresource Technology. 2011;**102**(3):3488-3497

[32] Matovic D. Biochar as a viable carbon sequestration option: Global and Canadian perspective. Energy. 2011;**36**(4):2011-2016

[33] FAOSTAT. FAOSTAT Crop Residues. Food and Agriculture Organization of the United Nations [Online]. 2017. Available from: http://www.fao.org/faostat/en/#data/GA [Accessed: 19 March 2020]

[34] Obia A, Mulder J, Martinsen V, Cornelissen G, Børresen T. In situ effects of biochar on aggregation, water retention and porosity in light-textured tropical soils. Soil and Tillage Research. 2016;**155**:35-44

[35] Major J. Guidelines on Practical Aspects of Biochar Application to Field Soil in Various Soil Management Systems. New York, United States: International Biochar Initiative; 2010

[36] Fryda L, Visser R. Biochar for soil improvement: Evaluation of biochar from gasification and slow pyrolysis. Agriculture. 2015;**5**(4):1076-1115

[37] Beeharry RP. Carbon balance of sugarcane bioenergy systems. Biomass and Bioenergy. 2001;**20**(12):361-370

[38] Lehmann J, Rillig MC, Thies J, Masiello CA, Hockaday WC, Crowley D. Biochar effects on soil biota—A review. Soil Biology and Biochemistry. 2011;**43**(9):1812-1836

[39] Cantrell KB, Hunt PG, Uchimiya M, Novak JM, Ro KS. Impact of pyrolysis temperature and manure source on physicochemical characteristics of biochar. Bioresource Technology. 2012;**107**:419-428

[40] Bekunda M, Ebanyat P, Nkonya E, Mugendi D, Msaky J. Soil fertility status, management, and research in East Africa. Eastern Africa Journal of Rural Development. 2005;**20**(1):1-23

[41] Berazneva J. Economic value of crop residues in African smallholder agriculture. In: AAEA and CAES Annual Meeting, August 4-6, 2013.; 2013, No. July. pp. 1-25

[42] Blein R et al. Agriculture in Africa, Transformation and Outlook. Johannesburg: NEPAD Transforming Africa; 2013. p. 72

[43] United Nations. Map of East African Community. United Nations Department of Field Support. p. 1

[44] CIA. The World Fact Book. The Centre of Intelligence Agency, Washington, DC, United States [Online]. 2020. Available from: https://www.cia.gov/library/publications/the-world-factbook/fields/214.html [Accessed: 03 March 2020]

[45] Mrema GC, Kienzle J, Mpagalile J. Current status and future prospects of agricultural mechanization in sub-Saharan Africa [SSA]. AMA, Agricultural Mechanization in Asia, Africa & Latin America. 2018;**49**(2):13-30

[46] Cornelissen G et al. Biochar effect on maize yield and soil characteristics in five conservation farming sites in Zambia. Agronomy. 2013;**3**:256-274

[47] McKee D. Maize Production Soaring in Eastern Africa. World Grain.com [Online]. 2016. Available from: https://www.world-grain.com/articles/7172-maize-production-soaring-in-eastern-africa [Accessed: 03 March 2020]

[48] USDA. Countries Agricultural Production Statistics, Index Mundi [Online]. 2020. Available from: https://www.indexmundi.com [Accessed: 03 March 2020]

[49] Bentsen NS, Felby C, Thorsen BJ. Agricultural residue production and potentials for energy and materials services. Progress in Energy and Combustion Science. 2014;**40**(1):59-73

[50] Koopmans A, Koppejan J. Agricultural and forest residues—Generation, utilization and availability. In: Regional Consultation on Modern

Applications of Biomass Energy; 1997, No. January, pp. 6-10

[51] Djurić SN, Brankov SD, Kosanić TR, Ćeranić MB, Nakomčić Smaragdakis BB. The composition of gaseous products from corn stalk pyrolysis process. Thermal Science. 2014;**18**(2):533-542

[52] Cornelissen G, Martinsen V, Obia A, Abiven S. Improving Crop Yield and Storing Carbon: Biochar in Conservation Farming in Zambia. Lusaka, Zambia: Norwegian Geotechnical Institute; 2014

[53] Berazneva J, Lee DR, Place F, Jakubson G. Allocation and valuation of smallholder maize residues in Western Kenya. Ecological Economics. 2018;**152**:172-182

[54] AF. Fertilizer Statistics. AfricaFertilizer.org [Online]. 2019. Available from: https://africafertilizer.org/statistics/ [Accessed: 05 March 2020]

[55] Ndhlovu M, Kiggundu N, Wanyama J, Banadda N. Effects of incorporating biochar into the soil using power tiller and. Sustainable Agricultural Resarch. 2017;**6**(4):93-103

Chapter 9

Enhancement of Soil Health Using Biochar

Oladapo T. Okareh and Alaba O. Gbadebo

Abstract

Over the years, the carbon-rich biochar has been used for the purpose of environmental conservation and reservation. Typically produced from varieties of materials ranging from sewage, farm produce, energy crops and agricultural waste or residues, the properties usually considered in the application of biochar include the chemical composition, porosity and stability. Contemporarily, the use of biochar has extended to its utilization in the industry, agriculture, forestry, and the natural environment. Soil fertility depends on the holistic process of managing the soil and likewise maintaining a robust biodiversity. This process involves the application of natural carbon-rich materials like biochar as soil amendments. The rapid absorption tendency of biochar, both on organic and inorganic materials has contributed immensely to the removal of excess antimicrobials from the environment. Biochar has been known to be a good enhancer of the soil due to its rich content of carbon and other nutrients good enough for the soil. Other striking effects of biochar on the soil and environment include the enhancement of the uptake of nitrogen, improvement of the soil structure, mitigation of greenhouse gases, conservation of the environment and enhancement of soil microbial community.

Keywords: biochar, soil fertility, soil enhancement, soil properties, wastes, soil carbon

1. Introduction

Biochar is a carbon-rich stable solid biomass in form of humus which is produced either from sustainable waste particles buried in the soil or through pyrolysis of plant or animal biomass under different temperature conditions. The nature and efficiency of the plant biomass used in biochar production depend on the type and the characteristics of the soil from which the plant biomass was developed [1]. The stability of organic matter in soils is determined by its ability to resist microbial and/or chemical decomposition, through chemical transformations and physical interactions with soil minerals. The sequestration of carbon and carbon dioxide reduction are being gradually achieved in recent years through the use of biochar. Biochar also help in the reduction of wastes and removal of the same from the environment. It actually comprise of organic carbon with low expectancy of degradation [1, 2].

Biochar has been in hunt for so many decades till now. Over the years, it has been used for the purpose of environmental conservation and reservation. Today, the use of biochar has extended to its utilization in the industry, agriculture, forestry, and the natural environment. It has also been used for the purpose of heavy metal digestion, pollutant immobilization as well as a supplement to composting and methane fermentation process [1, 3]. Recently, biochar is being used in pyrolytic filtering of tar and also in the production of hydrogen [3].

The presence of organic matter and nutrients in biochar is vital because the nutrients provide basic mineral supplements for the soil. Therefore, the amendment of soil with biochar increases the soil's pH, cation-exchange capacity (CEC), total nitrogen, organic carbon and conductivity. It has been reported that the fate and toxicity of heavy metals have been reasonably reduced due to the amendment of soil with biochar [4]. Furthermore, the porosity, large surface area and absorbance potential have made biochar a medium for soil nutrient improvement. Likewise, biochar also acts as a suitable habitat for soil microorganisms to thrive, alongside the support by roots of leguminous plants that aid the symbiotic relationships between the microorganisms and the plants [5].

Biochar has also been known to decrease the tensile strength of the soil because of its porosity and ability to create air pores within the soil, thus reducing soil compaction. With this development, biochar can be used as a sustainable tool for agricultural soil development. High crop yields can be produced with the use of biochar without necessarily depleting essential nutrients from the environment [6].

2. Biochar properties that enhance environmental safety

Biochar can be produced from varieties of materials ranging from sewage, farm produce, energy crops and agricultural waste or residues. Other materials used for biochar include forest wastes like conifer barks, sawdust pellets, paper and moss. After biochar is produced, the quality of the biochar is determined by the assessment of certain properties possessed by the biochar product [7, 8].

The properties usually considered at the post-production stage of biochar include the chemical composition, porosity and stability. However, at the point of production, the chemical composition can be manipulated by the type of substrate used in the biochar preparation and the composition of the substrate determines the chemical composition of the biochar itself [3, 9].

Typically, biochars contain stable organic compounds where the carbon content may be within the range of 50 to 90% and volatile content at 10 to 50% [2, 10]. Biochar seldom undergo microbial degradation because of their carbonized nature and their pH. Some of the properties of biochar are shown in **Table 1**. These properties determine the level of biochar quality. The temperatures of the pyrolytic processes at the point of production of the biochar are also indicated.

The pH is either neutral or alkaline. The alkaline pH sometimes results from the biomass pyrolysis of high temperature. The increase in the temperature of the soil containing the biomass has a resultant effect on the alkalinity of the biomass [10]. However, at lower temperatures, high ion-exchange biochars are produced. The cation exchange capacity which determines nutrient absorption indicates a rise in temperature at every rise in the cation exchange capacity of the soil biomass [11]. It is as a result of these physicochemical properties that make biochars qualify as good soil carbon sequesters and soil remediation agents [3, 12]. Furthermore, soil types that are supplemented with biochar have a very high degree of absorption.

Biochar has a predominantly condensed aromatic structure that is known to be highly resistant to microbial decomposition. The porosity of the biochar adds to the water retention capacity of the soil and increases the soils ability to form aggregates [13]. **Figure 1** describes the porous nature of a typical biochar.

DOI: http://dx.doi.org/10.5772/intechopen.92711

Feedstock for biochar production	pH	Carbon (g/kg)	Nitrogen (g/kg)	Carbon/ nitrogen ratio	Phosphorus (g/kg)	Potassium (g/kg)	Ash (%)	CEC (mEq/100 g)	Pyrolysis temperature (°C)
Acacia bark	7.4	398	10.4	38	—	—	—	350	260–360
Coconut bark		690	9.4	73	—	—	3.38	—	500
Corn	—	675	9.3	73	—	10.4	—	400	350
Corn	—	790	9.2	86	—	6.7	—	670	600
Green waste	6.2	680	1.7	400	0.2	1	—	228	450
Peanut shells	—	499	11.0	45	0.6	6.2	—	—	400
Pecan shells	7.6	834	3.4	245	—	—	3.8	640	700
Pecan shells	—	880	4.0	220	—	—	—	—	700
Rice slaw	—	490	13.2	37	—	—	9.54	—	500
Sewage sludge	—	470	64	7	56	—	35	—	450
Sugarcane bagasse	—	710	17.7	40	—	—	4.34	—	500
Eucalyptus wood	7.0	824	5.7	144	0.6	—	0.23	286	350
Oak wood	—	759	1.0	759	—	1.1	—	400	350
Oak wood	—	884	1.2	737	—	2.2	—	120	600

Source: Saletnik et al. [1].
NB: The biomass with suitable pH for biochar was highlighted. Other biomasses are more acidic, thus less suitable. Most emphasis was based on suitable pH and pyrolysis temperature while the CEC (cation exchange capacity) of the suitable biomass was determined.

Table 1.
Selected properties of biochar and the respective temperatures of pyrolytic processes.

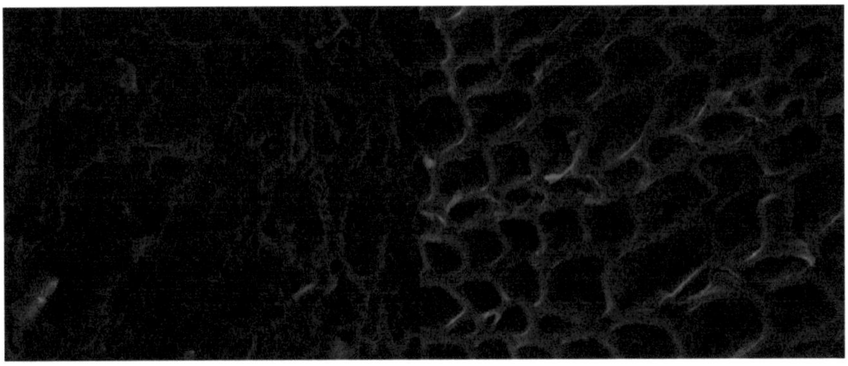

Figure 1.
The porous nature of biochar (Burrell et al. [14]).

3. Production processes of biochar and biochar efficiency

Biochars could be made either by burying organic waste particles in the soil or from pyrolysis of selected types of waste and fuel under different temperature conditions ranging from 260°C to about 600°C [4, 14]. When wastes are buried under the soil, for example wooden debris, sewage, sawdust and other similar wastes, they are placed in a shallow garden bed of 1–2 feet, smoked and covered with soil of up to 1 inch [12]. The waste material is thereafter left to char before the fire is put out. This leftover material referred to as biochar, is used in the improvement of soil composition. Likewise, the pyrolysis process involves the anaerobic digestion of biomass, usually between 260°C to about 600°C [4, 13–16]. These wastes could be regarded as raw materials for the pyrolytic process of making biochar. Thus, these waste materials are major constituents in the chemical and structural properties of biochars. Example of the raw materials for the pyrolytic process is categorized into three groups which include the following:

- Coal: charcoal can be made from coal to produce biochar. Apart from coal, charcoal can be made from peat, wood and petroleum. One of the common raw materials for charcoal is coal. The presence of solid coal increases the efficiency of the production of good biochar [17]. However, coal could also be used as a fuel material for the pyrolytic process.

- Biomass: example of biomass is wood, plant debris and organic matter. Biomass is one of the most important raw materials for pyrolytic production of biochar. It could be homogenous or heterogeneous in composition, and could either have a high or low humidity. The biomass could either be strongly bonded or loose in structure [4, 7].

- Sewage sludge: examples of these are tyres.

The composition of the raw materials determines the structural and chemical properties of the biochar. The processes involved in the biochar production through gasification procedures should be reviewed occasionally to avoid difficulty in the production of biofuel. The type of bioreactor and range of products should also be simplified enough to allow for the efficiency of the production process and technological reliability.

The high temperature enhances the production of hemicelluloses and depolymerization of celluloses in the biofuel production. The collisions alongside the high level of temperature are effective enough to prevent dehydration reactions; unlike the low temperature rates where the collisions always produce dehydration reactions [18]. Similarly, the fast and slow heating rate affects the product efficiency and volatility of the carbonated material. A slow heating rate has been shown to reduce the rate at which volatile substances from the production materials escape into the atmosphere causing more secondary reactions take place; while the fast heating rate increases volatility of volatile compounds thereby less secondary reactions take place [19]. Thus, more secondary reactions definitely result into high grade biochar production.

However, the most important properties are the chemical and physical properties. The raw materials with high lignin content have a very high tendency of producing efficient biochar but at relatively moderate temperatures. Furthermore, biomass materials that are relatively volatile produce efficient and large amount of pyrolitic gas and biofuel [20]. Furthermore, the amount of moisture in the raw materials for biochar determines the speed of the process of heat transfer and the distribution of products.

4. Soil fertility and biochar soil amendment

Soil fertility is a condition whereby a particular soil is physically, chemically and biologically stable with the required ecosystem intact without disrupting it. The state of complete functioning of the ecosystem is still intact in a soil termed to have the soil quality. There must be favorable interactions between the abiotic and biotic elements of the soil's ecosystem and still serve the purpose of crop production for sustainability [21]. A soil may have a complete ecosystem, yet not serving the purpose of crop production. Thus, soil fertility depends on the holistic process of managing the soil and at the same time keeping a robust biodiversity. Part of this process is through the application of natural carbon-rich materials like biochar as soil amendments [22]. The natural organic material added to the soil should fulfill the conditions of sustaining life, increasing biodiversity, keeping good water and air quality, beneficial to human health, as well as sequestering carbon.

5. Application of biochar and environmental safety standards

Legal standards, ab initio have been set on environmental matters, part of which includes biochar. Biochar in many countries is regarded as waste or fertilizers which have been guided by regulations and legal statements. Although, biochar may not appear in the legislation of many countries, even in the European Union (EU), but the law and regulations about waste control is in the offing of the constitutions of many nations across the world; though a few have gone beyond implementation.

In reality, biochar is considered as waste, but in several literatures, it is usually considered as a by-product of carbonization. Waste, according to the EU, is defined as any product that is supposed to be discarded or any substance the holder proposes to discard. The Waste Act of the EU therefore specifies that any waste that undergoes a recycling process and meet the following conditions may not be regarded as waste. These conditions are:

- If the specified waste is commonly used for certain purposes
- If the product is being traded in the market or has market value

- If the product meets the standard and specific requirement set out for the application of such product to the environment [23, 24]
- If the product itself or the use of it does no harm to the environment and to the people directly or indirectly affected by the product or its use
- If the use of the product does not cause long-term defects on the living and nonliving entities in the environment [6, 22]
  Biochar may also be classified as a by-product if it meets the above conditions.

Furthermore, other conditions whereby biochar cannot be regarded as waste are:

- If it is certain that the biochar can be reused
- If the biochar can be used without any further processing or manipulation
- If the biochar is an integral entity of the manufacturing process
- If the use of the biochar will not constitute an environmental nuisance or health hazard [25]

Apart from the existence of biochar as a by-product, it could also portray the picture of a product if it has been produced as the major product from the biomass raw material specifically designed for the process of that biochar production. Thus, the Waste Act is less concerned about the concept of using the biochar as product or by-product since biochar is not specifically mentioned in the Waste Act. The only country that has specific regulations about biochar is the Switzerland where biochar is permitted to be used in agriculture [26]. There is the European Biochar Certificate (EBC) which is recommended by the EU but only Switzerland has included the certification in their law in the use of biochar [7].

However, in Poland, the use of biochar as fertilizer or for the amelioration of soil must be preceded with the completed and signed approval form from the Ministry of Agriculture and Rural Development with the exception of the use of biochar for research purposes [25].

6. Applications of biochar for soil sustainability

In the face of declining soil infertility, climate change and human anthropogenic activities have made situation worse. However, several organic products have been introduced as part of the palliatives to lessen soil burden. Among these products is biochar which has been used as a popular choice for remediation of soil. Application of biochar to the environment is not detrimental because biochar samples have been found to contain several polycyclic aromatic hydrocarbons that are environmentally friendly. Pollutant compounds and toxic chemicals are yet to be found in the feedstocks used in the production of biochar [27].

However, acceptable proportions and amounts of biochar need to be determined through approved methods and environmental risk assessment so that the addition to land and aquatic bodies will be safe for the environment. The type of soil also needs to be considered to determine if specific physicochemical property of biochar is required for specific type of soil. The carbon sequestering potential of the biochar product is important for the determination of the amount of carbon sequestered for the purpose of evaluating the greenhouse gas effects and mitigating global warming [28, 29]. The Clean Development Mechanism (CDM) of the Kyoto Protocol is a model for the approval of materials involved in waste management. If biochar is approved by the CDM, then there is certainty of global commercial use. Fewer studies have been done to support the use of biochar when some studies have been skeptical about its use due to the ignorance of the interactions that take place between biochar and the soil. Therefore, there is need for innovations to create models that will be required for the assessing locations and soil types to determine the biochar-soil-climate interactions [19, 29].

7. Soil enhancement tendencies of biochar

Biochar has been effective on the soil in recent years through the evaluation of the soil quality after application to the soil. It has been observed that after application to the soil, the soil nutrients have been retained and soil quality improved. As earlier mentioned, the porous nature of the biochar, among other qualities has made the product to be a good soil conditioner and enhancer in recent years [30]. Some of these physicochemical properties exceptional of biochar will be discussed in details in this chapter.

7.1 Absorption tendencies of biochar

Biochar has been found to be a very good absorbent for the removal of excess pharmaceutical components from the environment. The kinds of pharmaceuticals that have been found to be removed by biochar are glyphosate, ibuprofen, atrazine, acetaminophen, and caffeine [9, 16, 31]. The excess release of pharmaceuticals into the environment has been a major risk factor for the resistance of microorganisms to antimicrobials. The rapid absorption tendency of biochar has therefore contributed immensely to the removal of these excess antimicrobials from the environment [10]. Toxins and unwanted drugs from the gastrointestinal tract have been shown to be reduced to reasonable amounts by biochar [11, 23, 29].

Furthermore, biochar has also been used to remediate waste fruit candy extract using deionized water as solvent with the aim of reusing the candy waste for the production of organic acids. The electrostatic interaction between polar and non-polar groups enhances the adsorption potential on organic substances by biochar. **Figure 2** reveals the mechanism of adsorption of organic substances by biochar. However, the adsorption of inorganic substances like heavy metals by biochar or activated carbon involves mechanisms like ion exchange, precipitation, cationic anionic metal attraction [29–31]. **Figure 3** shows the mechanisms of inorganic substance adsorption by biochar.

7.2 Soil amelioration of biochar

Biochar has been known to be a good enhancer of the soil due to its rich content of carbon and other nutrients good enough for the soil. Prior to the addition of biochar to soil, plant growth has not been optimum; but after the addition of biochar, the rate of growth of the plants on the same soil has been with optimum yield. A major advantage of biochar for soil amelioration is cost effectiveness and efficient process of application [32].

However, the feedstocks used in the production of biochar determine its efficiency and nutritive value. It also determines the proportion of the micro and macro-nutrients present in the biochar which resultantly affects the interactive patterns between the carboniferous biochar and organic as well as inorganic particles in the soil [29].

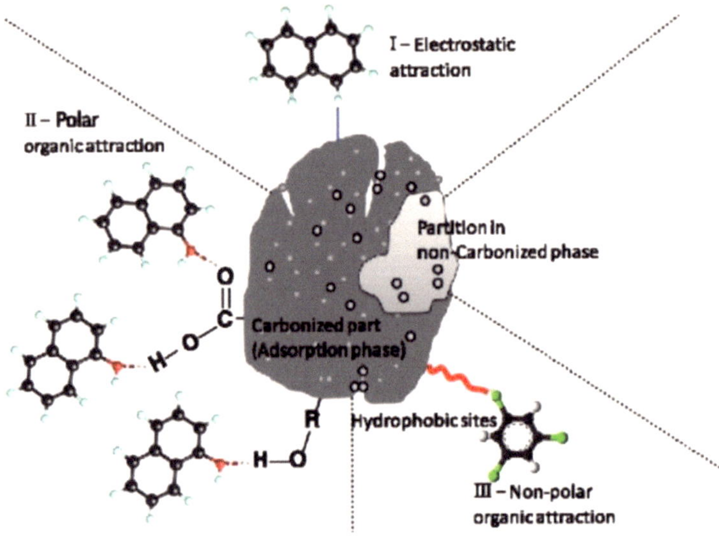

Figure 2.
Mechanism of adsorption of organic substances by biochar [31].

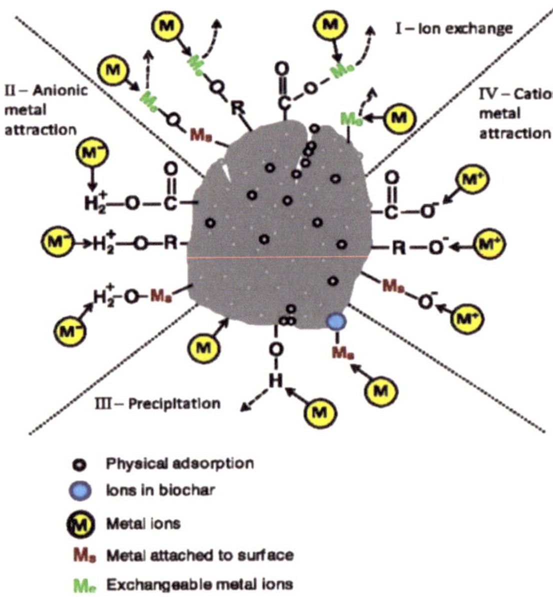

Figure 3.
Mechanism of inorganic adsorption by biochar [31].

Biochar from wood, poultry litter and cattle manure was discovered to produce the highest yield in cowpea, radish and maize plantations, respectively [28–31]. Certain characteristics of biochar that facilitates the enhancement of the soil includes its nature of rejecting biodegradation which makes it to spend more time within the soil for proper and effective carbon sequestration, thus making it an effective moderator for carbon dioxide in the soil. However, the presence of biochar in the soil tends to increase the amount of biogenic substances in the soil. Nonetheless, substances like phosphorus and potassium are bonded and chelated with biochars to avoid excess leaching of these substances into the environment [33]. The ion-exchange properties of biochars also enable the exchange and replacement of organic and inorganic substances that helps to complete the biogenic and geogenic cycles.

Additionally, biochar is a booster for the silicon cycle by supplying biogenic silica, also known as phytoliths into the soil. Likewise, it also helps in balancing the phosphorus and nitrogen cycle by the ion-exchange storage of phosphorus and nitrogen respectively in their compound forms [34]. Thus, when ammonia and ammonium compounds are trapped in the soil by biochars, it reduces the process of the formation of nitrous oxides, thereby reducing the emission of greenhouse gases into the atmosphere. Furthermore, by trapping more nitrogen in the soil, it increases the productivity rate of nitrogen-fixing bacteria in the soil, thereby increasing soil yield [35].

Biochars also increase the pH of the soil due to the presence of carbonized compounds like calcium carbonate [23, 30]. This may not be too beneficial in temperate regions where the soil pH is always tending toward the right side of the scale; but in the acid tropical soil regions, biochars are mostly used to regulate the pH of the soil. Due to the physical porous nature of biochars, they also affect the physical structure of soils thereby affecting the soil's retention capacity [36]. Therefore, such a soil is able to retain water at a very high rate but at the same time, forms aggregates and increases the capacity to resist erosion. **Figure 4** shows how soil enhancement can be achieved through biochar addition. Thus, with the combined effect of porosity and carbonization, biochar-in-soil is a favorable substrate for microorganisms to thrive on in the soil. *Nitrobacter* and *Nitrosomonas* spp. are suspected to thrive better in such biochar-laden soils due to the presence of trapped nitrogen by biochar [2, 4, 33]. Subsequently, there would be high microbial stimulation and efficient nitrogen mineralization in that soil.

7.3 Enhanced nitrogen fertilizer

Fertilizers and water need to be supplied increasingly as food and agricultural practices keep increasing. The rate at which population increases demands that more food should be supplied; hence, more fertilizer is needed. Over the years, the supply of nitrogen has been insufficient and mismanaged; thus, there is need to improve the supply of nitrogen and device affordable means of sustaining the

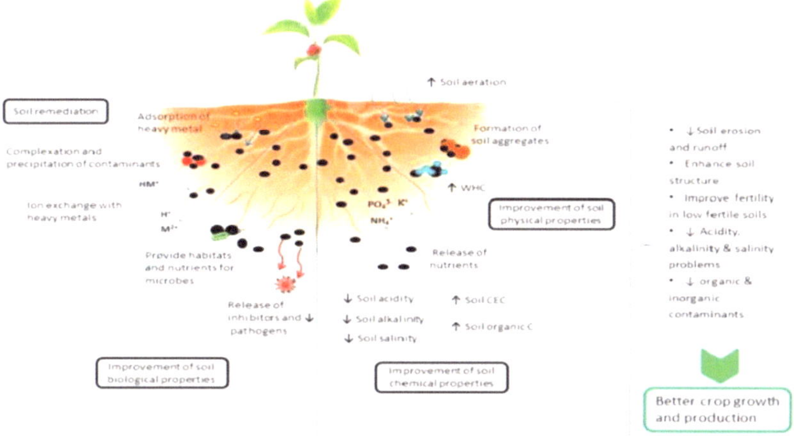

Figure 4.
Soil enhancement through biochar addition [35].

supply [15, 19]. Plants normally do not utilize more than 50% of the soil nitrogen; therefore, adequate technology should be invented in order to build the mechanism that would increase nitrogen uptake in plants. Part of these measures is the introduction of biochar into the soil to encourage the uptake of nitrogen from pure organic sources [18, 30].

The addition of biochar to soil increases the pool of nutrients and the volume of nitrogen available to plants. Organic amendment of soil provides the soil with nutrients like nitrogen, phosphorus and potassium [29, 33–35]. Therefore, biochar addition to the soil enhances improved supply of nutrients to plants and to the soil.

7.4 Soil structure improvements of biochar

Biochar amendments into the soil improves soil's nutrients and structure due to the high porosity of the biochar, leading to the formation of well-structured aggregates that enhances soil's water holding capacity. In tropical regions and areas with low rainfall, the addition of biochar to the soil improves the water retention capacity in the dry season [2, 13, 34]. The tensile strength of the soil increases as a result of accumulation of more organic matter. This forms a strong structure for the plant roots to adhere to. The strong adherence improves the soil ecosystem because the bonding of the soil aggregates prevents erosion and displacement of the top-most layers of the soil [11]. Thus, this overall soil structure improvement proves the evidence of the advantage of the physicochemical attributes of biochar to prevent adverse soil conditions caused by earth's geogenic and climate activities, thereby mitigating climate change [12, 36].

7.5 Biochar and mitigation of greenhouse gases

Biochar possess interesting properties like the presence of high-stable carbon necessary for the carbon capturing. The high stability of biochar carbon therefore gives biochar products an edge over other additives and conditioners. Thus, with the effective sinking of carbon from the soil, it regulates the presence of carbon dioxide in the atmosphere [29, 30]. Likewise, this property of biochar stability increases the turnover rate and half-life of biochar in the soil for more number of years. However, the mitigation of other greenhouse gases other than carbon dioxide by biochar has still not been investigated. Therefore, further research needs to be carried out to evaluate the effectiveness of biochar in regulating other greenhouse gases [33]. Also, effective measures of evaluating the emissions of these gases must be considered with the aim of saving up the greenhouse gases pertaining to the emission trading schemes [17, 18, 37].

Also, the use of plant and animal wastes for biochar is a means of preventing the emission of methane and nitrous oxide gas from landfills, which are potent greenhouse gases.

7.6 Biochar in environmental conservation

Biochars effect in environmental conservation deals with waste management, soil remediation and energy conservation. Animal wastes and biomass from agricultural produce can be processes under certain environmental conditions using pyrolysis [1, 10, 34]. The emission of greenhouse gases from wastes is reduced if the plant and animal biomass is processed into biochar [20].

Another major advantage of biochar for environmental conservation is the use of biochar as renewable fuel. The incineration of carbonized products can be processed into biochar which releases lesser amounts of inorganic materials into the environment because carbonized materials are purely organic [36].

The process of pyrolysis could however alter the extent of release of non-carbon materials like chlorine into the environment. However, the nature of the biomass determines the amount of chlorine released during the process of biomass conversion into biochar during pyrolysis. If the chlorine and other inorganic substances are formed continuously, it leads to the formation of sludge in fuel-burning boilers but with the carbonization of fuel, the sludge formation disappears. Thus, because the carbonization happens inevitably during biochar production, the process of production of biochar is a key process in solving the problem of sludge production [29].

Therefore, it is re-iterated that the type, properties and processes of feedstocks used for the formation of biochar is a potent determinant of the nature and effectiveness of the biochar [16, 36].

7.7 Biochar in enhancing soil microbial community

Biochar application to the soil has been found to increase the soil's physicochemical conditions, more especially rate of carbon sequestration and the overall soil fertility [22]. For example, biochar from wood waste was used to amend the soil on which wheat crops were planted in Tuscany, Italy. The soil samples were collected after 3 months of biochar amendment and was analyzed the pH, total organic carbon, microbial biomass, mean substrate-induced respiration and other parameters. It was observed that only the pH and the mean substrate-induced respiration had significant changes after 3 months of amendment, but after 14 months the effect was not pronounced [22, 37]. The result reflects the perking up of the activities of soil microorganisms after treatment with biochar.

Furthermore, biochar has an averagely neutral pH ranging from 6.2 to 7.6 (**Table 1**) and majority of microorganisms, especially soil microorganisms thrive well at this pH. Only a few microorganisms are acidophilic and alkanophilic. This therefore increases the chance of more microbial activities in a biochar-amended soil [24, 37].

Likewise, the porous nature of biochar is obviously an advantage for soil aeration, a good condition for soil aerobes. However, there are also anaerobic organisms (that do not require the presence of oxygen to survive) in the soil. In a condition where the soil aerobes use up the total oxygen in the soil, it provides a good ambience for soil anaerobes to thrive. The porous biochar also increases soil water retention capacity. However, chances of soil water-logging are possible, but the tendency of biochar to form aggregates reduces the chances of soil erosion [38].

The chemical properties of biochar also increase the microbial population in the rhizosphere region of the soil, which is popularly known as the rhizosphere effect [26, 36]. Furthermore, the overall effect of biochar enhances the growth of plants that release vitamins and amino acids from their roots and tissues. These nutrients subsequently increase microbial metabolism and increased microbial products yield.

In a biochar-amended tobacco-planting soil in China, rice-straw biochar was used to enhance the soil. Indigenous microorganisms present in the soil were bacteria of the phyla Proteobacteria, Actinomycetes and Acidobacteria; while the predominant fungal phyla were Ascomycota, Zygomycota and Basidiomycota [23, 35]. **Figures 5** and **6** shows that among these groups, only the Actinomycetes bacteria group and the Ascomycota fungal group respectively reduced in proportion after the addition of biochar while the remaining bacterial and fungal groups increased in proportion after the biochar treatment. Other soil fungal groups that exhibited increased proportion include the Zygomycota and Glomeromycota phyla [25, 38].

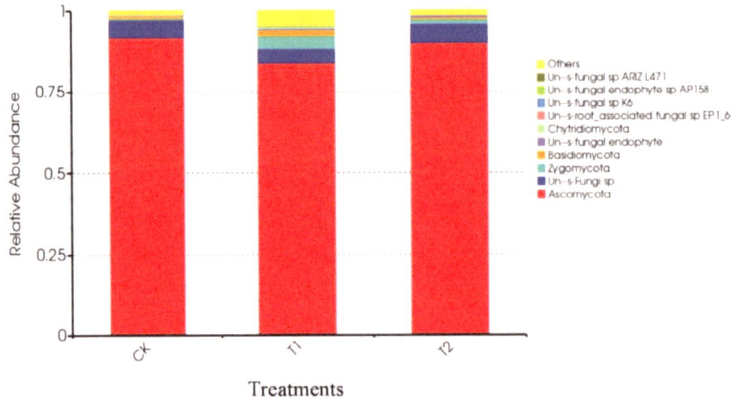

Figure 5.
Proportions of soil bacteria after biochar treatment [36]. CK = un-amended treatment; T1 = 2250 kg/ha biochar-amended treatment; T2 = 4500 kg/ha biochar-amended treatment.

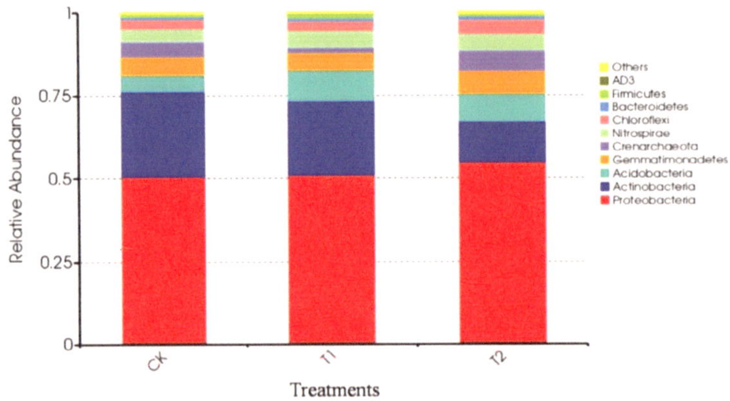

Figure 6.
Proportions of soil fungi after biochar treatment [36]. CK = un-amended treatment; T1 = 2250 kg/ha biochar-amended treatment; T2 = 4500 kg/ha biochar-amended treatment.

8. The implications of amending soils with biochar

Biochar is a natural soil booster and has recorded high success rates over the years in some parts of the world. In some regions including Africa, biochar has been found to yield positive results although some negative results were reported, for example due to improper incorporation of biochar into soil blends which fosters soil erosion [39]. Biochar has been found to balance soil pH, absorb pollutants, improve soil moisture and increases soil aeration. However, there are other economic importance attached with the use of biochar, these include the geographical distribution of the effective use of biochar [38, 39].

Biochar has not been known in some parts of the world due to lack of research, negligence and cultural practices. Another major implication of biochar application is contamination which is of high economic importance. The kinds of cattle manure used in biochar production could be contaminated with non-soil infectious microorganisms, heavy metals or may be denatured due to the high temperature of pyrolysis [40].

Excessive removal of feedstock from biochar raw material and production processes could lead to usage of biochar devoid of adequate organic nutrient constituents. Inadequate improvisation of biochar, for example the use of straw alone, or in inadequate amounts, will also have a negative effect on the soil. Earthworm survival is always put in doubts as well as the soil ecosystem niche that ensures overall soil health [40].

9. Discussion

Soil has been one major source of habitat for plants and animals. However, the anthropogenic activities of humans have contributed to the depletion of natural resources and nutrients from the soil. In a rescue mission to curtail the devastation of land and soil, biochar have been introduced for both agricultural and commercial purposes. Digestion of wastes in less amounts of oxygen at high temperature produces high quality biochar. The high temperature explains the pyrolysis process which is usually more than 200°C anaerobically [4, 41–43]. Biofertilizers and organic fertilizers have served as appetizers to the soil over the years. Recently, other forms of soil amendments have been procured for soil improvement. The soil amendments generally have a low cost and easy accessibility. An example of these soil amendments includes biochar.

Biochar is environmentally friendly and has a high binding capacity that allows the soil to adhere to biomolecules and absorb nutrients. Biochar is globally known to becoming a means of sustainable amendment of soil and means excavating heavy metals from the soil [44]. Although, biochar is produced from different biomass sources, but biochar produced from carbonization of organic wastes produces amended soil types with high carbon sequestration [45]. The quality of biochar depends on the pyrolysis conditions and raw materials, while the efficiency of biochar not only depends on the pre-production parameters but also the post-production and application procedure of the biochar; which include the type of soil under amendment and the amount of biochar applied [46]. In the process of pyrolysis of biomass, biochar is produced. There are new innovations that need to be applied when producing biochar. For example nanotechnologies and large scale model production are among the recent methods that need to be explored for adequate and mass production of biochar [15, 44, 47]. The temperature of pyrolysis, the particle size and residence time are factors that determine the end product of biomass conversion into biochar. These factors determine the state and half-life of the biochar product [48]. Biochar is a good soil and plant health promoter because it helps to retain the nutrients in the soil through its absorbent power thereby enhancing overall plant growth [49].

10. Conclusion

The physicochemical property of biochars is key to the enhancement and remediation of the soil. The porous structure and high carbon content are important factors for conservation and sustainability of the soil. A combination of the above properties and the ion-exchange characteristic is symbolic for carbon sequestration, immobilization of heavy metals, and removal of pollutants from the soil [28]. Wastes are used to generate biochars with specific processes like pyrolysis under certain parameters. Biochar is potent enough to trap excess carbon dioxide from the environment but studies have not been seen with regards to mitigating other greenhouse gases. Thus, further research needs to be carried out to evaluate the effectiveness of biochar in regulating other greenhouse gases. Also, effective measures of evaluating the emissions of these gases must be considered with the aim of saving up the greenhouse gases pertaining to the emission trading schemes.

Thus, as a substance of good physicochemical parameter, addition of the carbon-rich biochar to soil enhances the physical, chemical and biological properties of the soil. The enhancement includes the total organic and inorganic carbon, nitrogen, phosphorus and potassium. For agricultural purposes, the soil with added biochar has high crop yield. Therefore, biochar is recommended for most agricultural practices due to cost effectiveness and easy application techniques.

Furthermore, the effect of biochar on soil microbial community is significant considering the increase on the soil bacterial and fungal community with a higher increase in the later. Thus, biochar simultaneously enhances both the soil nutritional status and the soil microbial structure.

However, commercial use of biochar has not been widely accepted due to regulations backed by the law and bias about the fact that biochar is just a waste. Therefore, applications of biochar should be considered using efficient processes with the assurance of environmental safety so as to encourage the schools of thought that are tendentious about the safety of biochar application.

Author details

Oladapo T. Okareh* and Alaba O. Gbadebo
Department of Environmental Health Sciences, Faculty of Public Health, College of Medicine, University of Ibadan, Oyo State, Nigeria

*Address all correspondence to: dapsy2001@yaoo.co.uk

References

[1] Saletnik B, Zaguła G, Bajcar M, Tarapatskyy M, Bobula G, Puchalski C. Biochar as a multifunctional component of the environment—A review. Applied Sciences. 2019;**9**:1139. DOI: 10.3390/app9061139 www.mdpi.com/journal/applsci

[2] Lehmann J, Rilling MC, Thies J, Masiello CA, Hockaday WC, Crowley D. Biochar effects on soil biota—A review. Soil Biology and Biochemistry. 2011;**43**:1812-1836

[3] Malińska K. Biochar—A response to current environmental issues. Engineering and Environment Protection. 2012;**15**:387-403

[4] Popp J, Lakner Z, Harangi-Rákos M, Fári M. The effect of bioenergy expansion: Food, energy, and environment. Renewable and Sustainable Energy Reviews. 2014;**32**:559-578

[5] Montoya JI, Chejne-Janna F, Garcia-Pérez M. Fast pyrolysis of biomass: A review of relevant aspects. Part I: Parametric study. DYNA. 2015;**82**:239-248

[6] Ustawa z dnia 14 Grudnia 2012 o Odpadach [Waste Act of 14 December 2012]. Available from: http://isap.sejm.gov.pl/DetailsServlet?id=WDU20130000021

[7] Van Laer T, De Smedt P, Ronsse F, Ruysschaert G, Boeckx P, Verstraete W, et al. Legal constraints and opportunities for biochar: A case analysis of EU law. GCB Bioenergy. 2015;**7**:14-24

[8] Malińska K. Legal and quality aspects of requirements defined for biochar. Inżynieria i Ochrona Środowiska. 2015;**18**:359-371

[9] Tan X-F, Liu S-B, Liu Y-G, Gu Y-L, Zeng G-M, Hu X-J, et al. Biochar as potential sustainable precursors for activated carbon production: Multiple applications in environmental protection and energy storage. Bioresource Technology. 2017;**227**:359-372

[10] Hoegberg LC, Groenlykke TB, Abildtrup U, Angelo HR. Combined paracetamol and amitriptyline adsorption to activated charcoal. Clinical Toxicology. 2010;**48**:898-903

[11] Nanda S, Dalai AK, Berruti F, Kozinski JA. Biochar as an experimental bioresource for energy, agronomy, carbon sequestration, activated carbon and specialty materials. Waste and Biomass Valorization. 2016;**7**:201-235

[12] Ozsoy HD, van Leeuwen J. Removal of color from fruit candy waste by activated carbon adsorption. Journal of Food Engineering. 2010;**101**:106-112

[13] Burrell LD, Zehetner F, Rampazzo N, Wimmer B, Soja G Long-term effects of biochar on soil physical properties. Geoderma. 2016;**282**:96-102

[14] Saxena J, Rana G, Pandey M. Impact of addition of biochar along with *Bacillus* sp. on growth and yield of French beans. Scientia Horticulturae. 2013;**162**:351-356

[15] Cross A, Sohi S. The priming potential of biochar products in relation to labile carbon contents and soil organic matter status. Soil Biology and Biochemistry. 2011;**43**:2127-2134

[16] Glaser B, Lehmann J, Zech W. Ameliorating physical and chemical properties of highly weathered soils in the tropics with charcoal—A review. Biology and Fertility of Soils. 2002;**35**:1719-1730

[17] Tumuluru JS, Sokhansanj S, Hess JR, Wright CT, Boardman RD. A review on biomass torrefaction process and product properties for energy

applications. Industrial Biotechnology. 2011;7:384-402

[18] Sohi S, Loez-Capel S, Krull E, Bol R. Biochar's roles in soil and climate change: A review of research needs. CSIRO Land and Water Science Report. 2009;**5**(09):17-31

[19] Bogusz A, Cejner M. Biochar materials in adsorption of organic and inorganic contaminants. Inżynieria środowiska. 2016;**22**:9-33

[20] Sun K, Jin J, Keiluweit M, Kleber M, Wang Z, Pan Z, et al. Polar and aliphatic domains regulate sorption of phthalic acid esters (PAEs) to biochars. Bioresource Technology. 2012;**118**:120-127

[21] Liang B, Lehmann J, Solomon D, Kinyangi J, Grossman J, O'Neill B, et al. Black carbon increases cation exchange capacity in soils. Soil Science Society of America Journal. 2006;**70**:1719-1730

[22] Hammes K, Schmidt MWI. Changes in biochar in soils. In: Lehmann M, Joseph S, editors. Biochar for Environmental Management Science and Technology. London: Earthscan; 2009. pp. 169-182

[23] Brown TR, Wright MM, Brown RC. Estimating profitability of two biochar production scenarios: Slow pyrolysis vs fast pyrolysis. Biofuels, Bioproducts and Biorefining. 2011;**5**(1):54-68

[24] Blanco-Canqui H. Biochar and soil physical properties. Soil Science Society of America Journal. 2017;**81**:687-711

[25] Lehmann J, Gaunt J, Rondon M. Biochar sequestration in the terrestrial ecosystem—A review. Mitigation and Adaptation Strategies for Global Change. 2006;**11**:403-427

[26] Joseph SD, Camps-Arbestain M, Lin Y, Munroe P, Chia CH, Hook J, et al. An investigation into the reactions of biochar in soil. Australian Journal of Soil Research. 2010;**48**:501-515

[27] Dutta B. Assessment of pyrolysis techniques of lignocellulosic biomass for biochar production [dissertation—master's thesis]. Montreal, Quebec, Canada: McGill University; 2010

[28] Stoyle A. Biochar production for carbon sequestration [master's thesis]. Worcester, Massachusetts, USA: Worcester Polytechnic Institute, Shanghai Jiao Tong University; 2011

[29] Yuan J, Xu R, Zhang H. The forms of alkalis in the biochar produced from crop residues at different temperatures. Bioresource Technology. 2011;**102**:3488-3497

[30] Elmer W, White JC, Pignatello JJ. Impact of Biochar Addition to Soil on the Bioavailability of Chemicals Important in Agriculture. Report. New Haven: University of Connecticut; 2010

[31] Ahmad M, Rajapaksha AU, Lim JE, Zhang M, Bolan N, Mohan D, et al. Biochar as a sorbent for contaminant management in soil and water: A review. Chemosphere. 2014;**99**:19-33

[32] Elmer WH, Pignatello JJ. Effect of biochar amendments on mycorrhizal associations and Fusarium crown and root rot of asparagus in replant soils. Plant Disease. 2011;**95**:960-966

[33] Rutigliano FA, Romano M, Chan KY, Van Zwieten L, Meszaros I, Downie A, et al. Using poultry litter biochars as soil amendments. Australian Journal of Soil Research. 2008;**46**:437-444

[34] Uzoma KC, Inoue M, Andry H, Fujimaki H, Zahoor A, Nishihara E. Effect of cow manure biochar on maize productivity under sandy soil condition. Soil Use and Management. 2011;**27**:205-212

[35] Palansooriya KN, Ok YS, Awad YM, Lee SS, Sung JK, Koutsospyros A, et al. Impacts of biochar application on upland agriculture: A review. Journal of Environmental Management. 2019;**234**:52-64

[36] Cao Y, Gao Y, Qi Y, Li J. Biochar enhanced composts reduce the potential leaching of nutrients and heavy metals and suppress plant-parasitic nematodes in excessively fertilized cucumber soils. Environmental Science and Pollution Research International. 2018;**25**(8):7589-7599

[37] Rossana M, Castaldi S, et al. Effect of biochar addition on the soil microbial community in a wheat crop. European Journal of Soil Biology. 2013;**60**:9-15

[38] MicroDok. Factors That Affect Microbial Population in the Soil [Internet]. 2020. Available from: https://microdok.com/factors-that-affect-microbial-population-in-the-soil/ [Assessed: 17 February 2020]

[39] Slonczewski JL, Foster JW, Gillen KM. Microbiology: An Evolving Science. 2nd ed. New York, USA: W.W. Norton and Company, Inc.; 2011

[40] Gao L, Wang R, Shen G, Zhang J, Meng G, Zhang J. Effects of biochar on nutrients and the microbial community structure of tobacco-planting soils. Journal of Soil Science and Plant Nutrition. 2017;**17**(4):884-896

[41] Bonanomi G, Ippolito F, Cesarano G, Nanni B, et al. Biochar as plant growth promoter: Better off alone or mixed with organic amendments? Frontiers in Plant Science. 2017;**8**:1570

[42] Debela F, Thring RW, Arocena JM. Immobilization of heavy metals by co-pyrolysis of contaminated soil with woody biomass. Water, Air, and Soil Pollution. 2012;**223**:1161-1170

[43] Woolf D, Amonette J, Street-Perrott F, Lehmann J, Joseph S. Sustainable biochar to mitigate global climate change. Nature Communications. 2010;**1**:1-9

[44] Kammann C, Glaser B, Schmidt HP. Combining biochar and organic amendments. In: Shackley S, Ruysschaert G, Zwart K, Glaser B, editors. Biochar in European Soils and Agriculture: Science and Practice. New York: Routledge; 2016. pp. 136-164

[45] Spokas KA, Reikosky DC. Impacts of sixteen different biochars on soil greenhouse gas production. Annals of Environmental Science. 2009;**3**:179-193

[46] Houben D, Evrard L, Sonnet P. Beneficial effects of biochar application to contaminated soils on the bioavailability of Cd, Pb and Zn and the biomass production of rapeseed (*Brassica napus* L.). Biomass and Bioenergy. 2013;**57**:196-204. DOI: 10.1016/j.biombioe.2013.07.019

[47] Hu YL, Wu F-P, Zeng DH, Chang SX. Wheat straw and its biochar had contrasting effects on soil C and N cycling two growing seasons after addition to a black chernozemic soil planted to barley. Biology and Fertility of Soils. 2014;**50**:1291-1299. DOI: 10.1007/s00374-014-0943-6

[48] Mimmo T, Panzacchi P, Baratieri M, Davies CA, Tonon G. Effect of pyrolysis temperature on miscanthus (*Miscanthus × giganteus*) biochar physical, chemical and functional properties. Biomass and Bioenergy. 2014;**62**:149-157. DOI: 10.1016/j.biombioe.2014.01.004

[49] Mohan D, Sarswat A, Ok YS, Pittman CU Jr. Organic and inorganic contaminants removal from water with biochar, a renewable, low cost and sustainable adsorbent: A critical review. Bioresource Technology. 2014;**160**:191-202. DOI: 10.1016/j.biortech.2014.01.120

Chapter 10

Challenges of Biochar Usages in Arid Soils: A Case Study in the Kingdom of Saudi Arabia

Khalid A. Alaboudi

Abstract

Biochar is a carbon-rich material produced from the pyrolysis of organic biomasses in the absence of oxygen or under low-oxygen conditions. Biochar has received a great interest during the last few decades due to its beneficial roles for carbon dioxide capturing and soil fertility improvement. However, applications of biochar in arid soils are very limited, and there is a lack of knowledge on practical aspects of adding biochar to arid soils. In this chapter, we will focus on biochar applications in the Kingdom of Saudi Arabia soils as an example of arid soils. These soils are characterized by several marks, i.e., high soil pH, sand structures, high $CaCO_3$ contents, and low soil fertility. In addition, the unsuccessful recycling practices of agricultural and food wastes in the Kingdom of Saudi Arabia are also discussed. This chapter provides an overview of current biochar knowledge pertinent to its application to arid soils, summarizes what is known so far about biochar and its applications in arid regions, and demonstrates the possible strategies that can be used for enhancing the practices of biochar addition to these soils.

Keywords: biochar, carbon sequestration, soil fertility, applications, limitations

1. Introduction

Soil formation is a complex process resulting from long-term interactions among several environmental factors, i.e., climate, soil-forming processes, and land uses [1]. Such processes influence soil's physical, chemical, and biological characteristics and hence affect soil productivity [2]. In arid and semiarid regions, many challenges may face soil productivity; for example, many arid soils are of light texture with low organic matter and nutrient contents [3–5]. These soils exhibit low soil aggregation and can, therefore, be subjected easily to wind erosion [6]. Moreover, secondary minerals may dominate in such soils like calcite and gypsum [1], and these minerals can significantly diminish soil fertility [7, 8]. Another important threat that faces agricultural sustainability in arid sand semiarid soils is soil salinity [9]. Generally, the term arid or semiarid refers to the regions of limited rainfall and high evapotranspiration [10]. Areas with mean annual precipitation (MAP) ranging from 100 to 250 mm yr^{-1} are called arid climatic zones, while areas with MAP ranging between 250 and 600 mm yr^{-1} are called semiarid zones [11]. These two climatic regions cover approximately 30% or more of the total earth's surface [12]. To improve the productivity of low-fertility soils, organic applications should therefore be incorporated within the top soil [13, 14] to raise soil contents of both C and nutrients [14]. However, the negative implications of using easily decomposed organic amendments on the environment should be taken into account, e.g., emissions of greenhouse that possesses global warming [15]. Accordingly, biochar might be the appropriate organic amendments to improve the characteristics of such soils but decrease the emissions of greenhouse gases on the other hand. In the following section, we will discuss the distribution of arid and semiarid soil and the potentialities of using biochar to improve soil properties and attain sustainability in crop production

2. Distribution pattern of arid and semiarid soils

As shown in **Figure 1**, arid and semiarid soils are located in North and South Africa, the Middle East region, North and South America, and finally in Australia [16]. More than 95% of the total arid soils exist in Africa and Middle East regions. According to the UNEP [16], aridity index (AI) is commonly used to quantify the aridity of a specific region. Briefly, this index is estimated based on climate variability by calculating the ratio of annual average rainfall to potential evapotranspiration (P/PET). For this concern, lands are classified in the following ascending order based on the average precipitation: hyperarid, arid, semiarid, and dry subhumid, and their average precipitation rates are 0, 1–59, 60–119, and 120–179 mm yr^{-1}, respectively [16]. In this chapter we will focus on the Kingdom of Saudi Arabia lands as an example of arid lands and the major problems hindering the application of biochar technology in these soils. In the Kingdom of Saudi Arabia, almost 25% of the total land area is arable lands (52.7 million ha), in which 45% are calcareous, sandy textured soils, with very low contents of organic matter and nutrients [17].

3. Major characteristics of arid soils

Salt content, in many arid soils, is relatively high. These soils accumulate on the soil's surface because of the high evapotranspiration rates. In addition, these regions have dry climate with high temperature and high evaporation rate. The

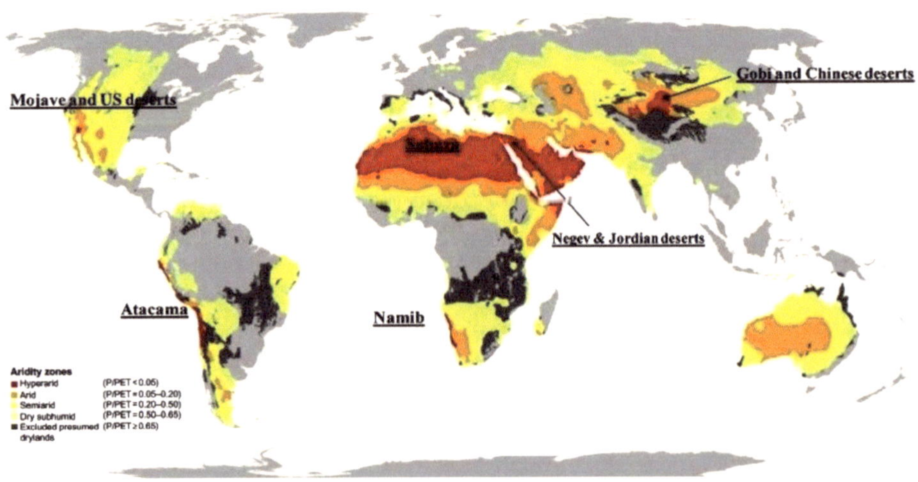

Figure 1.
Distribution pattern of arid and semiarid regions [16].

deeper soil layers are usually occupied by Ca. The arid soils can be applicable for cultivation in case proper water for irrigation becomes available. Due to high temperature, the degradation rate of soil's organic matter in arid soils is very high; consequently, these soils need further application rates. In the Kingdom of Saudi Arabia, most of agricultural soils are of coarse texture, with high $CaCO_3$ contents and high pH values. The lack of sufficient water in the Kingdom of Saudi Arabia led to increase the potentiality of soil salinization. Therefore, these soils are of poor fertility, in terms of physical (high infiltration rate, sand texture, and bad hydraulic properties), chemical (low organic matter contents, insufficient nutrients, and high soil pH), and biological (low microbial communities in soils due to the absence of organic residues and soil nutrients) characteristics [18]. Usually the pH value in the Kingdom of Saudi Arabia soils is greater than 8 with high $CaCO_3$ contents (>30%) [19]. In spite of the Kingdom of Saudi Arabia having the largest land mass among Middle East countries, it has the lowest arable land per capita worldwide [20]. A point to note is that the major water sources in the Kingdom of Saudi Arabia are groundwater and desalination of seawater [21]. Therefore, intensive studies are performed to overcome the infertility problems of arid soils in the Kingdom of Saudi Arabia by using organic and inorganic soil amendments, i.e., compost bio inoculums and mineral polymers [22–24]. However, these amendments need to be applied intensively because of their low nutritive contents and fast degradation rates [25].

4. Agricultural and food wastes in the Kingdom of Saudi Arabia

Due to the arid characteristics of Kingdom of Saudi Arabia lands, agricultural activities are also thought to be very low. According to the World Bank [20], the agricultural lands in the Kingdom of Saudi Arabia cover only 1736.472 km^2 [20]. On the other hand, the Kingdom of Saudi Arabia is considered the largest food and agricultural importer in the Gulf Cooperation Council with average imported food products of 80% in 2013 [26, 27]. These conditions hinder the development of the agricultural sector in the Kingdom of Saudi Arabia. A positive point to note is that such arid conditions are suitable for cultivation of date palm plant. According to the Food and Agricultural Organization (FAO), the Kingdom of Saudi Arabia has the highest harvested areas of date palm in 2016 with an average area of 145,516 ha. Palm trees generate huge wastes annually [28]; accordingly, proper recycling and management of these wastes can improve soil conditions.

5. Biochar

Biochar is an organic carbon-rich product, produced by burning agricultural and animal wastes in the absence of oxygen [29]. Several studies demonstrated its beneficial role for improving soil fertility and waste management, remediation of contaminated soils and water, and reducing greenhouse gas emissions [25, 30, 31]. In this chapter, we will discuss the potential benefits of biochar as a soil amendment for improving its fertility and productivity.

5.1 Ancient production of biochar

Biochar was initially produced by ancient Egyptians to produce liquid wood tar from charring processes in order to cover and preserve the dead bodies [32]. Thereafter, in South America (terra preta), 2500 years later, biochar is created both

Feedstock	Temperature	pH	%							CEC, cmolc kg^{-1}	C/N ratio	%		H/C ratio	O/C ratio	SSA, m^2 g^{-1}	Reference
			C	N	P	S	Ca	Mg	K			O.M	Ash				
Sugar cane bagasse	<500	8.63	74.02	1.00	0.24	—	0.17	0.32	2.00	69.62	74.02	87.80	12.21	0.42	0.23	92.30	[31]
Orange peel	<500	8.75	66.36	2.13	0.25	—	1.04	0.28	1.86	68.28	31.15	88.80	11.17	0.65	0.32	0.20	
Oak wood	600.00	6.38	87.50	0.20	—	—	—	—	—	75.70	489.00	—	0.01	0.33	0.07	642.00	[40]
Corn stover	350.00	9.39	60.40	1.20	—	—	—	—	—	419.30	51.00	—	11.40	0.75	0.29	293.00	[41]
	600.00	9.42	70.60	1.07	—	—	—	—	—	252.10	66.00	—	16.70	0.39	0.10	527.00	
Corn stalk	400.00	9.60	51.10	1.34	0.25	—	—	—	1.34	—	38.13	—	—	—	—	—	[42]
	500.00	10.30	48.40	0.55	0.44	—	—	—	2.65	—	88.00	—	—	—	—	—	
Wheat straw	425.00	10.40	46.70	0.59	—	—	1.00	0.60	2.60	—	79.15	—	20.80	—	—	—	[43]
Rice straw	400.00	—	71.30	1.46	—	—	—	—	24.60	—	48.84	—	36.20	—	—	—	[44]
Peanut hull	500.00	8.60	82.00	2.70	0.30	0.10	—	—	—	—	30.37	—	9.30	0.44	0.03	200.00	[45]
Coco peat	500.00	10.30	84.40	1.02	0.03	0.27	0.06	2.30	—	—	82.75	—	15.90	0.41	0.10	13.70	[46]
Coconut charcoal	<500	8.85	76.50	0.20	—	—	—	—	—	—	426.60	—	2.90	0.12	—	—	[47]
Pinewood	<500	8.47	53.20	0.40	—	—	—	—	—	—	143.40	—	65.70	0.35	—	—	
Eucalyptus deglupta	350.00	7.0[illegible]	82.40	0.57	0.06	0.03	—	—	—	4.69	144.56	—	0.20	—	0.12		[48]
Hardwood sawdust	500.00	—	63.80	0.22	—	0.01	—	—	—		290.00	—	22.80	0.60	0.14	1.00	[49]
Chinese pine	600.00	8.3[illegible]	66.67	2.21	—	—	—	—	—	31.58	30.17	—	12.50	0.58	0.31	—	[50]

Feedstock	Temperature	pH	%							CEC, cmolc kg^{-1}	C/N ratio	%		H/C ratio	O/C ratio	SSA, $m^2 g^{-1}$	Reference
			C	N	P	S	Ca	Mg	K			O.M	Ash				
Cattle waste	380.00	8.20	62.10	0.10	—	—	—	—	—	39.00	621.00	—	25.60	1.90	0.27	—	[51]
Sewage sludge	380.00	8.50	38.30	5.20	—	—	—	—	—	0.50	7.37	—	44.90	0.94	0.25	—	

Data obtained from Abdelhafez et al. [25].

Table 1.
Physicochemical characteristics of different types of biochar.

naturally by forest fires and by humans through burning bits for different practices, i.e., cooking and manufacturing [25]. In terra preta soils, the acidic conditions were the limiting factors affecting negatively crop production wherein these soils suffer severely from Al toxicity. To overcome this problem, the liming effect of biochar was an effective approach to overcome Al toxicity in soil [33].

5.2 Production technologies of biochar

All organic materials (feedstock, crop wastes, animal wastes manure) can be used for biochar production. Simply, biochar is a charcoal-like material that is produced in the absence of oxygen or limited oxygen conditions [25]. In this process organic wastes are burned at relatively low temperature < 700°C, and three main components are produced through the pyrolysis process, i.e., solid biochar (carbonized biomass with average C contents of >60), synthetic gas (which can be used as a power source), and bio oil (fuel material) [25]. Farmers in the past used to burn the agricultural wastes under limited oxygen conditions by covering the waste piles with soil dust. In this traditional method, approximately half the amount of organic C was lost into the atmosphere. Therefore, people have tried to develop the production technology through using pit kiln and brick kilns in order to eliminate the losses of C and other gas emission. After biochar technology has risen, non in situ equipment have been designed to maximize the biochar yield, eliminate the C lose and ash content and using syngas and bio oil as secondary products [25]. It is worthy to mention that organic materials start to decompose at low temperature (about 120°C), followed by hemicellulose and lignin compounds, which degrade at 200–260°C and 240–350°C, respectively [34].

5.3 Physical and chemical characteristics of biochar

Both pyrolysis conditions and the types of organic wastes identify the major characteristics of the produced biochar [25, 35]. Usually biochar (a carbon-rich product) is characterized by its high surface area and lower concentrations of hydrogen and oxygen [36, 37]. Thus, its application can improve soil characteristics (chemical, physical, and biological). Moreover, this organic product is considered relatively stable in soil because of its low availability of labile organic carbon [38] besides its low content of nutrients [39]. **Table 1** shows the main physiochemical characteristics of different types of biochars. For both physical and chemical characteristics, pyrolysis conditions and type of organic wastes are the main factors identifying them. Clearly, all biochars have the same characteristics, especially the high C contents and low N contents. Nitrogen usually starts to be volatile at 200°C; therefore, N contents are low in most types of biochars. The high pH of biochar might be attributed to the high content of alkaline metals, i.e., Ca, Mg, and K, which are stable during biochar production. Despite the low nutrient content of biochar, its application to soils improves its fertility because it is usually added at high rates as soil amendments. The pyrolysis conditions play an important role for identifying the physical characteristics of biochar. The higher surface area of biochar is a consequence of high temperature during the pyrolysis reaction [25].

6. Applications of biochar for soil fertility improvement

As mentioned above the porous structure of biochar facilitates its adsorption of water and, therefore, increases soil water holding capacity [52, 53]. This might increase the efficiency of water use in the arid zone soils [54]. The previous studies also demonstrated that the addition of biochar increases both soil aggregation and saturated hydraulic conductivity but decreases soil bulk density [53, 55, 56]. Therefore, application of biochar is a recommended practice to improve the physical characteristics of light textured soils [3]. For soil chemical characteristics, most studies showed that biochar has a negative effect on the availability of soil nutrients, i.e., its application increases soil pH [57, 58]. The liming effect of biochar can be attributed to the high concentrations of cationic metals, i.e., Ca^{2+}, Mg^{2+}, and K^{+}, which are stable and do not volatilize during the pyrolysis process [25]. In most cases, biochar has a relative high pH (within the range of 8–11.5) [57, 59, 60]. Therefore, addition of biochar is more favorable for acidic soils than the alkaline ones. The black earth (terra preta) was an acidic soil in the enteral Amazon basin, and AL toxicity and P deficiency were the main reasons hindering the agricultural activities. Continuous applications of biochar to these soils neutralized soil acidity but increased the available P fraction; consequently, biochar enhanced and sustained soil health of terra preta [25, 32, 59]. Moreover, high doses of biochar can increase soil salinity [61–65]. On the other hand, the addition of biochar can raise soil organic matter contents and elevate soil cation exchange capacity (CEC) [66]. For nutrients contents, it is worth mentioning that most biochar types are typically low in nutrient contents, especially N. As a result, applications of biochar only in agricultural is not adequate to supply the needed macro- and micronutrients [25, 67]. However, biochar plays an important role in mitigating soil nutrient losses by seepage or leaching [66]. Applications of biochar to soils increased its OC contents, which is suitable for soil organisms and provides more favorable habitats to microbes and, therefore, facilitates soil biological activities [68]. In addition, the release of organic molecules suppresses the activities of soil microbes [69].

7. Limitation of biochar application in the Kingdom of Saudi Arabia soils

As mentioned above the high pH of biochar limited its applications in arid soils. For this concern, application of biochar to agricultural soils in the Kingdom of Saudi Arabia is very limited due to many reasons as follows:

1. Low cultivated areas in the Kingdom of Saudi Arabia: as mentioned above, agricultural activities in the Kingdom of Saudi Arabia are very limited due to the arid conditions.

2. The chemical characteristics of soils in the Kingdom of Saudi Arabia; especially soil pH is one of the major factors hindering the application of biochar to agricultural soils. Our demonstration proves that most types of biochar are of alkaline nature and its application to agricultural soil may negatively affect the availability of soil nutrients due to increasing soil pH. The pH of the produced biochar is a function of pyrolysis temperature and time; by elevating the pyrolysis temperature and time, the pH of the produced biochars increased to reach 11.5 in some studies [53].

3. The lack of knowledge regarding biochar technology and its beneficial role for enhancing agricultural activates.

In a bibliometric study conducted by Arfaoui et al. [70], they have shown that Iran, the Kingdom of Saudi Arabia, and Egypt are the highest contributor countries for biochar studies and publications in the Middle East countries. As shown in **Figure 2** (biochar article number and geographic distribution according to the

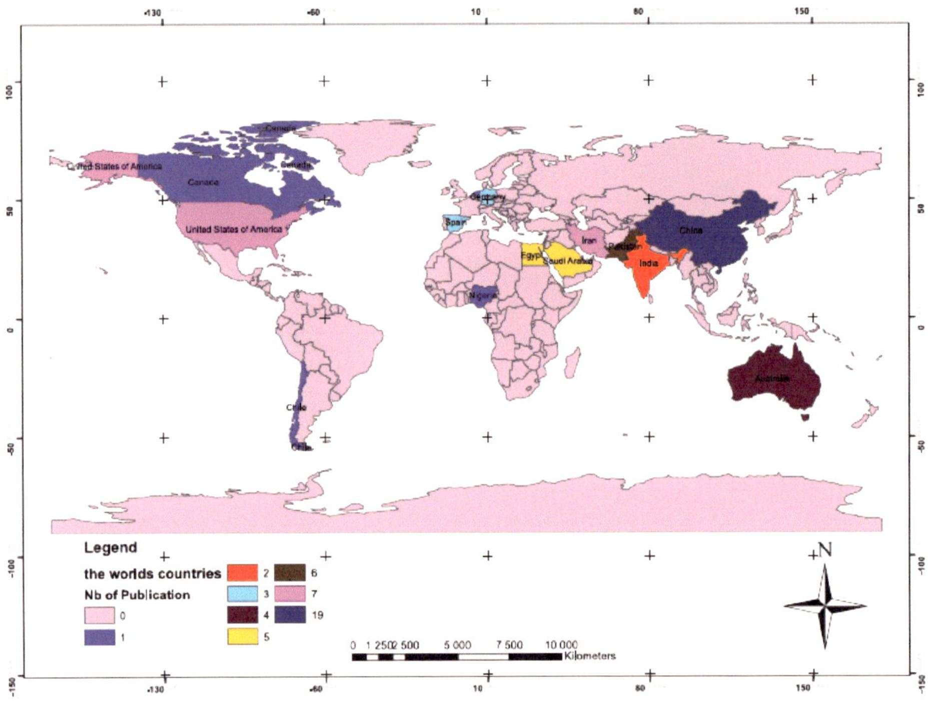

Figure 2.
Article number and geographic distribution according to the lead author's country of origin [70].

lead author's country of origin), China, the USA, and Iran are the leader countries for biochar studies and publications, followed by Pakistan, Middle East countries (Egypt and the Kingdom of Saudi Arabia), and to a lesser extent in Australia. In the Kingdom of Saudi Arabia, the Saudi Biochar Research Group in the King Saud University (Saudi Arabia) contributed to most publications in the Middle East countries.

We concluded that biochar is a promising soil amendment that can be used effectively for enhancing soil fertility. In arid regions like the Kingdom of Saudi Arabia, additional researches are needed to investigate the potential neutralization of biochar alkalinity; consequently, it can be added safely to agricultural soils. There are different sources of agricultural and food wastes that can be used for biochar production. In the case of date palm wastes, the average annual waste of one tree is about 40 kg; therefore, date palm wastes can be used effectively for biochar production in the Kingdom of Saudi Arabia. Therefore, the government of Kingdom of Saudi Arabia has to encourage the scientists for initiating intensive researches on biochar production and investigate its beneficial roles for improving soil fertility and agricultural production.

Author details

Khalid A. Alaboudi
National Centre for Biotechnology, King Abdulaziz City for Science and Technology, Kingdom of Saudi Arabia

*Address all correspondence to: khakha1979@gmail.com

References

[1] De Queiroz AF, Salviano AM, Da Cunha TJF, Olszevski N, DE Souza Júnior VS, De Oliveira Neto MB. Potentialities and limitations of agricultural use in soils of semi-arid region of the state of Bahia. Anais da Academia Brasileira de Ciências. 2018;**90**(4):3373-3387

[2] Ivanov IV, Prikhod'ko VE, Zamotaev IV, Manakhov DV, Novenko EY, Kalinin PI, et al. Synlithogenic evolution of floodplain soils in valleys of small rivers in the Trans-Ural Steppe. Eurasian Soil Science. 2019;**52**:593-609

[3] Elshony M, Farid I, Alkamar F, Abbas M, Abbas H. Ameliorating a sandy soil using biochar and compost amendments and their implications as slow release fertilizers on plant growth. Egyptian Journal of Soil Science. 2019;**59**(4):305-322

[4] Oliveira Filho JS, Vieira JN, da Silva EMR, de Oliveira JGB, Pereira MG, Brasileiro FG. Assessing the effects of 17 years of grazing exclusion in degraded semi-arid soils: Evaluation of soil fertility, nutrients pools and stoichiometry. Journal of Arid Environments. 2019;**166**:1-10

[5] Hammouda I, Elbaalawy A, Elf-fishy M. Effect of compost additives and application time of phosphorus in different methods on growth, productivity and quality of peanut in sandy soils. Egyptian Journal of Soil Science. 2019;**59**(4):339-352

[6] Sirjani E, Sameni A, Moosavi AA, Mahmoodabadi M, Laurent B. Portable wind tunnel experiments to study soil erosion by wind and its link to soil properties in the Fars province, Iran. Geoderma. 2019;**333**:69-80

[7] Bashir MA, Rehim A, Liu J, Imran M, Liu H, Suleman M, et al. Soil survey techniques determine nutrient status in soil profile and metal retention by calcium carbonate. Catena. 2019;**173**:141-149

[8] Marques MJ, Álvarez AM, Carral P, Esparza I, Sastre B, Bienes R. Estimating soil organic carbon in agricultural gypsiferous soils by diffuse reflectance spectroscopy. Water. 2020;**12**(1):261

[9] Mahdavi SM, Fujimaki H. Soil salinity resistance effect on evaporation. Eurasian Soil Science. 2019;**52**:526-534

[10] Food and Agrocultural Organization (FAO). Arid Zone Forestry: A Guide for Field Technicians. 1989. ISBN: 92-5-102809-5

[11] D'Odorico P, Porporato A, Runyan C. Ecohydrology of arid and semiarid ecosystems: An introduction. In: D'Odorico P, Porporato A, Wilkinson RC, editors. Dryland Ecohydrology. Cham: Springer; 2019. pp. 1-27. DOI: 10.1007/978-3-030-23269-6_1

[12] Shi WY, Zhu XC, Zhang FB, Wang KB, Deng L, Ma MG. Soil carbon biogeochemistry in arid and semiarid forests. In: Mazadiego LF, Garcia EDM, Barrio-Parra F, Izquierdo-Díaz M, editors. Applied Geochemistry with Case Studies on Geological Formations, Exploration Techniques and Environmental Issues. IntechOpen; 2019. DOI: 10.5772/intechopen.87951

[13] Barajas-Aceves M, Dendoovem L. Nitrogen, carbon and phosphorus mineralization in soils from semi-arid highlands of Central Mexico amended with tannery sludge. Bioresource Technology. 2001;**77**:121-130

[14] Farid I, Abbas M, El-Ghozoli A. Implications of humic, fulvic and

K-humate extracted from each of compost and biogas manures as well as their teas on faba bean plants grown on a typic Torripsamment soil and emissions of soil CO_2. Egyptian Journal of Soil Science. 2018;**58**(3):275-289

[15] Chang N, Zhai Z, Li H, Wang L, Deng J. Impacts of nitrogen management and organic matter application on nitrous oxide emissions and soil organic carbon from spring maize fields in the North China plain. Soil and Tillage Research. 2020;**196**:104441

[16] United Nations Convention to Combat Desertification (UNCCD). World Atlas of Desertification. 2nd ed. Nairobi: UNEP; 1997

[17] Mridha MAU, Al-Barakah FN. Green cultivation of moringa on arid agricultural land in Saudi Arabia. Acta Horticulturae. 2014:1158. ISHS 2017

[18] Ibrahim A, Usman ARA, Al-Wabel MI, Nadeem M, Ok YS, Al-Omran A. Effects of conocarpus biochar on hydraulic properties of calcareous sandy soil: Influence of particle size and application depth. Archives of Agronomy and Soil Science. 2017;**63**(2):185-197

[19] Alotaibi KD, Schoenau JJ. Addition of biochar to a sandy desert soil: Effect on crop growth. Water Retention and Selected Properties. Agronomy. 2019;**9**:327

[20] World Bank. Arable Land (hectares per person)-Saudi Arabia. 2017. Available from: http://data.worldbank.org/indicator/AG.LND.ARBL.HA.PC?end=2014&locations=SA&start=1961&view=chart

[21] Abdula'aly AI, Chammem AA. Groundwater treatment in the central region of Saudi Arabia. Desalination. 1997;**96**:203-214

[22] Mridha AU, Al-Qarawi AA. Bio-organic -an effective fertilizer for arid lands agriculture and date palm plantation in Saudi Arabia. Biosciences Biotechnology Research Asia. 2013;**10**(1):247-251

[23] Waqas M, Nizami AS, Aburiazaiza AS, Barakat MA, Rashid MI, Ismail IMI. Optimizing the process of food waste compost and valorizing its applications: A case study of Saudi Arabia. Journal of Cleaner Production. 2018;**176**(1):426-438

[24] Hozzein WN, Abuelsoud W, Wadaan MAM, Shuikan AM, Selim S, Al Jaouni S, et al. Exploring the potential of actinomycetes in improving soil fertility and grain quality of economically important cereals. Science of The Total Environment. 2019;**651**(2):2787-2798

[25] Abdelhafez AA, Abbas MHH, Li J. Biochar: The black diamond for soil sustainability, contamination control and agricultural production. In: Huang W-J, editor. Engineering Applications of Biochar. IntechOpen; 2017. DOI: 10.5772/intechopen.68803. Available from: https://www.intechopen.com/books/engineering-applications-of-biochar/biochar-the-black-diamond-for-soil-sustainability-contamination-control-and-agricultural-production

[26] USDA Foreign Agricultural Service. Report Number SA1417. Global Agricultural Information Network; 2014

[27] Lovelle M. Food and Water Security in the Kingdom of Saudi Arabia. Australia: Future Directions International Pty Ltd.; 2015, 2015. Available from: http://www.futuredirections.org.au/publication/food-and-water-security-in-the-kingdom-of-saudi-arabia/

[28] Alananbeh KM, Bouqellah NA, Kaff NS. Cultivation of oyster mushroom *Pleurotus ostreatus* on

date-palm leaves mixed with other agro-wastes in Saudi Arabia. Saudi Journal of Biological Sciences. 2014;**21**(6):616-625

[29] Lehmann J, Joseph S. Biochar for environmental management: An introduction. In: Lehmann J, Joseph S, editors. Biochar for Environmental Management: Science and Technology. London: Earthscan; 2009. pp. 1-12

[30] Major J, Rondon M, Molina D, Riha SJ, Lehmann J. Maize yield and nutrition during 4 years after biochar application to a Colombian savanna oxisol. Plant and Soil. 2010;**333**:117-128

[31] Abdelhafez AA, Li J. Removal of Pb(II) from aqueous solution by using biochars derived from sugarcane bagasse and orange peel. Journal of the Taiwan Institute of Chemical Engineers. 2016;**61**:367-375

[32] Emrich W. Handbook of Charcoal Making. The Traditional and Industrial Methods. Dordrecht, Holland: D. Reidel Publishing Company; 1985

[33] Verheijen FGA, Montanarella L, Bastos AC. Sustainability, certification, and regulation of biochar. Pesquisa Agropecuária Brasileira. 2012;**47**:649-653

[34] Sjöström E. Wood Chemistry: Fundamentals and Applications. 2nd ed. San Diego, CA: Academic Press; 1993

[35] Chen B, Yuan M. Enhanced sorption of polycyclic aromatic hydrocarbons by soil amended with biochar. Journal of Soils and Sediments. 2011;**11**:62-71

[36] Bird MI, Ascough PL, Young IM, Wood CV, Scott AC. X-ray microtomographic imaging of charcoal. Journal of Archaeological Science. 2008;**35**:2698-2706

[37] Abdullah H, Wu H. Biochar as a fuel: 1. Properties and grindability of biochars produced from the pyrolysis of mallee wood under slow-heating conditions. Energy and Fuels. 2019;**23**:4174-4181

[38] Farrell M, Kuhn TK, Macdonald LM, Maddern TM, Murphy DV, Hall PA, et al. Microbial utilisation of biochar-derived carbon. Science of the Total Environment. 2013, 2013;**465**:288-297

[39] Agblevor FA, Beis S, Kim SS, Tarrant R, Mante NO. Biocrude oils from the fast pyrolysis of poultry litter and hardwood. Waste Management. 2010;**30**:298-307

[40] Nguyen B, Lehmann J. Black carbon decomposition under varying water regimes. Organic Geochemistry. 2009;**40**:846-853

[41] Nguyen B, Lehmann J, Hockaday WC, Joseph S, Masiello CA. Temperature sensitivity of black carbon decomposition and oxidation. Environmental Science and Technology. 2010;**44**:3324-3331

[42] Feng Y, Xu Y, Yu Y, Xie Z, Lin X. Mechanisms of biochar decreasing methane emission from Chinese paddy soils. Soil Biology and Biochemistry. 2012;**46**:80-88

[43] Zhang A, Cui L, Pan G, Li L, Hussain Q, Zhang X, et al. Effect of biochar amendment on yield and methane and nitrous oxide emissions from a rice paddy from Tai Lake plain, China. Agriculture, Ecosystems and Environment. 2010;**139**:469-475

[44] Peng X, Ye LL, Wang C, Zhou H, Sun B. Temperature and duration-dependent rice straw-derived biochar: Characteristics and its effects on soil properties of an ultisol in southern China. Soil Tillage Research. 2011;**112**(2):159-166

[45] Novak JM, Lima I, Xing B, Gaskin JW, Steiner C, Das KC, et al. Characterization of designer biochar

produced at 18 engineering applications of biochar different temperatures and their effects on a loamy sand. Annals of Environmental Science. 2009;**3**(1):195-206

[46] Lee Y, Park J, Ryu C, Gang KS, Yang W, Park Y, et al. Comparison of biochar properties from biomass residues produced by slow pyrolysis at 500°C. Bioresource Technology. 2013;**148**:196-201

[47] Erin N, Yargicoglu EN, Sadasivam BY, Reddy KR, Spokas K. Physical and chemical characterization of waste wood derived biochars. Waste Management. 2014;**36**:256-268

[48] Rondon MA, Lehmann J, Ramírez J, Hurtado M. Biological nitrogen fixation by common beans (*Phaseolus vulgaris* L.) increases with biochar additions. Biology and Fertility of Soils. 2007;**43**(6):699-708

[49] Fabbri D, Torri C, Spokas KA. Analytical pyrolysis of synthetic chars derived from biomass with potential agronomic application (biochar): Relationships with impacts on microbial carbon dioxide production. Journal of Analytical and Applied Pyrolysis. 2012;**93**:77-84

[50] Liu XH, Han FP, Zhang XC. Effect of biochar on soil aggregates in the loess plateau: Results from incubation experiments. International Journal of Agriculture & Biology. 2012;**14**:975-979

[51] Shinogi Y, Yoshida H, Koizumi T, Yamaoka M, Saito T. Basic characteristics of low temperature carbon products from waste sludge. Advances in Environmental Research. 2003;7(3):661-665

[52] Dugan E, Verhoef A, Robinson S, Sohi S. Bio-char from sawdust, maize stover and charcoal: Impact on water holding capacities (WHC) of three soils from Ghana. In 19th World Congress of Soil Science, Symposium 2010;4(2):9-12

[53] Mukherjee A, Lal R. Biochar impacts on soil physical properties and greenhouse gas emissions. Agronomy. 2013;**3**:313-339

[54] Bassouny M, Abbas M. Role of biochar in managing the irrigation water requirements of maize plants: The pyramid model signifying the scoil hydro-physical and environmental markers. Egyptian Journal of Soil Science. 2019;**59**(2):99-115

[55] Asai H, Samson BK, Stephan HM, Songyikhangsuthor K, Homma K, Kiyono Y, et al. Biochar amendment techniques for upland rice production in Northern Laos: 1. Soil physical properties, leaf SPAD and grain yield. Field Crops Research. 2009;**111**:81-84

[56] Karhu K, Mattila T, Bergström I, Regina K. Biochar addition to agricultural soil increased CH_4 uptake and water holding capacity—Results from a short-term pilot field study. Agriculture, Ecosystems & Environment. 2011;**140**:309-313

[57] Yuan JH, Xu RK, Zhang H. The forms of alkalis in the biochar produced from crop residues at different temperatures. Bioresource Technology. 2011;**102**(3):3488-3497

[58] Abdelhafez AA, Abbas MHH, Hamed MH. Biochar: A solution for soil Pb pollution. In: The 8th International Conference for Development and the Environment in the Arab World Assiut University, Egypt. 2016. pp. 89-103

[59] Guo M, He Z, Uchimiya SM. Introduction to biochar as an agricultural and environmental amendment. In: Guo M, He Z, Uchimiya SM, editors. Agricultural and Environmental Applications of Biochar: Advances and Barriers. Madison, WI,

USA: Soil Science Society of America; 2016. pp. 1-14. SSSA Spec. Publ. 63

[60] Guo M, Xiao P, Li H. Valorization of agricultural byproducts through conversion to biochar and bio-oil. In: Simpson BK, Kwofie EM, Aryee AN, editors. Byproducts from Agriculture and Fisheries: Adding Value for Food, Feed, Pharma and Fuels. Somerset, NJ, USA: John Wiley & Sons, Inc.; 2020. pp. 501-522

[61] Sigua GC, Novak JM, Watts DW, Johnson MG, Spokas K. Efficacies of designer biochars in improving biomass and nutrient uptake of winter wheat grown in a hard setting subsoil layer. Chemosphere. 2016;**142**:176-183

[62] Zhang Y, Idowu OJ, Brewer CE. Using agricultural residue biochar to improve soil quality of desert soils. Agriculture. 2016;**6**(1):10

[63] Blok C, van der Salm C, Hofland-Zijlstra J, Streminska M, Eveleens B, Regelink I, et al. Biochar for horticultural rooting media improvement: Evaluation of biochar from gasification and slow pyrolysis. Agronomie. 2017;**7**:6

[64] Luo X, Liu G, Xia Y, Chen L, Jiang Z, Zheng H, et al. Use of biochar-compost to improve properties and productivity of the degraded coastal soil in the YellowRiver Delta, China. Journal of Soils and Sediments. 2017;**17**:780-789

[65] Zheng H, Wang X, Chen L, Wang Z, Xia Y, Zhang Y, et al. Enhanced growth of halophyte plants in biochar-amended coastal soil: Roles of nutrient availability and rhizosphere microbial modulation. Plant Cell Environment. 2018;**41**(3):517-523

[66] Igalavithana AD, Ok YS, Usman ARA, Al-Wabel MI, Oleszczuk P, Lee SS. The effects of biochar amendment on soil fertility. In: Guo M, He Z, Uchimiya SM, editors. Agricultural and Environmental Applications of Biochar: Advances and Barriers. Madison, WI, USA: Soil Science Society of America; 2016. pp. 123-144. SSSA Spec. Publ. 63

[67] Tian J, Miller V, Chiu PC, Maresca JA, Guo M, Imhoff PT. Nutrient release and ammonium sorption of poultry litter and wood biochars in stormwater treatment. Science of The Total Environment. 2016;**553**:596-606

[68] Paz-Ferreiro J, Mendez A, Gasco G. Application of biochar for soil biological improvement. In: Guo M, He Z, Uchimiya SM, editors. Agricultural and Environmental Applications of Biochar: Advances and Barriers. Madison, WI, USA: Soil Science Society of America; 2016. pp. 145-174. SSSA Spec. Publ. 63

[69] Lehmann J, Rillig MC, Thies J, Masiello CA, Hockaday WC, Crowley D. Biochar effects on soil biota—A review. Soil Biology and Biochemistry. 2011;**43**:1812-1836

[70] Arfaoui A, Ibrahimi K, Trabelsi F. Biochar application to soil under arid conditions: A bibliometric study of research status and trends. Arabian Journal of Geosciences. 2019;**12**:45

Section 4

Role of Biochar for Soil Remediation and Ameliorating Salinity Effects

Chapter 11

The Use of Biochar as a Soil Amendment to Reduce Potentially Toxic Metals (PTMs) Phytoavailability

João Arthur Antonangelo and Hailin Zhang

Abstract

The contamination of potentially toxic metals (PTMs) is widespread in the world and has negatively affected plants, humans, soil health, and environmental quality. Some metals are essential plant nutrients but they are also toxic to vegetation and aquatic live when present in high concentrations, such as Cu, Mn and Zn. Others (e.g., Pb, Cd, Cr, and As) are potential toxic metals for all organisms, and are not needed (or are toxic) for plant growth. This chapter summarizes the use of readily available biochars (BCs) to reduce PTMs phytoavailability in soils thus improving crop yields and to minimize its impact on the environment. The physico-chemical and morphological properties of BCs as affected by feedstock sources and pyrolysis temperatures are discussed. The effectiveness of biochar rates on plant growth and metal fractions are also highlighted. Biochar has the potential to be used as a viable bioproduct for the remediation of contaminated soils since it reduces the phytoavailability of PTMs pollutants. Biochars produced from different feedstocks and at different pyrolysis temperatures present highly heterogeneous physico-chemical and morphological properties, which can affect the effectiveness in the remediation of PTMs contaminated soils. Therefore, potential technologies need to be developed and research gaps still need to be overcome to optimize the use of BCs as a feasible alternative for remediation of metal contaminated soils.

Keywords: soil remediation, biochar amendment, potentially toxic metals, phytoavailability, immobilization, soil health

1. Introduction

Biochar is a carbon-rich by-product produced from the thermochemical conversion of biomass feedstock under partial or total absence of oxygen [1, 2]. Feedstocks used in biochar (BC) production are mostly wood, municipal and agriculture wastes [3–6]. Amending soil with biochar has received increased attention as a method for carbon sequestration in soils, thereby reducing carbon dioxide (CO_2) emissions [7–9] and improving soil quality due to the vital role of carbon (C) in soil physical, chemical and biological processes [10].

IntechOpen

Amending soil with biochar has been practiced for a long time. The high fertility of anthropogenic dark earth soils known as 'Terra Preta de Indio' in the Amazon basin has been related to the high content of charred materials [11–13]. Historically, the source of char in these soils has been considered as a disposal of charcoal from domestic fires and the practice of slash and char agriculture by Pre-Columbian Amazonian Indians [11, 14]. Hence, these soils have remained fertile and rich in biochar derived C stock for hundreds to thousands of years after they were abandoned.

In addition to the role of biochar in increasing the C sequestration and influencing the reduction of CO_2 emissions, biochar has been shown to enhance soil quality and to stabilize PTMs [15]. Biochar has a potential benefit for improving soil fertility [16, 17], improving soil properties such as pH [11–13, 18], cation exchange capacity (CEC) and water holding capacity [19], enhancing plant growth [20], and reducing nutrient leaching losses [21]. The significant amount of calcium (and magnesium) carbonate ($Ca/MgCO_3$) in BCs enables them to function as lime materials providing Ca and Mg to plants and neutralizing acidity when applied to acid soils.

The role of biochar in improving soil pH, organic carbon (OC), and CEC was also highlighted by [16]. Moreover, biochar can immobilize PTMs (immobilization is the reduction of the potential migration of PTMs to plants, or reduction of phytoavailability) such as cadmium (Cd), lead (Pb), and zinc (Zn) and thereby to reduce the phytoavailability of PTMs (concentration of PTMs in plant parts, or contents of PTMs in soils available to plants) to plants in contaminated soils, notably because it raises the soil pH [18, 22] and increases CEC and OC [23]. Many studies also found biochar application promotes the ability to remove organic contaminants [24, 25]. Because of its porous structure and diverse functional groups [26], biochar has been widely used in the field of agriculture and environmental protection [27] due to its ability to improve soil health and crop yields, and sequestering carbon, immobilizing PTMs and adsorbing organic pollutants such as polycyclic aromatic hydrocarbons (PAHs).

For these reasons, studies on biochar land-application have exponentially increased in the last 20 years (**Figure 1**). During the same period (1999–2018), the word 'potentially toxic metal' or 'heavy metal' places itself in the **top 5** within the 25 keywords used in biochar researches, numbering **308** publications [6]. Therefore, this chapter is to provide a summary of the most recent studies on biochar use to improve soil quality and to immobilize the phytoavailability of PTMs to plants of agricultural importance. The main goal is to improve our

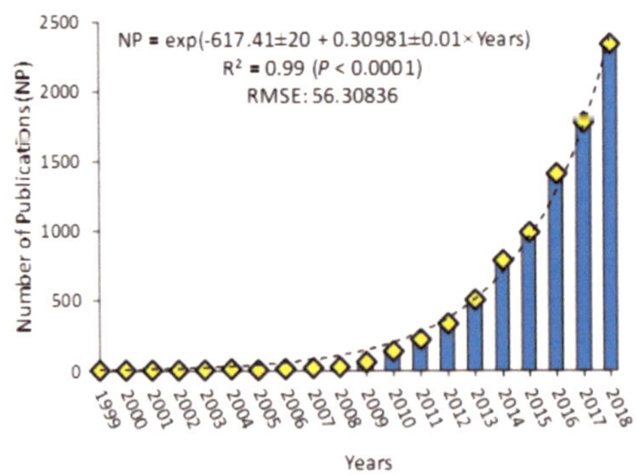

Figure 1.
Number of publications (NP) of biochar studies since 1999 (adapted from [6]). RMSE: Root mean square error value of the exponential model adopted.

understanding of biochar production and application as a soil remediation technique and to serve as the basis for future research work.

2. Biochar increases soil pH and soil organic carbon content and affects PTMs phytoavailability

The effects of pH rising in soils are significantly influenced by biochar addition more than by other practices such as liming [28]. **Table 1** shows some of the main characteristics of BCs (pH included) as affected by feedstock sources and pyrolysis temperature. Biochar is superior to lime to remediate PTMs-polluted areas, mainly because acidic conditions can lead to the leaching of metals and threatening of groundwater [28]. Biochars can supply OC and raise soil pH, but lime only increases soil pH. Hence, poultry litter-derived biochar (PLB) proves itself as very effective in immobilizing Cd, even under strong acidic conditions, thus preventing Cd leaching

Biochar feedstock	Pyrolysis temperature (°C)	Water (wt%)	TC (wt%)	TOC (wt%)	TN (wt%)	pH	Reference
Wood of 'Quaresmeira' (*Tibouchina arborea*)	330	55	82	80	—	9.7	[13]
Poultry litter	450	—	38	—	2	9.9	[16]
	550	—	33	—	0.85	13	
Miscanthus (Miscanthus × giganteus) straw	600	—	53.5	—	0.31	10	[18]
1-Switchgrass straw	500	—	23.7	—	0.8	9.4	[19]
2-Anaerobically digested fiber		—	25	—	0.85	9.3	
3-Softwood bark		—	28.4	—	0.14	7.6	
4-Wood pellets		—	28.2	—	0.05	7.2	
1-Swicthgrass	350	1.4	42.6	—	0.9	5.2	[23]
	700	1.7	31.4	—	0.7	10	
2-Poultry litter	350	2.7	38.4	—	4.1	7.4	
	700	3.9	27.8	—	1.6	10	
1-Pig manure	350	—	31.6	—	3.8	8.3	[25]
	700	—	25.2	—	2.1	9.5	
2-Deashed pig manure-biochar	350	—	66.7	—	7.6	6.2	
	700	—	74.2	—	5	6.4	
1-Poultry litter	400	—	16.8	—	1.4	10	[28]
2-Eucalyptus	600	—	81	—	1.1	10.4	

TC: total carbon. TOC: total organic carbon. TN: total nitrogen. "—": not given.

Table 1.
Summary of some biochar properties as affected by feedstock sources and pyrolysis temperature.

to the groundwater [28]. Besides raising the soil pH, the enhanced OC provided by biochar addition contributes to a decrease in the phytoavailability of PTMs by reducing metal mobility due to bonding metals into more stable fractions [29, 30], such as organic matter-bound and/or highly stable organic complexes which are not readily dissolved by water. The increase of both pH and OC also contributes to a higher CEC, then resulting in a higher PTMs adsorption [31].

The application of orchard prune-derived biochar (OPBC) to mine tailings reduced phytoavailable (DTPA-extractable) concentrations of Pb, Cd, and Zn [32]. Rice straw-derived biochar reduced Cd concentration in the plant available soil fraction grown in a greenhouse [30]. Biochar addition also showed a potential in reducing Cd and Pb accumulation in ryegrass (*Lolium perenne*) shoots, thus presenting a viable option for safe cultivation in PTMs-polluted soils [29]. Recent studies [22, 33] have found the ability of biochar to reduce the phytoavailability of PTMs (Cu, Zn, Pb, Cd, Mn, and Ni) to ryegrasses. Particularly, the study conducted by [22] elucidated that soil pH and OC increases as a function of biochar application rates played a big role in PTMs (Zn, Pb, and Cd) immobilization, mainly by forming stable (and undissolved) complexes with hydroxyls (OH^-) ($Zn/Pb/Cd_x(OH)_y$) and surface functional groups, respectively. **Figure 2** summarizes their findings and illustrates the importance of biochar application as a soil remediation technique. In another study, the concentrations of Cu, Pb, and Zn decreased as the rates of BCs applied increased, with a significant effect for amendments >1% w/w applied. Especially, the phytoavailability of PTMs decreases gradually with time when the soil is amended by 5% or 10% w/w of biochar [18].

It has been shown that the addition of sugar cane bagasse-derived BC (SCBC) decreased the phytoavailable Pb fraction by 97%, and that the PTM uptake by maize plants decreased with increasing the level of applied BC [34]. The authors attributed such results to an enhancement of soil pH and soil organic carbon (SOC) because of BC addition. In addition, an unpublished work conducted by Antonangelo and Zhang has also revealed how impacting is the increase of soil pH and SOC, as biochar application rates increase, on PTMs (Zn, Pb and Cd) mobility. The study was carried out with no plants and the application rates of poultry litter- (PLB) and switchgrass-derived biochar (SGB) ranged from 0 to 8% w/w

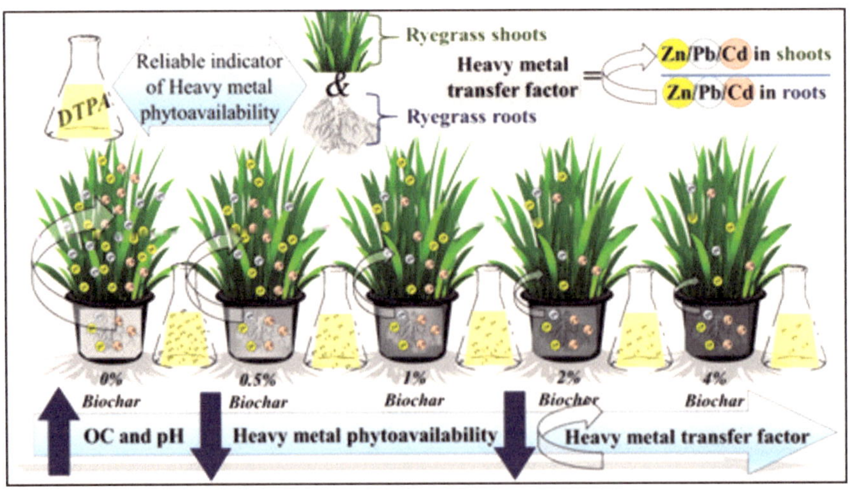

Figure 2.
Potentially toxic metal (heavy metal) immobilization to ryegrass shoots and roots as a function of biochar application rates. Diagram was created from the work of [22].

in multi-metal contaminated soil. Potentially toxic metal phytoavailability was assessed by using two extraction methods (NH_4NO_3 and DTPTA) after 10 weeks of soil + biochar incubation under laboratory room temperatures. The results of PTMs contents in the extracts and SOC (%) as a function of biochar application rates are summarized in **Figures 3** and **4**, and **Table 2**. The pH increase from 6.5 to 8.0, as shown in **Figure 3**, is a consequence of the increased BCs (SGB and PLB) application rates from 0 to 8%, so is the SOC increase, as shown in **Figure 4**. **Table 2** highlights the significant negative correlations (inverse relationship) of pH and SOC (independent variables) with phytoavailable PTMs in the filtered extracts (dependent variables).

The immobilization of Cu, Pb, Cd, and Ni by BCs was attributed to the quantity of surface oxygen-functional groups, which is directly related to the amount of carbon (C) present in the biochar composition [35]. Uchimiya et al. [36] reported that biochar enhanced Cu sorption in a sandy loam soil primarily by cation exchange mechanism, enhanced by the soil C increase. Hence, biochar addition increased the sorption capacity of the soil matrix for both organic and inorganic contaminants [37]. However, BCs may favor the availability of some PTMs such

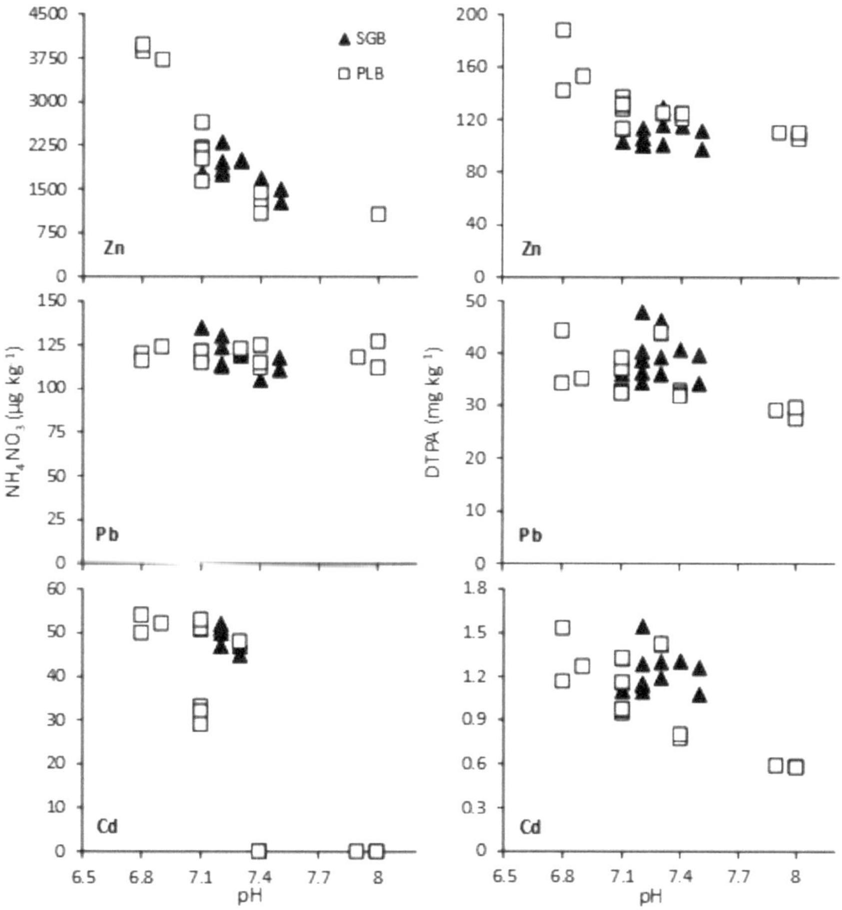

Figure 3.
Relationship between pH and metal concentrations in NH_4NO_3 and DTPA extracts after 10 weeks of biochar incubation. Points were plotted from the whole dataset of measurements. One outlier from three replicates (n = 3) was removed, when detected, by using IML and UNIVARIATE (ROBUSTSCALE) procedures of the SAS program. pH ranged from 6.5 to 8.0 as biochars (SGB and PLB) application rates increased from 0 to 8% w/w of biochar.

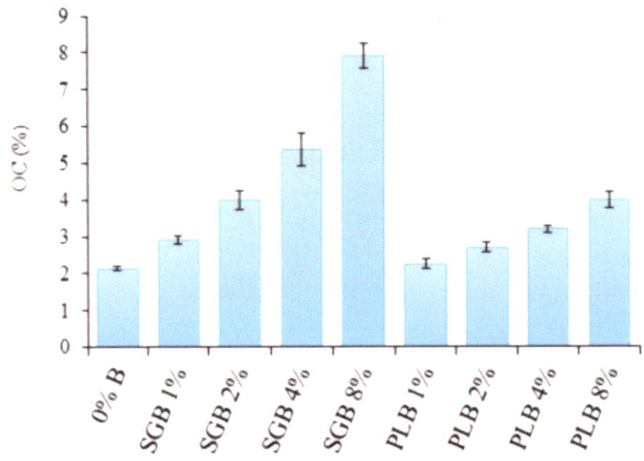

Figure 4.
Soil organic carbon (OC%) changes due to biochar (SGB and PLB) application rates after 10 weeks of incubation. Bars represent the standard deviation of the mean (n = 3) results are significant at P < 0.01 *(Tukey test).*

Soil attribute	SGB	PLB	SGB + PLB	SGB	PLB	SGB + PLB
		Zn (NH_4NO_3)			Zn (DTPA)	
pH	−0.75**	−0.82***	−0.81***	−0.15^{NS}	−0.76***	−0.55**
OC	−0.79**	−0.74**	−0.43*	−0.31^{NS}	−0.60*	−0.47*
		Pb (NH_4NO_3)			Pb (DTPA)	
pH	−0.55*	−0.05^{NS}	−0.16^{NS}	0.03^{NS}	−0.63*	−0.44*
OC	−0.64*	−0.11^{NS}	−0.46*	−0.12^{NS}	−0.77**	0.10^{NS}
		Cd (NH_4NO_3)			Cd (DTPA)	
pH	−0.91***	−0.82***	−0.72***	0.04^{NS}	−0.83***	−0.66***
OC	−0.91***	−0.91***	−0.01^{NS}	−0.11^{NS}	−0.92***	0.06^{NS}

**: P < 0.05; **: P < 0.01; ***: P < 0.001; NS: non-significant (P > 0.05). The R-values were calculated from the whole dataset of measurements. SGB + PLB: two biochar treated soils. One outlier from three replicates (n = 3) was removed, when detected, by using IML and UNIVARIATE (ROBUSTSCALE) procedures of SAS program.*

Table 2.
Pearson correlation coefficient (R) between metal extracted from NH_4NO_3 or DTPA and soil attributes (pH and organic carbon-OC) after 10 weeks of soil+biochar (SGB and PLB) incubation.

as arsenic (As). The addition of sugar cane bagasse-derived BC (SCBC) stabilized Pb but accelerated the desorption of arsenic (As); consequently, increased its availability [38]. That is probably a result of charge repulsion between the negatively charged SCBC and the arsenate anion (AsO_4^{3-}).

3. The effect of biochar feedstock sources on PTMs phytoavailability

While investigating the effects of chicken manure (CMB) and greenwaste biochar (GWB), both produced at 550°C (pyrolysis temperature), on the immobilization and phytoavailability of Cd, Cu and Pb in metal-spiked and multi-metal contaminated soils, [39] found that both BCs significantly decreased Cd and Pb mobility, mostly by modifying those PTMs from the easily exchangeable soil fraction to less available organic bond fraction. Additionally, they reported that the application of the two feedstock-derived BCs increased the root and shoot dry biomass and decreased the accumulation of Cd, Cu, and Pb in Indian mustard (*Brassica juncea*), thus illustrating the role of biochar in reducing metal phytoavailability while supplying plant nutrients, regardless of the feedstock source. However, according to the authors, the CMB was more effective in metal immobilization and plant growth than the GWB.

A significant decrease in the transfer factor values (TF) of PTMs (Zn, Pb, and Cd) from ryegrass roots to ryegrass shoots when evaluating PLB and SGB additions to a multi-metal contaminated soil was found by [22], and that the PLB was more efficient in such reduction than SGB. This was probably a consequence of their higher pH, CEC, specific surface area (SSA), and stronger buffering capacity as reported by [23], which resulted in the higher efficiency of PLB in decreasing PTMs uptake, as highlighted by the higher decrease of the bioconcentration factor (BCF = [PTMs in shoots/PTMs concentration in soil]) as PLB application rates increased. **Figure 5** (and **Figure 3**) briefly summarizes such findings and highlights the PTMs immobilization as a function of two feedstocks derived-biochar application rates.

When comparing other soil amendments or feedstocks (raw material) with a feedstock-derived biochar, [40] showed that mussel shells, cow manure, and oak wood biochar application reduced Pb phytoavailability and phytoavailability in a highly contaminated military shooting range soil in Korea. Their study also showed increases in germination percentage and root elongation of lettuce (*Lactuca sativa*) in soil treated with the tested amendments, indicating a reduction of Pb accessibility. Outstandingly, biochar was more effective in decreasing Pb availability than the other tested soil amendments [40]. The application of BCs derived from animal wastes (pig manure biochar, and PLB) reduced the mobility of Cu, Cd, Pb and Zn from 28 to 69%, 77 to 100%, 94 to 99%, and 15 to 97%, respectively, in a multi-metal contaminated soil [41].

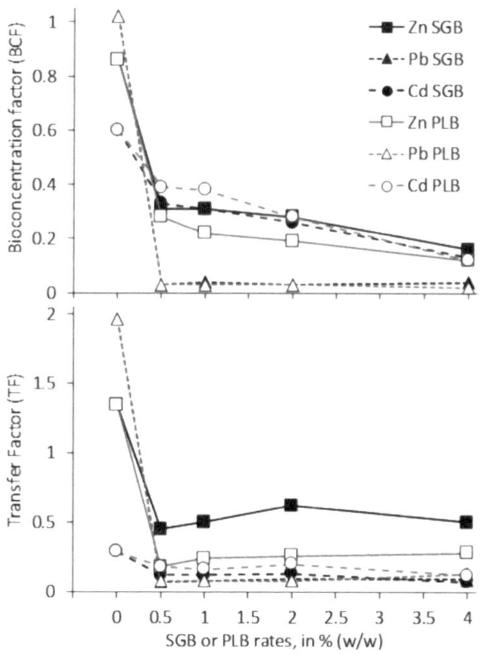

Figure 5.
Bioconcentration factor (BCF) and transfer factor (TF) of potentially toxic metals (Zn, Pb, and Cd) from ryegrass roots to ryegrass shoots as a function of biochar application rates. Biochars were either derived from switchgrass (SGB) or poultry litter (PLB) feedstocks. Graphs were modified from [22].

A study conducted by [42] observed that the addition of hardwood-derived biochar (HWB) to a PTMs contaminated mine soil reduced pore water solubility of Pb concentrations and ryegrass Pb levels. On the other hand, the combination of biochar with greenwaste compost (GWC) was more effective in reducing Pb in soil pore water and uptake by ryegrass. However, the biochar itself was more effective in reducing pore water Cu than GWC. Additionally, [43] reported that the addition of HWB and GWC to a multi-element contaminated soil significantly reduced concentrations of Cd and Zn in pore water during a 60 days exposure to field conditions and reduced phytoavailability of these elements resulting in increased shoot emergence of ryegrass. In contrast, concentrations of Cu and As in pore water increased with amendment applications [38, 43]. In a laboratory column study, [44] reported that HWB reduced the concentrations of Cd and Zn in leachate obtained from a multi-metal polluted soil with evidence of surface retention of both metals on biochar.

The work of [45] compared the impacts of broiler litter-derived biochar and pecan shell-derived steam activated carbon amendments on PTMs (Cu, Ni, and Cd) immobilization and the effects of oxidation on mineral retention in synthetic rainwater leaching experiments. Conversely, their study found that biochar was most effective in immobilizing Cu, whereas activated carbon immobilized Ni and Cd to a larger extent than biochar.

Contrarily, some BCs might only slightly decrease or even significantly increase extractable PTMs depending on the feedstock and pyrolytic temperature [33, 42]. Overall, the influence of biochar on PTMs extractability varies depending on the feedstock, application rate, and BCs particle size [46]. Generally speaking, biochar is a promising tool to reduce the mobility of PTMs in mining areas [22].

4. The effects of biochar conversion processes on PTMs phytoavailability

Biochars are effective in the immobilization of PTMs and this effect varies depending on biochar nature and pyrolysis conditions. [47] investigated the impact of pyrolysis temperature on BCs ability to stabilize PTMs in Small Arms Range soil using broiler litter BCs produced at 350 and 650°C. They found that both BCs were effective in stabilizing Pb and Cu at application rates of ≤5% without releasing Sb. In other experiments, [48, 49] suggested using BCs prepared at high temperatures, 650–800°C, for remediation purposes. Additionally, the uptake of PTMs by ryegrass planted in biochar-treated soils generally decreased with increasing pyrolytic temperature [33]. It was also pointed out by [50] that two different feedstocks-derived BCs were more effective in chromium (Cr) adsorption when pyrolyzed at higher temperatures. However, low-temperature biochar was more effective in stabilizing Pb than high-temperature biochar [17]. Such a result was attributed to the higher soluble P concentration of low-temperature biochar, which resulted in a greater Pb immobilization by the formation of lead-phosphate precipitates. In similar experiments using oxidized and unoxidized plant-derived BCs, [49] observed that oxidized BCs rich in carboxyl functional groups had greater ability for Pb, Cu, and Zn immobilization than unoxidized BCs. Therefore, the effect of BCs on the mobility of PTMs in soil is not only a function of the pyrolysis temperature, but also the feedstock used, as previously mentioned, soil properties, and surface functional groups. Indeed, the ability of BCs in reducing the phytoavailability of PTMs in soil depends on its surface functional groups, specific surface area, and porosity [23].

Figure 6 was adapted from the recent work of [51] and clearly shows that simply increasing pyrolysis temperature of BCs to enhance PTMs immobilization is not a pragmatism. The recovery efficiency (RE, in %) of PTMs (in a multi-metal contamination scheme: Pb + Cu + Zn) from soils amended with mesquite BCs (MBC) pyrolyzed at four different temperatures (300, 400, 500, and 600°C), have decreased as the initial concentration of added PTMs increased (**Figure 6**). However, MBC pyrolyzed at the highest temperature (600°C) has shown an overall higher RE% of Pb, Cu, and Zn (**Figure 6**). The authors also emphasized that there was a competitive adsorption among the PTMs into BCs exchangeable sites with a preferable affinity for Pb sorption. That would probably favor other PTMs to be phytoavailable in the medium. According to [22, 23, 51], surface functional groups responsible for metals retention are prone to change when pyrolysis temperature increases, which changes metals' sorption effectiveness to the same extent. Phenolic groups (OH) decrease followed by the increase of aromatic carbon contents (C=C stretching) in produced BCs are attributed to the depolymerization and dehydration of materials as pyrolysis temperature increases, then resulting in the formation of C=C double bonds, carbonyl, and carboxylic functional groups [23, 51–53]. Those functional groups are also responsible for PTMs adsorption and complexation.

Biomass gasification has also been demonstrated as an alternative method of pyrolysis to produce BCs [54, 55], although to a much lesser extent. It is a

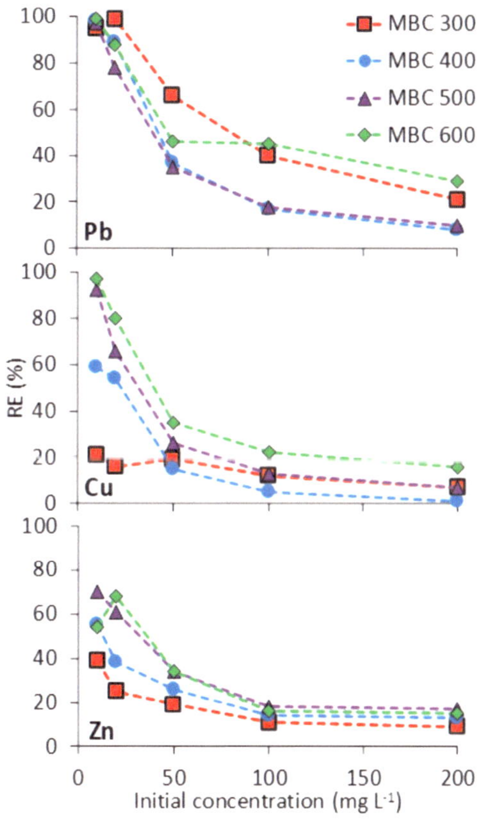

Figure 6.
Lead (Pb), copper (Cu), and zinc (Zn) removal efficiency (RE%) (PTM adsorption) in a multi-metal contaminated soil amended with biochars derived from mesquite and pyrolyzed at 300°C (MBC 300), 400°C (MBC 400), 500°C (MBC 500), and 600°C (MBC 600). The graph is modified from [51] published work.

technology that uses a controlled process involving heat, steam, and oxygen to convert biomass to hydrogen (and other products), in the absence of combustion. A recent study of [55] have demonstrated that the SSA, CEC, and basic functional groups of the pine woodchips-derived BCs (PWC) increased as the rate of airflow increased during the BCs conversion process. Therefore, such improved properties would favor PTMs immobilization in contaminated soils if a proper rate of PWCs were applied. More studies on different gasification processes of applied BCs affecting PTMs mobility in soils are encouraged to broaden the BCs options for remediation purposes.

5. Mechanisms of PTMs adsorption into biochars

The mechanisms proposed for PTMs immobilization are explored in [56, 57]. In summary, three mechanisms are mostly responsible for PTMs retention into BCs, among them: (1) PTMs exchange with calcium (Ca^{2+}), magnesium (Mg^{2+}), and other cations associated with BCs (**Figure** 7); (2) complexation of PTMs into different surface functional groups, as previously highlighted (see Section 4); and (3) the physical adsorption followed by surface precipitation contributing to PTMs immobilization.

In the first case, the PTMs/cations exchange is further attributed to the co-precipitation of PTMs and their innersphere complexation with complex humic matter and mineral oxides contained in the biochar [57]. First, the strengths of PTMs adsorption into BCs surface are low due to the presence of water molecules surrounding the ion (oscillation in the distance of the electrostatic retention). Latterly, water molecules are released and the affinity to complexation enhances depending on the composition and structure of the biochar reactive surface, thus a much stronger inner-sphere complex is formed disfavoring the PTMs release back to the soil solution. In the second case, the PTMs inner-sphere complexation is with the free hydroxyl of mineral oxides surface (OH) and other surface precipitation [57]. In the third case, the surface precipitation of PTMs occurring is designated as *amorphorse*, since biochars present an amorphous although highly reactive structure, similarly as the organic matter compounds [58].

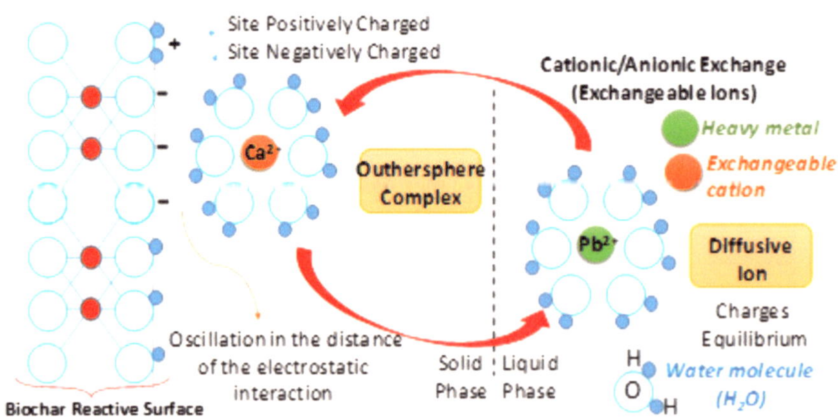

Figure 7.
Mechanisms of cation exchange between positively charged ions contained in the reactive surface of biochars and potentially toxic heavy metals ('heavy metal') dissolved in soil solution. The graph was made by the authors.

6. Conclusions and final considerations

Most studies from the last decade have demonstrated the strong potential of BCs in reducing leachability and immobilizing PTMs pollutants in soils that were previously phytoavailable. By increasing soil pH, BCs act as liming materials and PTMs precipitate as insoluble PTMs hydroxides in a high pH environment. Additionally, the greater SOC attributed to the BCs addition to multi-metal contaminated soil contributes to the formation of PTMS-organic anion complexes that are precipitated out from the system, since the previous bioavailable PTMs contents were much higher than usual. The characteristics of BCs vary widely with different feedstocks (biomass materials) and pyrolysis conditions (low and high temperatures). Generally, there are three mechanisms related to the direct removal of PTMs (and other pollutants) towards BCs reactive surface, which are related to a strong sorption and weak desorption of cationic PTMs then indicating that BCs sequesters pollutants in itself.

The current works emphasizing the use of biochar for soil remediation purposes have mainly been conducted in laboratories and/or greenhouses on a small scale with controlled conditions. As pointed out by [57], large-scale field trials are essential before operational scale remediation projects are implemented. Since the BC properties are largely varying, and sometimes contrasting, it is important to design BC products for every specific remediation project. The BCs ability to sequestrate may lead to the accumulation of PTMs contaminants in the amended soils in the long-term, and yet the pollutant environmental fate over time is not well elucidated. It is well known that the capacity of BCs to adsorb and/or complex PTMs decreases with time as a consequence of the aging process and saturation. Therefore, it is strongly encouraged to research on the aging process of BCs activity in the future because such information would help in the decision-making of the BC application rate and frequency to improve soil PTMs remediation efficiency.

Author details

João Arthur Antonangelo* and Hailin Zhang
Department of Plant and Soil Sciences (PSS), Oklahoma State University (OSU), Stillwater, OK, United States of America

*Address all correspondence to: joao.antonangelo@okstate.edu

References

[1] International Biochar Initiative (IBI) (Hg.). Standardized Product Definition and Product Testing Guidelines for Biochar That Is Used in Soil [Internet]. Climate Engineering - Kiel Earth Institute. 2012. Available from: https://www.climate-engineering.eu/single/international-biochar-initiative-ibi-hg-2012-standardized-product-definition-and-product-testing-guidelines-for-biochar-that-is-.html [cited 13 March 2020]

[2] Lehmann J, Joseph S. Biochar for Environmental Management: Science, Technology and Implementation. London: New York; 2015

[3] Yuan Y, Bolan N, Prévoteau A, Vithanage M, Biswas JK, Ok YS, et al. Applications of biochar in redox-mediated reactions. Bioresource Technology. 2017;**246**:271-281

[4] Chen B, Zhou D, Zhu L. Transitional adsorption and partition of nonpolar and polar aromatic contaminants by biochars of pine needles with different pyrolytic temperatures. Environmental Science & Technology. 2008;**42**(14):5137-5143

[5] Jang HM, Kan E. Engineered biochar from agricultural waste for removal of tetracycline in water. Bioresource Technology. 2019;**284**:437-447

[6] Li D, Zhao R, Peng X, Ma Z, Zhao Y, Gong T, et al. Biochar-related studies from 1999 to 2018: A bibliometrics-based review. Environmental Science and Pollution Research. 2019;**27**(3):2898-2908

[7] Lehmann J, Gaunt J, Rondon M. Bio-char sequestration in terrestrial ecosystems—A review. Mitigation and Adaptation Strategies for Global Change. 2006;**11**(2):403-427

[8] Lehmann J. Bio-energy in the black. Frontiers in Ecology and the Environment. 2007;**5**(7):381-387

[9] Woolf D, Amonette JE, Street-Perrott FA, Lehmann J, Joseph S. Sustainable biochar to mitigate global climate change. Nature Communications. 2010;**1**(1):1-9

[10] Stevenson FJ. Humus Chemistry: Genesis, Composition, Reactions. New York: John Wiley; 1994

[11] Glaser B, Lehmann J, Zech W. Ameliorating physical and chemical properties of highly weathered soils in the tropics with charcoal—A review. Biology and Fertility of Soils. 2002;**35**(4):219-230

[12] Petter FA, Madari BE, Silva MASD, Carneiro MAC, Carvalho MTDM, Júnior BHM, et al. Soil fertility and upland rice yield after biochar application in the Cerrado. Pesquisa Agropecuária Brasileira. 2012;**47**(5):699-706

[13] Gunten KV, Hubmann M, Ineichen R, Gao Y, Kurt KO, Alessi DS. Biochar-induced changes in metal mobility and uptake by perennial plants in a ferralsol of Brazil's Atlantic forest. Biochar. 2019;**1**(3):309-324

[14] Lehmann J. Amazonian Dark Earths: Origin Properties Management. Dordrecht: Kluwer Academic Publishers; 2003

[15] Ahumada I, Sepúlveda K, Fernández P, Ascar L, Pedraza C, Richter P, et al. Effect of biosolid application to Mollisol Chilean soils on the bioavailability of heavy metals (Cu, Cr, Ni, and Zn) as assessed by bioassays with sunflower (*Helianthus annuus*) and DGT measurements. Journal of Soils and Sediments. 2014;**14**(5):886-896

[16] Chan KY, Zwieten LV, Meszaros I, Downie A, Joseph S. Using poultry litter biochars as soil amendments. Soil Research. 2008;**46**(5):437

[17] Novak JM, Busscher WJ, Laird DL, Ahmedna M, Watts DW, Niandou MAS. Impact of biochar amendment on fertility of a Southeastern coastal plain soil. Soil Science. 2009;**174**(2):105-112

[18] Houben D, Evrard L, Sonnet P. Mobility, bioavailability and pH-dependent leaching of cadmium, zinc and lead in a contaminated soil amended with biochar. Chemosphere. 2013;**92**(11):1450-1457

[19] Streubel JD, Collins HP, Garcia-Perez M, Tarara J, Granatstein D, Kruger C. Influence of contrasting biochar types on five soils at increasing rates of application. Soil Science Society of America Journal. 2011;**75**(4):1402-1413

[20] Peng X, Ye L, Wang C, Zhou H, Sun B. Temperature- and duration-dependent rice straw-derived biochar: Characteristics and its effects on soil properties of an Ultisol in southern China. Soil and Tillage Research. 2011;**112**(2):159-166

[21] Laird D, Fleming P, Wang B, Horton R, Karlen D. Biochar impact on nutrient leaching from a Midwestern agricultural soil. Geoderma. 2010;**158**(3-4):436-442

[22] Antonangelo JA, Zhang H. Heavy metal phytoavailability in a contaminated soil of northeastern Oklahoma as affected by biochar amendment. Environmental Science and Pollution Research. 2019;**26**(32):33582-33593

[23] Antonangelo JA, Zhang H, Sun X, Kumar A. Physicochemical properties and morphology of biochars as affected by feedstock sources and pyrolysis temperatures. Biochar. 2019;**1**(3):325-336

[24] Devi P, Saroha AK. Simultaneous adsorption and dechlorination of pentachlorophenol from effluent by Ni–ZVI magnetic biochar composites synthesized from paper mill sludge. Chemical Engineering Journal. 2015;**271**:195-203

[25] Zhang W, Zheng J, Zheng P, Qiu R. Atrazine immobilization on sludge derived biochar and the interactive influence of coexisting Pb(II) or Cr(VI) ions. Chemosphere. 2015;**134**:438-445

[26] Ahmad I, Maathuis FJ. Cellular and tissue distribution of potassium: Physiological relevance, mechanisms and regulation. Journal of Plant Physiology. 2014;**171**(9):708-714

[27] Wang J, Wang S. Preparation, modification and environmental application of biochar: A review. Journal of Cleaner Production. 2019;**227**:1002-1022

[28] Lu H, Li Z, Fu S, Méndez A, Gascó G, Paz-Ferreiro J. Effect of biochar in cadmium availability and soil biological activity in an anthrosol following acid rain deposition and aging. Water, Air, & Soil Pollution. 2015;**226**(5):1-11

[29] Xu P, Sun C-X, Ye X-Z, Xiao W-D, Zhang Q, Wang Q. The effect of biochar and crop straws on heavy metal bioavailability and plant accumulation in a Cd and Pb polluted soil. Ecotoxicology and Environmental Safety. 2016;**132**:94-100

[30] Zhang R-H, Li Z-G, Liu X-D, Wang B-C, Zhou G-L, Huang X-X, et al. Immobilization and bioavailability of heavy metals in greenhouse soils amended with rice straw-derived biochar. Ecological Engineering. 2017;**98**:183-188

[31] Bolan N, Kunhikrishnan A, Thangarajan R, Kumpiene J, Park J, Makino T, et al. Remediation of heavy metal(loid)s contaminated soils—To mobilize or to immobilize? Journal of Hazardous Materials. 2014;**266**:141-166

[32] Fellet G, Marchiol L, Vedove GD, Peressotti A. Application of biochar on mine tailings: Effects and perspectives for land reclamation. Chemosphere. 2011;**83**(9):1262-1267

[33] Zhang G, Guo X, Zhao Z, He Q, Wang S, Zhu Y, et al. Effects of biochars on the availability of heavy metals to ryegrass in an alkaline contaminated soil. Environmental Pollution. 2016;**218**:513-522

[34] Abdelhafez AA, Abbas MH, Hamed MH. Biochar: A solution for soil Lead (Pb) pollution. In: The 8th International Conference for Development of the Environment in the Arab World; 2016. pp. 22-24

[35] Uchimiya M, Wartelle LH, Klasson KT, Fortier CA, Lima IM. Influence of pyrolysis temperature on biochar property and function as a heavy metal sorbent in soil. Journal of Agricultural and Food Chemistry. 2011;**59**(6):2501-2510

[36] Uchimiya M, Chang S, Klasson KT. Screening biochars for heavy metal retention in soil: Role of oxygen functional groups. Journal of Hazardous Materials. 2011;**190**(1-3):432-441

[37] Uchimiya M, Klasson KT, Wartelle LH, Lima IM. Influence of soil properties on heavy metal sequestration by biochar amendment: 1. Copper sorption isotherms and the release of cations. Chemosphere. 2011;**82**(10):1431-1437

[38] Abdelhafez AA, Li J, Abbas MH. Feasibility of biochar manufactured from organic wastes on the stabilization of heavy metals in a metal smelter contaminated soil. Chemosphere. 2014;**117**:66-71

[39] Park JH, Choppala GK, Bolan NS, Chung JW, Chuasavathi T. Biochar reduces the bioavailability and phytotoxicity of heavy metals. Plant and Soil. 2011;**348**(1-2):439-451

[40] Ahmad M, Lee SS, Yang JE, Ro H-M, Lee YH, Ok YS. Effects of soil dilution and amendments (mussel shell, cow bone, and biochar) on Pb availability and phytotoxicity in military shooting range soil. Ecotoxicology and Environmental Safety. 2012;**79**:225-231

[41] Gondek K, Mierzwa-Hersztek M. Effect of low-temperature biochar derived from pig manure and poultry litter on mobile and organic matter-bound forms of Cu, Cd, Pb and Zn in sandy soil. Soil Use and Management. 2016;**32**(3):357-367

[42] Karami N, Clemente R, Moreno-Jiménez E, Lepp NW, Beesley L. Efficiency of green waste compost and biochar soil amendments for reducing lead and copper mobility and uptake to ryegrass. Journal of Hazardous Materials. 2011;**191**(1-3):41-48

[43] Beesley L, Moreno-Jiménez E, Gomez-Eyles JL. Effects of biochar and greenwaste compost amendments on mobility, bioavailability and toxicity of inorganic and organic contaminants in a multi-element polluted soil. Environmental Pollution. 2010;**158**(6):2282-2287

[44] Beesley L, Marmiroli M. The immobilisation and retention of soluble arsenic, cadmium and zinc by biochar. Environmental Pollution. 2011;**159**(2):474-480

[45] Uchimiya M, Lima IM, Klasson KT, Chang S, Wartelle LH, Rodgers JE. Immobilization of heavy metal ions (CuII, CdII, NiII, and PbII) by broiler litter-derived biochars in water and

soil. Journal of Agricultural and Food Chemistry. 2010;**58**(9):5538-5544

[46] Yang X, Liu J, Mcgrouther K, Huang H, Lu K, Guo X, et al. Effect of biochar on the extractability of heavy metals (Cd, Cu, Pb, and Zn) and enzyme activity in soil. Environmental Science and Pollution Research. 2015;**23**(2):974-984

[47] Uchimiya M, Bannon DI, Wartelle LH. Retention of heavy metals by carboxyl functional groups of biochars in small arms range soil. Journal of Agricultural and Food Chemistry. 2012;**60**(7):1798-1809

[48] Uchimiya M, Klasson KT, Wartelle LH, Lima IM. Influence of soil properties on heavy metal sequestration by biochar amendment: 2. Copper desorption isotherms. Chemosphere. 2011;**82**(10):1438-1447

[49] Uchimiya M, Bannon DI, Wartelle LH, Lima IM, Klasson KT. Lead retention by broiler litter biochars in small arms range soil: Impact of pyrolysis temperature. Journal of Agricultural and Food Chemistry. 2012;**60**(20):5035-5044

[50] Rafique MI, Usman ARA, Ahmad M, Sallam A, Al-Wabel MI. In situ immobilization of Cr and its availability to maize plants in tannery waste-contaminated soil: Effects of biochar feedstock and pyrolysis temperature. Journal of Soils and Sediments. 2019;**20**(1):330-339

[51] Abdin Y, Usman A, Ok YS, Tsang YF, Al-Wabel M. Competitive sorption and availability of coexisting heavy metals in mining-contaminated soil: Contrasting effects of mesquite and fishbone biochars. Environmental Research. 2020;**181**:108846

[52] Bavariani MZ, Ronaghi A, Ghasemi R. Influence of pyrolysis temperatures on FTIR analysis, nutrient bioavailability, and agricultural use of poultry manure biochars. Communications in Soil Science and Plant Analysis. 2019;**50**(4):402-411

[53] Lange J-P. Lignocellulose conversion: An introduction to chemistry, process and economics. Biofuels Bioproduct Biorefining. 2007;**1**(1):39-48

[54] Rivas AMJ, Yuan W, Boyette MD, Wang D. Airflow and insulation effects on simultaneous syngas and biochar production in a top-lit updraft biomass gasifier. Renewable Energy. 2018;**117**:116-124

[55] Rivas AMJ, Yuan W, Wang D, Wang D, Kumar A. The effect of gasification conditions on the surface properties of biochar produced in a top-lit updraft gasifier. Applied Sciences. 2020;**10**(2):688

[56] Lu H, Zhang W, Yang Y, Huang X, Wang S, Qiu R. Relative distribution of Pb2 sorption mechanisms by sludge-derived biochar. Water Research. 2012;**46**(3):854-862

[57] Zhang X, Wang H, He L, Lu K, Sarmah A, Li J, et al. Using biochar for remediation of soils contaminated with heavy metals and organic pollutants. Environmental Science and Pollution Research. 2013;**20**(12):8472-8483

[58] Uchimiya M, Lima IM, Klasson KT, Wartelle LH. Contaminant immobilization and nutrient release by biochar soil amendment: Roles of natural organic matter. Chemosphere. 2010;**80**(8):935-940

Chapter 12

Biochar Effects on Amelioration of Adverse Salinity Effects in Soils

Ana Carolina Feitosa de Vasconcelos

Abstract

Biochar is the term given to biomass subjected to the process of change in the composition by the action of high temperatures. Advantages of biochar in soil quality have been reported, including amelioration of salinity effects. Salinity has a negative effect on soil physical properties and plant production by adversely affecting the process of plant growth, hence seed germination, nutrient uptake, and yield. Moreover, salt stress causes oxidative stress in plant and the reduction in antioxidant enzyme activities. Biochar is an amendment, which could decrease the negative effect of salt stress on crop growth and production. Application of biochar enriches mineral nutrients; improves the soil's physical, chemical, and biological characteristics such as bulk density, hydrological properties, aggregate structure, ion exchange capacity, and microbial activity; and consequently enhances plant growth. Enhancing physical properties, biochar balances water holding capacity and air porosity in soils. Biochar promotes benefits in plant growth in saline soils through reduction in oxidation stress and in osmotic stress, lower production of phytohormones, improvement in stomatal density and conductance, improvement in seed germination, and the promotion of microbial activities. Biochar amendment can contribute to reduce salt stress in plants under saline condition due to its high salt adsorption capability.

Keywords: organic residues, pyrolysis, soil quality improvement, plant growth, carbon sequestration

1. Introduction

Biochar (charcoal) is the term given to biomass subjected to the process of decomposition or change in the composition by the action of heat at high temperatures. Biochar is obtained by pyrolysis of biomass at temperatures of 300–600°C and has a great potential to mitigate possible impacts of climate change, such as periods of excessive rain or severe droughts. Because it is a thermally altered material, it degrades much more slowly, creating a large long-term carbon stock in the soil, being about 1500–2000 times more stable than non-pyrolyzed organic matter [1].

Good-quality biochar has an internal structure similar to graphite, which preserves (sequester) the carbon in the soil for hundreds and even thousands of years [1], in addition to having a reactive peripheral structure, which acts as the natural organic matter of the environment. The presence of internal organic structures similar to that of graphite contributes to the biochar recalcitrance, i.e., biochar stays for a longer period in soil, characterizing a more efficient negative carbon sequestration system [1].

IntechOpen

The interest in using biochar as a soil amendment is increasing in the years 2000 because of the work of researchers who study the production of an organic soil-conditioning fertilizer trying to be similar to the black lands of Amazonian Indians [2]. Pre-Columbian peoples produced the black lands of Indians, but it is not known exactly whether it was an intentional process of soil improvement or a by-product of the agricultural and housing activities of these peoples. Human activity in the pre-Columbian past resulted in the accumulation of plant and animal waste, as well as large amounts of ash and coals with various chemical elements, such as P, Mg, Zn, Cu, Ca, Sr, and Ba [1].

The tailings produced by the industrialization of plant-origin products and the tailings originated from animal production can be used as inputs for agriculture and can be potential environmental liabilities, such as animal residues, rests of wood, and crop residues. The establishment of mechanisms, by which external environmental benefits can be monetized or internalized, may be important for the adoption of biochar production technologies [3].

Thus, biochar of vegetable or animal source is an alternative to act in carbon sequestration and as an organic soil conditioner. In addition, in many agricultural and forestry production systems, there is an expressive amount of produced waste, such as cut waste, dead wood, surplus seedlings, and sawmill and crop residues left in the field after harvest. Many of these residues can be used to produce biochar, which can be applied to agricultural soil both to sequester carbon and to improve crop production potential [3].

When applied to the soil, biochar acts as a soil conditioner promoting plant growth by retaining the nutrients and enhancing the physical and chemical soil properties [3, 4]. Experiments carried out in field with biochar application in the soil have presented benefits to agricultural productivity.

In this way, many functions of the biochar stand out, such as promote the structure of the soil with chemical connections between the biochar and the inorganic macromolecular structures, thus avoiding landslides during the rainy periods; retention of rainwater and irrigation to be released during dry periods; retention and release of H^+ and OH^- ions in the action of controlling soil pH; retention of nutrient metal ions from plants such as Ca, Fe, Cu, or toxic to them (e.g., Al); increase in plant growth and agricultural productivity; decrease in N_2O emissions; reduction in the need for mineral fertilizers; and increase in the organic carbon stock in the soil by sorption of labile soil organic matter onto biochar particles, thus decreasing its mineralization, for instance [5].

The uses of biochar cannot be limited only to increasing crop productivity and carbon sequestration, as biochar also reduces other important greenhouse gas emissions. Thus, studies reported by [6] show that a decrease in methane emissions was verified with additions of 30 g kg^{-1} of biochar to the soil, as well as a significant reduction in nitrous oxide emissions. According to the author, such facts may be due to the improvement of soil aeration, reducing the occurrence of anaerobic conditions and possibly decreasing the nitrogen cycle by increasing the soil's C/N ratio. In addition, biochar can alter the rates of nitrogen cycling in soil systems by influencing nitrification and denitrification, which are key sources of the greenhouse gas nitrous oxide. According to Liu et al. [7], biochar can potentially reduce N_2O emission in soil by affecting ammonia- and nitrite-oxidizing bacteria, and these effects depended on the biochar application rate in soil.

It is estimated that the world population is expected to increase to 9.7 billion by 2050, and in 2100, it is expected to reach 10.9 billion people, which will inevitably lead to an increasing demand for food [8]. According to [9], to feed more people and better feed them, in a scenario with higher prices, inputs that are more expensive, and increasingly limited resources, and, at the same time, fight against climate change is an unprecedented challenge for humankind.

In this sense, the application of biochar as a sustainable soil corrective has been proposed as an attractive approach to mitigate greenhouse gas emissions due to its contribution to carbon sequestration and to improve crop productivity [3]. Thus, Woolf et al. [10] estimated that the annual net emissions of CO_2, N_2O, and CH_4 could be reduced by 12% with the implementation of biochar, without endangering food security, habitat, or soil conservation.

The potential of biochar in increasing crop productivity has been demonstrated in a large number of studies on tropical agricultural products [11–13]. It has been found that treatments with biochar increased crop yields averaged 10%, with larger effects observed in acid soils and thick texture [14].

Although the detailed physiological mechanisms remain unclear [15], the favorable effects of biochar on crop productivity are due to the high specific surface area and cation exchange capacity and depend on pyrolysis conditions and microporosity [13]. In addition to improving water and nutrient retention in the soil, these properties shown by biochar also allow adsorbing a wide range of potentially toxic materials, including heavy metals [16], pesticides [17], and other contaminants [16, 18, 19]. In saline conditions of soils, biochar improved soil conditions for plant growth [20].

Soil salinity is one of the factors that affect the crop yield. In arid and semiarid regions, salinity constitutes a serious problem, limiting agricultural production and reducing crop productivity to uneconomic levels. In these regions that are characterized by low rainfall and high evapotranspiration, inadequate irrigation management, quality of irrigation water, and conditions of insufficient drainage contribute to accelerating the soil salinization process [21].

Soils affected by salts, also known as soils halomorphic or saline and sodium soils, are soils developed in imperfect drainage conditions, which are characterized by the presence of soluble salts, exchangeable sodium, or both, in horizons or layers close to the surface. Salt-affected soils are generally classified as saline, sodic, or saline-sodic, which is mainly based on their electrical conductivity (EC), sodium adsorption ratio (SAR), and exchangeable sodium percentage (ESP) of the saturated paste extracts [21].

Among saline soils, saline-sodic soils are highly degraded and least productive which is mainly due to the effect of both salinity and sodicity on soil properties, and if these soils are dispersed, then water infiltration and hydraulic conductivity are reduced, which negatively affect the plant growth [22]. Salinity stress negatively correlates with soil properties such as organic matter and C/N ratio [23]. Soil salinity reduces the microbial activity and biomass and alters the microbial community structure in the soil [24].

Increasing the concentration of soluble salts in the soil affects plant growth due to increased osmotic tension of the soil solution, which reduces the absorption of water by plants, the accumulation of toxic amounts of various ions, and disturbances in the ion balance. On the other hand, the saturation of the exchange complex for Na^+ results in physical conditions highly unfavorable to plant growth, besides causing nutritional disturbances [20].

The high capacity to activate carbons in order to adsorb a variety of salts has been observed, and for this reason, the biochar has been used in industrial processes such as desalination [25]. However, the potential use of biochar as a soil amendment to mitigate the stress induced by salt in the plant has received little attention [26].

In addition, some studies have also shown that the application of organic amendments improved the physicochemical properties of saline soil; however, little data are available on the effect of biochar on the saline soil properties [27–30].

Thus, the objective of this chapter was to approach the role of biochar on amelioration of adverse salinity effects in soils.

2. General considerations about biochar production

Biochar is produced by heating any kinds of organic waste materials (crop residue, animal, or poultry manure) at high temperature through the process of pyrolysis (fast and slow). In addition, rotating kilns, vertical silo-type reactors, gasification, hydrothermal carbonization, and pyrolysis are common techniques used for biochar production [31].

A great amount of waste such as forest waste and crop residues is left in the field after harvesting in several agricultural and forest production. Many of the agricultural and forestry waste can be used to produce biochar, a product that when applied to agricultural land can both sequester carbon and improve crop production potential. Moreover, animal wastes can also be converted to biochar [3, 32–34]. In many cases, these residues have little value, and their disposal incurs costs.

The characterization of biochars is hard to be specified due to the large variety of potential biomass to be used for its production. In addition, the carbonization conditions applied for the conversion of biomass into biochar also interfere in the final product characterization [25].

Some studies have reported the differences in results of biochar applications in saline soils according to the method used for obtaining it [31]. In a study carried out by [35], reduction in plant growth was observed when eucalyptus wood-derived biochar produced at high temperature(800°C) was applied in a sandy ultisol, whereas biochar produced at lower temperature (350°C) enhanced plant growth. According to Almaroai et al. [30], characteristics such as duration and temperature of pyrolysis as well as the application rate of biochar are highly variable according to soil characteristics (fertility level, for instance) and soil biological activities.

According to Qayyum et al. [36], the use of biochar is a practice for achieving multiple benefits of sustainable agriculture. Biochars are characterized with a high concentration of total organic carbon (30–70%) depending on the pyrolysis conditions (temperature, aeration, and time), high mineral contents (Na, K, Mg, Fe, etc.), high pH, high electrical conductivity, and a low concentration of ash and volatile matter [36]. However, the feedstock type significantly affects the biochar properties [37].

The sophisticated techniques of characterization include the quantification and identification of surface functional groups, aromatic compounds, polycyclinc aromatic hydrocarbon, active surface area, and scanning electron microscopy. Thus, biochars should be carefully analyzed prior to their utilization [38].

In a study carried out by Hansen et al. [39], the authors applied two kinds of biochar: straw gasification biochar and wood gasification biochar in sandy soil for evaluating the shoot and root growth of barley (*Hordeum vulgare* L.). They observed that straw gasification biochar was more effective compared to the application of wood gasification biochar in soil, since straw gasification biochar presented considerable potential for enhancing crop productivity in coarse sandy soils by increasing soil water retention and improving root development.

Two biochar materials produced from maize using two different pyrolysis techniques, heating at 600°C for 30 min and batch-wise hydrothermal carbonization at 210°C, were used in a study carried out by Almaroai et al. [30]. The results of this study demonstrated positive synergistic effects of biochar amendments on plant growth, plant nutrient uptake, soil nutrient contents, and soil biological properties in sandy loam soil, with a more significant effect on the measured biological indicators for the biochar produced by batch-wise hydrothermal carbonization at 210°C. According to these authors, different methods of producing biochar from the same source (maize) play a critical role in the expression of soil ecological effects, which underpin the assumption of a link between chemical and physical properties of biochar and enhanced plant nutrient acquisition, symbiotic performance, and plant stress tolerance.

3. Application of biochar in saline soils and improvement of plant growth

Mineral salts are an important plant stress factor, having adverse impacts on crops particularly in arid and semiarid regions. High soil salinity and/or sodicity affects an estimated 1.1 Gha or more than 7% of the world's total land area [40]. Salinity changes the water absorption and uptake of nutrients, as well as the permeability of membranes. These changes reflect in the water and nutrient balance of the plant and cause changes in metabolism, hormonal balance, gas exchange, and production of reactive oxygen species (ROS) [41]. All these changes compromise the growth and division of cells, vegetative and reproductive growth and acceleration of leaves senescence, resulting in the eventual death of the plant [42].

Salt stress adversely affects the process of plant growth, since seed germination, nutrient uptake, and yield. Moreover, salt stress causes oxidative stress in plant and the reduction in antioxidant enzyme activities [43]. Salinity reduces crop growth by affecting several processes that depend on salt accumulation in shoots [44]. The independent processes reduce shoot biomass predominately by closing stomata and inhibiting leaf expansion. Plant initial responses to salt stress are generally the reduction in leaf expansion and partial/full closure of stomata to conserve water resource. These responses are coordinated by an increased accumulation of stress hormones, particularly abscisic acid (ABA). An increased level of ABA in xylem stream is an indication of plant roots facing osmotic stress [43].

The application of biochar in saline soils favors the increase of soil organic matter and nutrients, also increasing the cation exchange capacity and replacing Na from exchange sites by providing Ca in soil solution, improving the stabilization of soil structure. Therefore, by enhancing physical properties, biochar balances water holding capacity and air porosity in soils. In addition, biochar works as habitat for many soil microorganisms that can help improve salt-affected soils [45, 46].

Biochar can also hasten salt leaching and thus decrease the time required for reducing salt concentration to a level suitable for growing plants [47]. Moreover, biochar adds soil organic C and increases the stability of organic molecules that would help bind soil aggregates for long periods compared to easily degradable molecules from other organic amendments [48].

Several studies observed that the application of biochar has been shown to be effective in reducing salinity stress by improving soil physicochemical [49, 50] and biological [50, 51] properties directly related to Na removal such as Na leaching, Na adsorption ratio, and electrical conductivity, as reported by Saifullah et al. [52].

The interaction effects of biochar and salinity on growth and yield of wheat significantly decreased wheat production through increase in soil salinity as the produced biochar while the grain yield and straw dry weight were declined by application of biochar. Therefore, under dry condition, biochar can be used as an appropriate level, as it could store more water compared to treatments without biochar [42].

It was reported that incorporation of biochar into salt-affected soil could alleviate salinity stress in potatoes [48] mainly because of its high salt (Na^+) adsorption potential. In another study, Akhtar et al. [53] examined the effect of different levels of salinity and biochar on wheat yield. The results showed that biochar application positively influenced growth and yield of wheat under saline condition. However, Thomas et al. [28] noticed high salt adsorption potential of biochar, some studies have reported negative effect of biochar on crop productivity, but these were generally restricted to specific type of biochar [54].

Biochar could improve the soil physicochemical and biological properties under conditions of abiotic stresses [55]. Biochar poultry manure compost (BPC) with pyroligneous solution (PS) in the saline soil increased microbial biomass carbon and the activities of urease, invertase, and phosphatase in bulk soils and rhizosphere soils under maize cultivation, according to Lu et al. [56]. Similarly, Bhaduri et al. [57] concluded that the effects of biochar on soil enzyme activities in saline soil vary with the applied rate of biochar, incubation time, and soil enzyme types.

The biochar application in salt-stressed soil (30 g m^{-2}) did not affect the soil pH but increased the soil electrical conductivity as compared to the control [28]. Similarly, a biochar produced by furfural (a colorless liquid used in synthetic resin manufacture, originally obtained by distilling bran) in saline soil decreased pH, while increasing the soil organic carbon, cation exchange capacity (CEC), and available P in the soil [31].

When applied in saline soils, composted biochar increased the soil organic matter content and CEC and decreased the exchangeable Na and soil pH [58]. These studies showed that biochar addition in saline soils could improve the plant growth by improving the soil biological activity and physicochemical properties.

The accumulation of Na and impairment of K nutrition are major characteristics of salt-stressed plants [59]. Thus, improved K/Na ratio through enhancing K availability is considered a useful tool to increase plant growth and yield under saline soils [60, 61]. Biochar, depending upon feedstock, may increase K concentration in soils, and this increase in salt-affected soils counteracts the adverse impacts of Na, being considered one of the major benefits associated with biochar application in saline soils [52].

Corroborating [52], a study carried out by Lin et al. [62] reported that the biochar application in saline soil improved wheat and soybean yields by increasing the exchangeable K concentration (by 44% over control) and increasing the K/Na ratio in plants, improving plant salt tolerance. According to Lashari et al. [63], a considerable increase in K concentration and K/Na ratio in the leaf sap of corn under salt stress and an increasing supply of K were suggested as major mechanisms responsible for the alleviation of salt stress to plants.

The benefits of biochar in plant growth in saline soils observed in several studies cited by Saifullah et al. [52] also include reduction in oxidation stress through degradation of O_2^- and H_2O_2 concentration reduction in osmotic stress through improving water holding capacity and thus availability of water; lower production of phytohormones; improvement in stomatal density and conductance; improvement in seed germination and the promotion of microbial activities; and a bacterial community shift toward the beneficial taxa in the rhizosphere.

Plants under salinity stress produce abscisic acid (ABA), and it is a good indicator of the osmotic stress, acting as a long-distance signal molecule to close stomata under water deficit conditions [64]. Thus, decreased production of ABA could be attributed to a biochar-induced improvement in water availability to plants, which would result ultimately in increased stomatal conductance. Further, enhanced availability of water and nutrients with biochar application under saline conditions could improve seed germination.

However, Thomas et al. [28] affirm that the biochar impact on the growth of plants in salt-affected soils is species dependent. Biochar application significantly improved the growth of salt-sensitive plant species; however, salt-tolerant species did not show any growth improvement with biochar amendment.

On the other hand, Luo et al. [58] reported significant improvement in the growth and yield of two salt-tolerant species grown in biochar manure compost-amended salt-affected soils.

4. Conclusions

The interaction of biochar with soils with salinity conditions is essential for determining any contrasting effects, which also depend on the physicochemical properties of biochar and the raw material used for biochar production. Elucidating the effect of biochar type on plant growth and development and soil biochemical properties provides important guidance on the selection of feedstock type and production technology, which could be applied under specific environmental conditions.

Different methods of producing biochar from the same source play a critical role in the expression of soil ecological effects, which underpin the assumption of a link between chemical and physical properties of biochar and enhanced plant nutrient acquisition, symbiotic performance, and plant stress tolerance.

Although there is an increasing number of studies about biochar and its effects in saline soils for improvement of plant growth, the results obtained until the present are still not conclusive given the diversity of raw material and methods for biochar production. It is still necessary to conduct more investigations in order to better use biochar for ameliorating the adverse salinity effects in soils.

Author details

Ana Carolina Feitosa de Vasconcelos
CNPq, Campina Grande, Brazil

*Address all correspondence to: ana3carol@yahoo.com.br

References

[1] Rezende EIP, Angelo LC, Santos SS, Mangrich AS. Biocarvão (Biochar) e Sequestro de Carbono. Revista Virtual Química. 2011;**3**:426-433. DOI: 10.5935/1984-6835.20110046

[2] Novotny EH, Hayes MHB, Madari BE, Bonagamba TJ. Lessons from the Terra Preta de Índios of the Amazon region for the utilization of charcoal for soil amendment. Journal of the Brazilian Chemical Society. 2009;**20**:1003-1010

[3] Lehmann J, Silva JP, Steiner C, Nehls T, Zech W, Glaser B. Nutrient availability and leaching in an archaeological anthrosol and a ferralsol of the Central Amazon basin: Fertilizer, manure and charcoal amendments. Plant and Soil. 2003;**249**:343-357. DOI: 10.1590/S0100-204X2012000500010

[4] Glaser B, Lehmann J, Zech W. Ameliorating physical and chemical properties of highly weathered soil in the tropics with charcoal—A review. Biology and Fertility of Soils. 2002;**35**:219-230. DOI: DOI 10.1007/s00374-002-0466-4

[5] Hammond J, Shackley S, Sohi S, Brownsort P. Prospective life cycle carbon abatement for pyrolysis biochar systems in the UK. Energy Policy. 2011;**39**:2646-2655. Available from: https://EconPapers.repec.org/RePEc:eee:enepol:v:39:y:2011:i:5:p:2646-2655 [Accessed: 21 March 2020]

[6] McHenry MP. Soil organic carbon, biochar, and applicable research results for increasing farm productivity under Australian agricultural conditions. Communications in Soil Science and Plant Analysis. 2011;**42**:1187-1199. DOI: 10.1080/00103624.2011.566963

[7] Liu L, Shen G, Sun M, Cao X, Shang G, Chen P. Effect of biochar on nitrous oxide emission and its potential mechanisms. Journal of the Air & Waste Management Association. 2014;**64**:894-902. DOI: 10.1080/10962247.2014.899937

[8] United Nations, Department of Economic and Social Affairs, Population Division. World Population Prospects 2019: Highlights (ST/ESA/SER.A/423) [Internet]. 2019. Available from: https://population.un.org/wpp/Publications/Files/WPP2019_Highlights.pdf [Accessed: 12 March 2020]

[9] Diouf J. La crisis alimentaria actual: desafios y oportunidades para El desarrollo agrícola. Cuba: Conferencia Magistral em La Habana; 2008. Available from: http://www.fao.org/fileadmin/user_upload/foodclimate/HLCdocs/HLC08-Rep-S.pdf [Accessed: 21 March 2020]

[10] Woolf D, Amonette JE, Stree-Perrott FA, Lehmann J, Joseph S. Sustainable biochar to mitigate global climate change. Nature Communications. 2010;**1**:1-8. DOI: 10.1038/ncomms1053

[11] Rajkovich S, Enders A, Hanley K, Hyland C, Zimmerman AR, Lehmann J. Corn growth and nitrogen nutrition after additions of biochars with varying properties to a temperate soil. Biology and Fertility of Soils. 2012;**48**:271-284. DOI: 10.1007/s00374-011-0624-7

[12] Major J, Steiner C, Downie A, Lehmann J. Biochar effects on nutrient leaching. In: Lehmann J, Joseph S, editors. Biochar for Environmental Management: Science and Technology. London: Earthscan; 2009. pp. 271-287

[13] Thies JE, Rillig MC. Characteristics of biochar: Biological properties. In: Lehmann J, Joseph S, editors. Biochar for Environmental Management: Science

and Technology. London: Earthscan; 2009. pp. 85-105

[14] Jeffery S, Verheijen FG, van der Velde M, Bastos AC. A quantitative review of the effects of biochar application to soils on crop productivity using meta-analysis. Agriculture, Ecosystems and Environment. 2011;**144**:175-187. DOI: 10.1016/j.agee.2011.08.015

[15] Atkinson CJ, Fitzgerald JD, Hipps NA. Potential mechanisms for achieving agricultural benefits from biochar application to temperature soils: A review. Plant and Soil. 2010;**337**:1-18. DOI: 10.1007/s11104-010-0464-5

[16] Beesley L, Moreno-Jimenez E, Gomez-Eyles JL. Effects of biochar and green waste compost amendments on mobility, bioavailability, and toxicity of inorganic and organic contaminants in a multi-element polluted soil. Environmental Pollution. 2010;**158**:2282-2287. DOI: 10.1016/j.envpol.2010.02.003

[17] Kookana RS. The role of biochar in modifying the environmental fate, bioavailability, and efficacy of pesticides in soils: A review. Australian Journal of Soil Research. 2010;**48**:627-637. DOI: 10.1071/SR10007

[18] Rhodes AH, Carlin A, Semple KT. Impact of black carbon in the extraction and mineralization of phenanthrene in soil. Environmental Science &Technology. 2008;**42**:740-745. DOI: 10.1021/es071451n

[19] Buss W, Kammann C, Koyro HW. Biochar reduces copper toxicity in *Chenopodium quinoa* Willd in a sandy soil. Journal of Environmental Quality. 2011;**40**:1-9. DOI: 10.2134/jeq2011.0022

[20] Vasconcelos ACF, Chaves LHG, Gheyi HR, Fernandes JD, Tito GA. Crambe growth in a soil amended with biochar and under saline irrigation. Communications in Soil Science and Plant Analysis. 2017;**48**(11):1291-1300. DOI: 10.1080/00103624.2017.1341911

[21] Agricultural Handbook No. 60Richards LA. Diagnosis and improvement of saline and alkali soils. In: Riverside, California: U. S. Salinity Laboratory; 1954. Available from: https://www.ars.usda.gov/ARSUserFiles/20360500/hb60_pdf/hb60complete.pdf [Accessed: 23 March 2020]

[22] Suarez D, Wood J, Lesch S. Effect of SAR on water infiltration under a sequential rain-irrigation management system. Agricultural Water Management. 2006;**86**:150-164. DOI: 10.1016/j.agwat.2006.07.010

[23] Morrissey EM, Gillespie JL, Morina JC, Franklin RB. Salinity affects microbial activity and soil organic matter content in tidal wetlands. Global Change Biology. 2014;**20**:1351-1362. DOI: 10.1111/gcb.12431

[24] Yan N, Marschner P, Cao W, Zuo C, Qin W. Influence of salinity and water content on soil microorganisms. International Soil and Water Conservation Research. 2015;**3**:316-323. DOI: 10.1016/j.iswcr.2015.11.003

[25] Zou L, Morris G, Qi D. Using activated carbon electrode in electrosorptive deionization of brackish water. Desalination. 2008;**225**:329-340. DOI: 10.1016/j.desal.2007.07.014

[26] Lehmann J, Joseph S. Biochar for environmental management: An introduction. In: Lehmann J, Joseph S, editors. Biochar for Environmental Management: Science and Technology. London: Earthscan; 2009. pp. 1-12

[27] Huang M, Zhang Z, Zhu C, Zhai Y, Lu P. Effect of biochar on sweet corn and soil salinity under conjunctive irrigation with brackish water in coastal saline soil. Scientia Horticulturae.

2019;**250**:405-413. DOI: 10.1016/j.scienta.2019.02.077

[28] Thomas SC, Frye S, Gale N, Garmon M, Launchbury R, Machado N, et al. Biochar mitigates negative effects of salt additions on two herbaceous plant species. Journal of Environmental Management. 2013;**129**:62-68. DOI: 10.1016/j.jenvman.2013.05.057

[29] Wu Y, Xu G, Shao HB. Furfural and its biochar improve the general properties of a saline soil. Solid Earth. 2014;**5**:665-671. DOI: 10.5194/se-5-665-2014

[30] Almaroai YA, Usman AR, Ahmad M, Moon DH, Cho JS, Joo YK, et al. Effects of biochar, cow bone, and eggshell on Pb availability to maize in contaminated soil irrigated with saline water. Environment and Earth Science. 2014;**71**:1289-1296. DOI: 10.1007/s12665-013-2533-6

[31] Egamberdieva D, Li L, Ma H, Wirth S, Bellingrath-Kimura SD. Soil amendment with different maize biochars improves chickpea growth under different moisture levels by improving symbiotic performance with *Mesorhizobium ciceri* and soil biochemical properties to varying degrees. Frontiers in Microbiology. 2019;**10**:2423. DOI: 10.3389/fmicb.2019.02423

[32] Chan KY, Van Zwieten L, Meszaros I, Dowie A, Joseph S. Using poultry litter biochars as soil amendments. Australian Journal of Soil Research. 2008;**46**:437-444. DOI: 10.1071/SR08036

[33] Lehmann J, Silva JP, Steiner C, Nehls T, Zech W, Glaser B. Nutrient availability and leaching in an archaeological Anthrosol and a Ferralsol of the Central Amazon basin: Fertilizer, manure and charcoal amendments. Plant and Soil. 2003;**249**:343-357. DOI: 10.1023/A:1022833116184

[34] Lehmann J, Gaunt J, Rondon M. Biochar sequestration in terrestrial ecosystems—A review. Mitigation and Adaptation Strategies for Global Change. 2006;**11**:403-427. DOI: 10.1007/s11027-005-9006-5

[35] Butnan S, Deenik JL, Toomsan B, Antal MJ, Vityakona P. Biochar characteristics and application rates affecting corn growth and properties of soils contrasting in texture and mineralogy. Geoderma. 2015;**23**:105-116. DOI: 10.1016/j.geoderma.2014.08.010

[36] Qayyum MF, Abid M, Danish S, Saeed MK, Ali MA. Effects of various biochars on seed germination and carbon mineralization in an alkaline soil. Pakistan Journal of Agricultural Sciences. 2014;**51**:977-982

[37] Ronsse F, van Hecke S, Dickinson D, Prins W. Production and characterization of slow pyrolysis biochar: Influence of feedstock type and pyrolysis conditions. Global Change Biology. Bioenergy. 2013;**5**:104-115. DOI: 10.1111/gcbb.12018

[38] Kuppusamy S, Thavamani P, Megharaj M, Venkateswarlu K, Naidu R. Agronomic and remedial benefits and risks of applying biochar to soil: Current knowledge and future research directions. Environment International. 2016;**87**:1-12. DOI: 10.1016/j.envint.2015.10.018

[39] Hansen V, Hauggaard-Nielsen H, Petersen CT, Mikkelsen TN, Müller-Stöver D. Effects of gasification biochar on plant-available water capacity and plant growth in two contrasting soil types. Soil and Tillage Research. 2016;**161**:1-9. DOI: 10.1016/j.still.2016.03.002

[40] Wicke B, Smeets E, Dornburg V, Vashev B, Gaiser T, Turkenburg W, et al. The global technical and economic potential of bioenergy from salt-affected soils. Energy & Environmental

Science. 2011;**4**:2669-2681. DOI: 10.1039/c1ee01029h

[41] Munns R. Comparative physiology of salt and water stress. Plant, Cell & Environment. 2002;**25**:239-250. DOI: 10.1046/j.0016-8025.2001.00808.x. PubMed

[42] Rezaei N, Razzaghi F. Effect of different levels of water salinity and biochar on wheat yield under greenhouse conditions. Acta Horticulturae. 2018;**1190**:83-88. DOI: 10.17660/Acta Horticulturae.2018.1190.14

[43] Ali S, Rizwan M, Qayyum MF, Ok YS, Ibrahim M, Riaz M, et al. Biochar soil amendment on alleviation of drought and salt stress in plants: A critical review. Environmental Science and Pollution Research. 2017;**24**:12700-12712. DOI: 10.1007/s11356-017-8904-x

[44] Mohamed AK, Qayyum MF, Abdel-Hadi AM, Rehman RA, Ali S, Rizwan M. Interactive effect of salinity and silver nanoparticles on photosynthetic and biochemical parameters of wheat. Archives of Agronomy and Soil Science. 1736-1747;**63**:12. DOI: 10.1080/03650340.2017.1300256

[45] Amini S, Ghadiri H, Chen C, Marschner P. Salt-affected soils, reclamation, carbon dynamics, and biochar: A review. Journal of Soils and Sediments. 2016;**16**:939-953. DOI: 10.1007/s11368-015-1293-1

[46] Zheng H, Wang X, Chen L, Wang Z, Xia Y, Zhang Y, et al. Enhanced growth of halophyte plants in biochar-amended coastal soil: Roles of nutrient availability and rhizosphere microbial modulation. Plant, Cell & Environment. 2017;**41**: 517-532. DOI: 10.1111/pce.12944

[47] Yue Y, Guo WN, Lin QM, Li GT, Zhao XR. Improving salt leaching in a simulated saline soil column by three biochars derived from rice straw (*Oryza sativa* L.), sunflower straw (*Helianthus annuus*), and cow manure. Journal of Soil and Water Conservation. 2016;**71**:467-475. DOI: 10.2489/jswc.71.6.467

[48] Yeboah E, Ofori P, Quansah GW, Dugan E, Sohi SP. Improving soil productivity through biochar amendments to soils. African Journal of Environmental Science and Technology. 2009;**3**(34):2009. DOI: 41.0.1105/tpc.112.107227

[49] Gul S, Whalen JK, Thomas BW, Sachdeva V, Deng H. Physico-chemical properties and microbial responses in biochar-amended soils: Mechanisms and future directions. Agricultural Ecosystem and Environment. 2015;**206**:46-59. DOI: 10.1016/j.agee.2015.03.015

[50] Cornelissen G, Martinsen V, Shitumbanuma V, Alling V, Breedveld GD, Rutherford DW, et al. Biochar effect on maize yield and soil characteristics in five conservation farming sites in Zambia. Agronomy. 2013;**3**:256-274

[51] Hammer EC, Forstreuter M, Rillig MC, Kohler J. Biochar increases arbuscular mycorrhizal plant growth enhancement and ameliorates salinity stress. Applied Soil Ecology;**2015, 9**(6):114-121. DOI: 10.1016/j.apsoil.2015.07.014

[52] Saifullah SD, Naeem A, Rengel Z, Naidu R. Biochar application for the remediation of salt-affected soils: Challenges and opportunities. Science of The Total Environment. 2018;**625**:320-335. DOI: 10.1016/j.scitotenv.2017.12.257

[53] Akhtar SS, Andersen MN, Liu F. Residual effects of biochar on improving growth, physiology and yield of wheat under salt stress. Agricultural

Water Management. 2015;**158**:61-68. DOI: 10.1016/j.agwat.2015.04.010

[54] Liu X, Zhang A, Ji C, Joseph S, Bian R, Li L, et al. Biochar's effect on crop productivity and the dependence on experimental conditions—A meta-analysis of literature data. Plant and Soil. 2013;**373**:583-594. DOI: 10.1007/s11104-013-1806-x

[55] Rizwan M, Ali S, Qayyum MF, Ibrahim M, Rehman MZ, Abbas T, et al. Mechanisms of biochar-mediated alleviation of toxicity of trace elements in plants: A critical review. Environmental Science and Pollution Research. 2016;**23**:2230-2248. DOI: 10.1007/s11356-015-5697-7

[56] Lu H, Lashari MS, Liu X, Ji H, Li L, Zheng J, et al. Changes in soil microbial community structure and enzyme activity with amendment of biochar-manure compost and pyroligneous solution in a saline soil from Central China. European Journal of Soil Biology. 2015;**70**:67-76. DOI: 10.1016/j.ejsobi.2015.07.005

[57] Bhaduri D, Saha A, Desai D, Meena HN. Restoration of carbon and microbial activity in salt-induced soil by application of peanut shell biochar during short-term incubation study. Chemosphere. 2016;**148**:86-98

[58] Luo X, Liu G, Xia Y, Chen L, Jiang Z, Zheng H, et al. Use of biochar-compost to improve properties and productivity of the degraded coastal soil in the Yellow River Delta, China. Journal of Soils and Sediments. 2017;**17**:780-789. DOI: 10.1007/s11368-016-1361-1

[59] Liu J, Zhu JK. Proline accumulation and salt-stress-induced gene expression in a salt-hypersensitive mutant of Arabidopsis. Plant Physiology. 1997;**114**:591-596. DOI: 10.1104/pp.114.2.591

[60] Cakmak I. The role of potassium in alleviating detrimental effects of abiotic stresses in plants. Journal of Plant Nutrition and Soil Science. 2005;**168**:521-530. DOI: 10.1002/jpln.200420485

[61] Chakraborty K, Bhaduri D, Meena HN, Kalariya K. External potassium (K^+) application improves salinity tolerance by promoting Na^+-exclusion, K^+-accumulation and osmotic adjustment in contrasting peanut cultivars. Plant Physiology and Biochemistry. 2016;**103**:143-153. DOI: 10.1016/j.plaphy.2016.02.039

[62] Lin XW, Xie ZB, Zheng JY, Liu Q, Bei QC, Zhu JG. Effects of biochar application on greenhouse gas emissions, carbon sequestration and crop growth in coastal saline soil. European Journal of Soil Science. 2015;**66**:329-338. DOI: 10.1111/ejss.12225

[63] Lashari MS, Ye Y, Ji H, Li L, Kibue GW, Lu H, et al. Biochar–manure compost in conjunction with pyroligneous solution alleviated salt stress and improved leaf bioactivity of maize in a saline soil from Central China: A 2-year field experiment. Journal of the Science of Food and Agriculture. 2015;**95**:1321-1327. DOI: 10.1002/jsfa.6825

[64] Duan L, Dietrich D, Ng CH, Chan PMY, Bhalerao R, Bennett MJ, et al. Endodermal ABA signaling promotes lateral root quiescence during salt stress in Arabidopsis seedlings. The Plant Cell. 2013;**25**:324-341. DOI: 10.1105/tpc

Section 5

Applications of Biochar for Water Treatment

Chapter 13

Sorption of Heavy Metals onto Biochar

Ramalingham Senthilkumar
and Donipathi Mogili Reddy Prasad

Abstract

Biochar is a stable carbon-rich product synthesized from biological materials through different heating methods above the decomposition temperature. The potential uses of biochar in various fields include soil fertility improvement, C sequestration, pollutant removal and waste minimization/reuse. In recent years, large number of research has confirmed that biochar can be used successfully for the removal of heavy metal ions from aqueous solutions. The main aim of this chapter is to summarize and assess the sorption capacity of biochar toward various heavy metal ions. Considering that sorption is a surface phenomenon, the key parameters controlling the formation of biochar including pyrolysis temperature, residence time, and feedstock type will be discussed in detail. In addition, the mechanism associated with remediation of heavy metal ions and the physicochemical factors affecting the sorption potential will be discussed. Mathematical models employed in the sorption studies will be given special importance. The modification procedures used to enhance the sorption capacity of biochar will also be highlighted.

Keywords: biochar, adsorption, heavy metals, water quality, pyrolysis

1. Introduction

Water usage has been rising immensely with growing population and industrial activities in both developed and developing countries [1]. This resulted in deterioration of water sources as various contaminants such as dyes [2], toxic heavy metals [3], organic compounds like detergents, phenols, dyes, pesticides in addition to the other persistent organic pollutants [4] are increasingly being dumped into the water bodies [5]. Among these contaminants, heavy metals are of high priority because they persist in soils and do not undergo biodegradation [6]. This might affect significantly the suitability and sustainability of the water resources [7]. These contaminants reach water bodies through various industrial activities, including mining, electrolysis, metallurgy, battery manufacture, metal finishing, electroplating, electro-osmosis, pigment manufacture, tanneries, etc. [8, 9]. Heavy metals are then taken up by the biological systems through food intake and thereby cause major hazardous health impacts [10, 11]. Owing to this, different biological and physico-chemical treatment techniques have been proposed to remediate heavy metal-bearing contaminated waters. These remediation technologies include adsorption, biosorption, ion-exchange, electrocoagulation, membrane technologies and precipitation [12]. Adsorption is one of the widely used remediation approaches, which has its unique advantages including cost-effectiveness, high performance toward metal ion of interest, flexibility and ease in operation. [13]. Some of the widely used adsorbents for heavy metal remediation include fly ash, activated carbon, sorbents prepared from agricultural, industrial and biological waste materials [14]. Recently, char derived from biological materials under oxygen free condition, popularly known as "biochar" has been recently been introduced as an effective sorbent for various toxins [4, 15].

Biochar is a stable carbon-rich product synthesized from biological materials through different heating methods above the decomposition temperature. Biochar is produced through thermal degradation of organic components in absence of O_2 or under limited oxygen conditions (pyrolysis) [16]. In recent years, owing to the inherent biochar properties such as surplus surface binding sites (hydroxyl, carboxyl, phenolic hydroxyl and carbony groups), porous surface, high cation exchange capacity and its surface area, this organic amendment be utilized as an efficient and practical sorbent for remediation [17]. Biochars from different feed stock materials were prepared and successfully examined for their adsorption potential toward various metal ions [3] and nutrients [18]. The pyrolysis conditions such as temperature, rate of heating and residence time are all critical factors influenceing the potentiality of metal sorption on biochar. In addition, changes in sorption potential can also be obtained through physico-chemical techniques. Thus through this chapter, we critically explore the state of knowledge on different methods to modify biochar with preferred properties. We focused on various aspects including biochar preparation methods, factors affecting the biochar sorption potential and metal removal mechanisms.

2. Production of biochar and the factors influencing the properties

Biochar has been receiving considerable attention in recent years as an important material for sorption of contaminants from polluted waters [16, 19]. Carbon-rich materials are established adsorbents for inorganic and organic contaminants. For instance, activated carbon (AC) is a well-known adsorbent extensively used for removal of heavy metal ions [20]. It is worth noting that the preparation of AC requires high temperature, costly and an additional activation step (activated carbon, sorbents prepared from agricultural, industrial and biological waste materials) [14]. Hence the cost of remediation is high. Alternatively, biochar is a recommended cost-effective alternative sorbent because of its binding sites, high permeable structure and extensive surface area [21]. Its preparation is comparatively cheaper while considering its fewer requirements for energy [21, 22]. In addition, biochar consist of non-carbonized portion that may react with soil pollutants [19]. Uchimiya et al. [23] also indicated that degree of O-comprising hydroxyl, phenolic and carboxyl binding sites on the biochar surface can play crucial roles to immobilize soil pollutants.

The sorption potential of biochar is critically dependent on the pyrolysis conditions (temperature and residence time) as well as chemical composition of the feedstock. Overall, pyrolysis temperature markedly affects the functional groups and surface area of biochar. Jung et al. [24] recorded that the pores of biochar were blocked at pyrolysis temperatures greater than 400°C. Similarly, Gai et al. [25] examined the effects of feedstock (wheat-straw, corn-straw and peanut-shell) and pyrolysis temperature (400–700°C) on the properties and adsorption potential of biochar and found that yield of biochar as well as composition of H_2, O_2 and N_2 decreased as pyrolytic temperature surges from 400 to 700°C. Xiao et al. [26] studied influence of pyrolysis temperature during synthesis of biochar from crayfish shell for adsorption of Pb(II) ions from aqueous solutions. The authors observed that pyrolysis temperature exhibited varied effects on active functional groups, surface area and elemental composition of the produced biochar. Also, the sorption performance of biochar increased with increase in pyrolysis temperature, with biochar produced at 600°C exhibited maximum Pb(II) uptake of 190.7 mg/g.

Biochar are generally synthesized from cheap and copiously available waste biomaterials [27]. To be specific, biochar feedstocks are primarily produced from solid wastes and biomasses of agricultural activities. The agricultural residues are generally available in vast quantities and frequently pose disposal challenges [3]. For instance, preparing biochar from invasive plant can resolve challenges posed during disposal as well as help in waste management. Similarly, marine algae are generally abundant and can clog waterways; therefore other usages such as biochar synthesis can be beneficial to the local people. In recent years, a wide range of biomaterials were proposed as feedstocks for biochar, including animal manure, plant waste, seaweed, municipal solid waste and wood chip [27]. The type of feed stock strongly influences the biochar attributes and subsequently its further applications. Sohi et al. [28] indicated that the size of pores, surface area and functional groups in biochars are strongly influenced by the feedstock type. Hodgson et al. [29] examined the feasibility of different feedstocks (*Salix viminalis*, *Lolium perenne* fiber, *Lolium perenne*, *Picea sitchensis, Fraxinus excelsior* and *Miscanthus x giganteus*) as adsorbents for zinc. Their results pointed out to the efficiency of the synthesized bioach from *L. perenne* to remove Zn from water and it removal potential ranged from 83.27 to 92.96%. Zhao et al. [30] tested the impacts of feedstock type on properties of biochar by assessing several organic wastes, including various aquatic plant, waste wood, municipal waste, crop residue, animal manures and food wastes. The results indicated that type of feedstock affected the biochar properties, such as carbon sequestration capacity, cation exchange capacity, biochar C-content, fixed C, ash content and mineral composition. Therefore, preparing biochar with desired characteristic for a specific application may require careful screening and correct choice of a feedstock.

3. Sorption of heavy metals and associated mechanisms

Biochar has been investigated for adsorption of pesticides, heavy metals, nutrients, and organic compounds. Several researchers explored the adsorption capacity of biochar and confirmed favorable results for heavy metal ions [3, 31], nutrients [15], and organic pollutants [4]. Shakya and Agarwal [32] derived biochar from pineapple peel at different pyrolysis temperatures and investigated its efficiency for Cr(VI) sorption from aqueous solution. The results indicated that biochar synthesized at 350°C exhibited maximum sorption potential of 41.7 mg/g. Liu et al. [33] prepared biochar from corn stalk to test its capability of removing Pb(II) from aqueous solutions. Using X-ray diffraction, X-ray photoelectron spectroscopy, Fourier transform infrared spectroscopy and Scanning electron microscopy with energy dispersive spectrometer analyses, the authors identified combined complexation, mineral precipitation and ion exchange mechanisms contributed to Pb(II) sorption onto corn stalk-derived biochar. The maximum Pb(II) adsorption capacity of biochar was identified to be 49.7 mg/g.

Biochar sorption experimental trials are generally performed in continuous/batch operational modes. In most batch trials, the researchers aim at examining the effects of initial metal concentration, adsorbent dosages, temperatures, and solution pH. On the other hand, continuous trials aimed at understanding the continuous contaminant removal potential of biochar.

3.1 Solution pH

Solution pH is a vital operating factor influencing the adsorption process and usually plays a critical part in overall success of adsorption. Precisely, the solution pH influences the surface properties of the sorbent, as well as the metal speciation and finally the extent of metal sorption. The pH also decides the extent of adsorbent protonation, thereby affecting the specific charge of functional groups and finally the adsorption capacity of adsorbent [13]. In general, under acidic (low pH) conditions, the uptake of cationic metal ions is low owing to strong competition from H^+ ions. As the pH surges, the amount of H^+ ions declines and sorption of cationic metal species increases [13]. In order to endorse the effect of pH on the adsorption potential, few researchers investigated the impact of pH on sorption capacity of biochar. Liu et al. [33] witnessed that Pb(II) sorption capacity of corn stalk derived biochar surged as the solution pH increased. The removal performance was improved within the pH ranges of 4–6. The authors suggested that under acidic conditions, the existence of H^+ inhibited the sorption of Pb cations. On the other hand, Senthilkumar et al. [34] observed that remediation of As(V) by *Ulva reticulata* derived biochar enhanced from 55 to 93% as the pH surged from 2 to 4. Further increase in pH decreased the adsorption potential of biochar. The authors indicated that As(V) oxyanion mostly occurs as $HAsO_4^{2-}$ and $H_2AsO_4^-$ species under acidic conditions (pH 4 to 6). Thus, relatively high As(V) sorption in low pH conditions by seaweed derived biochar was due to high protonated positively charged binding sites on biochar surface owing to saturation of excess H^+ ions, thereby enhancing the sorption of As(V) through electrostatic attraction.

3.2 Temperature

Temperature tends to affect the kinetics rate and adsorption capacity of any adsorbent. The increase or decrease of the adsorption capacity upon varying the temperature will be useful to establish the type of the sorption process. On the basis of change in temperature, the process is identified to be endothermic when the adsorption capacity rises with the increase in temperatures; whereas the process is exothermic when the sorptional capacity decreases with temperature [13]. Several research studies have confirmed that temperature plays a critical part during adsorption of heavy metal ions by biochar [33, 35].

3.3 Biochar dosage

In an attempt to determine the optimum adsorbent dose essential to attain maximum adsorption, many researchers have performed adsorbent dosage optimization experiments during metal removal studies [36, 37]. In general, the % metal removal is directly linked with the adsorbent dosage. Precisely, the increase in adsorbent dosage generally increases the % metal removal of the adsorbent. This general trend can be explained as follows: as the sorbent dose increases, the total number of binding groups present on the surface of the adsorbent increases which, in turn, increases the overall binding of metal ions [38]. On the other hand, the sorptional capacity decreases with increasing adsorbent doses [39]. This is due to nature of interaction between sorbent and sorbate. The important factor being at high biochar dose, the metal ions in the solution are less compared to the exchangeable groups on the biochar, typically results in in less metal uptakes [13].

3.4 Initial solute concentration

Initial solute concentration is a critical parameter that influences the adsorption potential of any adsorbent. Past studies have shown that increase in initial metal concentration generally resulted in decline in the % metal removals [33, 34]. However, the sorptional uptake normally improves with the increase in the initial metal concentration. This was because at lower initial metal concentration, the ratio of the initial moles of metals in the solution to the biochar surface area was low and consequently, the adsorption became independent of initial concentration. Nevertheless, at higher metal concentration the accessible binding groups of sorbent become fewer in comparison to the moles of metal ions available in solution and hence, the percentage metal removal would be severely impacted by the initial metal ion concentration. During adsorption of arsenic(V) by *Ulva reticulata* derived biochar, Senthilkumar et al. [34] observed that augmentation of initial concentration of arsenic(V) from 10 to 25 mg/L produced enhancement of As(V) uptakes from 4.65 to 7.40 mg/g, whereas % removal decreased from 93.0 to 59.2, respectively.

3.5 Parameters influencing column sorption of metals

In comparison to batch sorption research, very little background literature is available about the possibility of utilizing column sorption in the removal of metal ions from aqueous solutions. Packed column sorption refers to feeding contaminated solution into the column packed with sorbent for continuous treatment. Of these little continuous-flow studies, it was identified that column adsorption potential strongly depends on operational parameters such as flow rate, influent metal concentration and bed depth [13]. The batch experimental trials are helpful in elucidating the fundamental information about the characteristics of adsorbent and the factors affecting the adsorption process [38]. Nevertheless, the batch experimental results cannot be utilized for accurate scale-up in real industrial wastewater systems [40]. This is due to the fact that in industrial wastewater systems, continuous adsorption column setup are generally used [13]. For cyclic adsorption/elution processes, packed columns are effective and practical arrangement, as they efficiently utilizes the concentration difference which is known to be the driving force for sorption of heavy metals [41]. Also, the column assembly allows more efficient utilization of the adsorbent capacity and generally results in superior effluent quality. Thus, adsorption using packed columns has important advantages including fast and high yield operations as well as easy scaling up [42]. Additionally, packed columns permit large amount of wastewater to be continuously remediated using a small amount of sorbent loaded inside the column [43]. Regeneration and subsequent reuse of sorbent is also possible using appropriate elutant. After adsorption, metal ions loaded-adsorbent can be eluted using suitable desorbent, or otherwise can be contained/disposed [44].

Vilvanathan and Shanthakumar [45] conducted continuous column adsorption experiments using biochar prepared from *Tectona grandis* leaves to remediate Co(II) and Ni(II) ions from aqueous solutions. The breakthrough curves were generated by fluctuating the inlet metal ion concentration, flow rate and bed depth. The results confirmed that the column exhaustion time prolonged with increasing bed depth and/or reducing each of the metal ion concentrations and flow rate. The metal-loaded column was desorbed using HCl, which indicates the possible regenerated and reuse of column bed for subsequent sorption cycles. Senthilkumar et al. [34] utilized 2 cm internal diameter and 35 cm depth column loaded with *U. reticulata* biochar to perform column experiments for arsenic(V) remediation from aqueous solutions. At a flow rate of 0.3 L/h, initial arsenic(V) concentration of 25 mg/L and bed depth of 25 cm, the column recorded breakthrough and exhaustion times of 3.25 and 13 h, respectively. Around 3.9 L of arsenic(V) solution was remediated by the column. The %As(V) removal and adsorption capacity of column were calculated as 59.5% and 8.12 mg/g, respectively. The bed was successfully eluted using 0.01 M sodium hydroxide with 99.5% elution efficiency.

As indicated before, very limited research studies focused on column applications compared to batch applications. Thus, serious efforts ate needed to explore the adsorption capacity of adsorbent in continuous operational mode to elucidate the adsorbent compatibility in real wastewater plants.

3.6 Biochar modification

Although biochar exhibits good sorption properties; however, it can be additionally altered to improve its sorption efficiency. The modification procedures employed include acid/base modification, functional group modification and impregnation with mineral oxides.

Through acid/base modification, alteration of surface acidities and porous nature of biochar can be obtained [15]. After exposure to chemicals including HNO_3, H_2SO_4, HCl, KOH and NaOH, El-Hendawy [46] identified that HNO_3 exposure resulted in improved adsorption and pore diffusion of hydrated Pb^{2+} with O_2 groups, and therefore improved the hydrophilic nature of biochar. Li et al. [31] evaluated lead adsorption capacity of two biochar materials (low mesopore char (AC1) and high mesopore char (AC2)) derived from bagasse modified using nitric acid. The results indicated that the adsorption capacities of AC2 and AC1 toward lead ions were recorded by 27 and 15 mg/g, respectively, due to high mesopore volume of AC2. Precisely, the lead removal rate of by AC1 surged from 46 to 99% after treatment with HNO_3. Liu et al. [47] investigated the influence of KOH and H_2SO_4 modifications onto biochar during sorption of tetracycline. The results indicated that the KOH-exposed biochar showed high porosity, larger specific surface area, and high C and O composition than the H_2SO_4-exposed and virgin biochars. The remediation of inorganic constituents during alkali treatment allowed the biochar to sorb more pollutant.

The biochar hydrophilicity and surface functional sites can be chemically modified for remediation of specific pollutants at a specific rate from solutions [48]. It is well-known that carboxyl, amine, hydroxyl, phosphonate, and phenolic groups are functional groups often responsible for adsorption of different dyes/metals [49]. The biochar material exhibit low pollutant uptake capacities if the amount of these binding sites is low. Nevertheless, several modification techniques are present to improve the number of these functional sites on the surface of biochar. Xue et al. [50] highlighted that modification using H_2O_2 for peanut hull-derived biochar enhanced the oxygen comprising functional groups particularly carboxyl groups on surface of biochar, which caused enhanced Pb(II) adsorption potential of over 20 times compared to raw biochar.

Biochar can also be prepared for particular applications through mineral impregnation methods. Yao et al. [51] improved the biochar functionality by distributing clay particles in biochar matrix. The authors mixed the biomaterial (bamboo, bagasse and hickory chips) with clay and consequently pyrolysed at 600°C without O_2 for 1 h. The adsorption potential of clay-biochar composite was enhanced five times compared to virgin biochar due to highly porous structure and presence of clay. Magnetic biochar can be prepared through chemical co-precipitation of Fe^{2+}/Fe^{3+} onto biomass and subsequent pyrolysis [17]. The hybrid nature of magnetic biochar permits enhanced adsorption of various organic and inorganic toxins. Through exposure of peanut hull biochar to $FeCl_3$, Han et al. [52] synthesized magnetic biochar for removal of Cr(VI) ions. The prepared magnetic biochar showed improved adsorption potential toward Cr(VI), around 1–2 times compared to raw biochar. The study also identified the removal mechanism through XPS, XRD and SEM and revealed that Cr(VI) was interacted electrostatically to the protonated -OH onto the surface of γ-Fe_2O_3.

4. Mathematical modeling

Adsorption isotherm is the mathematical representation of adsorption capacity (Q) versus equilibrium concentration of the solute (C_e). Modeling adsorption isotherm data is important for prediction/comparison among adsorption performances. Two, three and four-parameter isotherm models are suggested to model the sorption data. Some of the important sorption isotherm models used in the sorption studies include, the Langmuir, Freundlich, Toth and Sips models.

The Langmuir model [53] was fundamentally derived to define the sorption (gas-solid phase) of activated carbon. However, in later years, it was employed to assess and calculate the adsorption behavior of various adsorbents. In its formulation, binding to the surface was primarily by physical forces and implicit in its derivation was the assumption that all sites possess equal affinity for the sorbate. Its use was extended to empirically describe equilibrium relationships between a bulk liquid phase and a solid phase [53]. The model can be expressed as

$$Q = \frac{Q_{max} b_L C_e}{1 + b_L C_e} \tag{1}$$

where Q is the sorptional capacity (mg/g); C_e is the equilibrum concentration (mg/L); Q_{max} is the maximum uptake of toxin by the adsorbent (mg/g) and b_L is the equilibrium coefficient of the Langmuir model (L/mg).

The Freundlich model [54] was empirically derived equation; however it can be applied to adsorption onto diverse surfaces or surfaces with sites of varied affinities. It is assumed that the stronger binding sites are occupied first and that the binding strength decreases with increasing degree of site occupation. It can expressed as,

$$Q = K_F C_e^{1/n_F} \tag{2}$$

where n_F is the exponent of the Freundlich model and K_F is the Freundlich model coefficient $(L/g)^{1/n}{}_F$,

The Sips model [55] is based on the assumption that binding sites on the adsorbent have varied strengths and each active binding site interact with one sorbate molecule. The constant K_s represents sorptional uptake of the adsorbent, whereas a_S denotes affinity of adsorbent toward metal ions. At high metal ion concentrations, the model ultimately takes the Langmuir form, whereas at low metal concentrations reduces to the Freundlich model [56]. The model can be expressed as

$$Q_e = \frac{K_S C_e^{\beta_S}}{1 + a_S C_e^{\beta_S}} \tag{3}$$

where a_S is the Sips model coefficient $(L/mg)^{\beta}{}_S$, β_S is the Sips model exponent and K_S is the Sips model isotherm coefficient $(L/g)^{\beta}{}_S$.

The Toth model [57] is the other three parameter model frequently employed to describe metal-adsorent isotherms. The model assumes quasi-Gaussian energy distribution and is derived from the potential theory. The Toth model can be expressed as

Toth model:

$$Q = \frac{Q_{max} b_T C_e}{\left[1 + (b_T C_e)^{1/n_T}\right]^{n_T}} \tag{4}$$

where b_T is the Toth model constant (L/mg) and n_T is the Toth model exponent.

For any practical applications, the process design, operation control and sorption kinetics are very important [13]. The sorption kinetics can be described using several models.

The most commonly used method to identify the contribution of intraparticle diffusion during adsorption is through fitting the kinetic data to an intraparticle diffusion plot, as presented by Weber and Morris [58] as below:

$$Q_t = k_i t^{1/2} \tag{5}$$

where Q_t is the sorptional capacity at any time t (mg/g) and k_i is the intraparticle diffusion constant. This involves plotting the sorptional capacity at a given time vs. the square root of that time. If the plot passes through the origin, then intraparticle diffusion is the rate determining step.

The pseudo-first-order model assumes that the rate of change of solute uptake with time is directly proportional to the difference in saturation concentration and the amount of solid uptake with time. The model can be expressed as,

$$Q_t = Q_e(1 - \exp(-k_1 t)) \tag{6}$$

where Q_e is the equilibrium uptake (mg/g) and k_1 is the pseudo-first-order constant (1/min). In an attempt to understand the Cd(II) adsorption mechanism of rice straw-derived biochar from aqueous solutions, Fan et al. [59] fitted Cd(II) kinetics data using pseudo-first-order and pseudo-second-order kinetic models. The results indicated that Cd(III) adsorption kinetics by rice straw biochar was better described by the pseudo-first-order kinetic model.

The pseudo-second-order kinetics is framed to predict adsorption capcity over entire experimental conditions (ranges) as the model based on the adsorption capacity of the solid phase.

$$Q_t = \frac{Q_e^2 K_2 t}{1 + Q_e K_2 t} \tag{7}$$

where k_2 is pseudo-second-order constant. Xiao et al. [60] prepared biochar using cow bone meal for adsorption of Cd(II), Pb(II) and Cu(II) ions. On analyzing the kinetics data, the authors identified that pseudo-second-order model fitted the kinetics data well compared to the pseudo-first-order model based on the correlation coefficients and calculated equilibrium uptake values.

5. Conclusions

Biochar represents an effective class of sorbent for remediation of heavy metals from solutions. Several studies recognized superior adsorption potential of biochar compared to other established sorbents. The pyrolysis temperature and feedstock type strongly influences the sorption capacity of biochar. In addition, the process operating parameters such as pH, temperature, initial solute concentration and biochar dosage strongly influences the extent of metal sorption by biochar. Despite the application of biochar as sorbents is increasing as indicated through published literatures, more knowledge needed especially in the area of column sorption and real effluent clean-up.

Author details

Ramalingham Senthilkumar[1]* and Donipathi Mogili Reddy Prasad[2]

1 Department of Engineering, College of Applied Sciences, Sohar, Sultanate of Oman

2 Faculty of Engineering, Universiti Teknologi Brunei, Gadong, Brunei Darussalam

*Address all correspondence to: kumar.soh@cas.edu.om

References

[1] Arfanuzzaman M, Atiq Rahman A. Sustainable water demand management in the face of rapid urbanization and ground water depletion for social–ecological resilience building. Global Ecology and Conservation. 2017;**10**:9-22

[2] Rangabhashiyam S, Balasubramanian P. The potential of lignocellulosic biomass precursors for biochar production: Performance, mechanism and wastewater application—A review. Industrial Crops and Products. 2019;**128**:405-423

[3] Inyang MI, Gao B, Yao Y, Xue Y, Zimmerman A, Mosa A, et al. A review of biochar as a low-cost adsorbent for aqueous heavy metal removal. Critical Reviews in Environmental Science and Technology. 2016;**46**(4):406-433

[4] Dai Y, Zhang N, Xing C, Cui Q, Sun Q. The adsorption, regeneration and engineering applications of biochar for removal organic pollutants: A review. Chemosphere. 2019;**223**:12-27

[5] Abdolali A, Guo WS, Ngo HH, Chen SS, Nguyen NC, Tung KL. Typical lignocellulosic wastes and by-products for biosorption process in water and wastewater treatment: A critical review. Bioresource Technology. 2014;**160**:57-66

[6] Anyanwu BO, Ezejiofor AN, Igweze ZN, Orisakwe OE. Heavy metal mixture exposure and effects in developing nations: An update. Toxics. 2018;**6**(4):65

[7] Tchounwou PB, Yedjou CG, Patlolla AK, Sutton DJ. Heavy metal toxicity and the environment. Experientia Supplementum. 2012;**101**:133-164

[8] Vareda JP, Valente AJM, Durães L. Assessment of heavy metal pollution from anthropogenic activities and remediation strategies: A review. Journal of Environmental Management. 2019;**246**:101-118

[9] Vijayaraghavan K, Ashokkumar T. Plant-mediated biosynthesis of metallic nanoparticles: A review of literature, factors affecting synthesis, characterization techniques and applications. Journal of Environmental Chemical Engineering. 2017;**5**(5):4866-4883

[10] Jan AT, Azam M, Siddiqui K, Ali A, Choi I, Haq QM. Heavy metals and human health: Mechanistic insight into toxicity and counter defense system of antioxidants. International Journal of Molecular Sciences. 2015;**16**(12):29592-29630

[11] Vijayaraghavan K, Raja FD. Design and development of green roof substrate to improve runoff water quality: Plant growth experiments and adsorption. Water Research. 2014;**63**:94-101

[12] Senthilkumar R, Reddy Prasad DM, Lakshmanarao G, Krishnan S, Naveen Prasad BS. Ocean-based sorbents for decontamination of metal-bearing wastewaters: A review. Environmental Technology Reviews. 2018;**7**(1):139-155

[13] Vijayaraghavan K, Yun Y-S. Bacterial biosorbents and biosorption. Biotechnology Advances. 2008;**26**:266-291

[14] Aeisyah A, Ismail MHS, Lias K, Izhar S. Adsorption process of heavy metals by low-cost adsorbent: A review. Research Journal of Chemistry and Environment. 2014;**18**(4):91-102

[15] Rajapaksha AU, Chen SS, Tsang DCW, Zhang M, Vithanage M, Mandal S, et al. Engineered/designer biochar for contaminant removal/immobilization from soil and water: Potential and implication of

biochar modification. Chemosphere. 2016;**148**:276-291

[16] Wang J, Wang S. Preparation, modification and environmental application of biochar: A review. Journal of Cleaner Production. 2019;**227**:1002-1022

[17] Li L, Zou D, Xiao Z, Zeng X, Zhang L, Jiang L, et al. Biochar as a sorbent for emerging contaminants enables improvements in waste management and sustainable resource use. Journal of Cleaner Production. 2019;**210**:1324-1342

[18] Yao Y, Gao B, Chen J, Yang L. Engineered biochar reclaiming phosphate from aqueous solutions: Mechanisms and potential application as a slow-release fertilizer. Environmental Science and Technology. 2013;**47**:8700-8708

[19] Beesley L, Moreno-Jiménez E, Gomez-Eyles JL, Harris E, Robinson B, Sizmur T. A review of biochars' potential role in the remediation, revegetation and restoration of contaminated soils. Environmental Pollution. 2011;**159**(12):3269-3282

[20] Saleem J, Shahid UB, Hijab M, Mackey H, McKay G. Production and applications of activated carbons as adsorbents from olive stones. Biomass Conversion and Biorefinery. 2019;**9**(4):775-802

[21] Tan X, Liu Y, Zeng G, Wang X, Hu X, Gu Y, et al. Application of biochar for the removal of pollutants from aqueous solutions. Chemosphere. 2015;**125**:70-85

[22] Cao X, Ma L, Gao B, Harris W. Dairy-manure derived biochar effectively sorbs lead and atrazine. Environmental Science and Technology. 2009;**43**(9):3285-3291

[23] Uchimiya M, Wartelle LH, Klasson KT, Fortier CA, Lima IM. Influence of pyrolysis temperature on biochar property and function as a heavy metal sorbent in soil. Journal of Agricultural and Food Chemistry. 2011;**59**(6):2501-2510

[24] Jung KW, Kim K, Jeong TU, Ahn KH. Influence of pyrolysis temperature on characteristics and phosphate adsorption capability of biochar derived from waste-marine macroalgae (*Undaria pinnatifida* roots). Bioresource Technology. 2016;**200**:1024-1028

[25] Gai X, Wang H, Liu J, Zhai L, Liu S, Ren T, et al. Effects of feedstock and pyrolysis temperature on biochar adsorption of ammonium and nitrate. PLOS One. 2014;**9**(12):e113888

[26] Xiao Y, Xue Y, Gao F, Mosa A. Sorption of heavy metal ions onto crayfish shell biochar: Effect of pyrolysis temperature, pH and ionic strength. Journal of the Taiwan Institute of Chemical Engineers. 2017;**80**:114-121

[27] Vijayaraghavan K. Recent advancements in biochar preparation, feedstocks, modification, characterization and future applications. Environmental Technology Reviews. 2019;**8**(1):47-64

[28] Sohi SP, Krull E, Lopez-Capel E, Bol R. A review of biochar and its use and function in soil. Advances in Agronomy. 2010;**105**:47-82

[29] Hodgson E, Lewys-James A, Rao Ravella S, Thomas-Jones S, Perkins W, Gallagher J. Optimisation of slow-pyrolysis process conditions to maximise char yield and heavy metal adsorption of biochar produced from different feedstocks. Bioresource Technology. 2016;**214**:574-581

[30] Zhao L, Cao X, Wang Q, Yang F, Xu S. Mineral constituents profile of

biochar derived from diversified waste biomasses: Implications for agricultural applications. Journal of Environmental Quality. 2013;**42**(2):545-552

[31] Li K, Jiang Y, Wang X, Bai D, Li H, Zheng Z. Effect of nitric acid modification on the lead(II) adsorption of mesoporous biochars with different mesopore size distributions. Clean Technologies and Environmental Policy. 2016;**18**(3):797-805

[32] Shakya A, Agarwal T. Removal of Cr(VI) from water using pineapple peel derived biochars: Adsorption potential and re-usability assessment. Journal of Molecular Liquids. 2019;**293**:111497

[33] Liu L, Huang Y, Zhang S, Gong Y, Su Y, Cao J, et al. Adsorption characteristics and mechanism of Pb(II) by agricultural waste-derived biochars produced from a pilot-scale pyrolysis system. Waste Management. 2019;**100**:287-295

[34] Senthilkumar R, Reddy Prasad DM, Govindarajan L, Saravanakumar K, Naveen Prasad BS. Synthesis of green marine algal-based biochar for remediation of arsenic(V) from contaminated waters in batch and column mode of operation. International Journal of Phytoremediation. 2020;**22**:279-286. DOI: 10.1080/15226514.2019.1658710

[35] Ullah A, Kaewsichan L, Tohdee K. Adsorption of hexavalent chromium onto alkali-modified biochar derived from *Lepironia articulata*: A kinetic, equilibrium, and thermodynamic study. Water Environment Research. 2019;**91**(11):1433-1446

[36] Cobbina SJ, Duwiejuah AB, Quainoo AK. Single and simultaneous adsorption of heavy metals onto groundnut shell biochar produced under fast and slow pyrolysis. International Journal of Environmental Science and Technology. 2019;**16**(7):3081-3090

[37] Peng C, Xiao T, Li Z. Effects of pyrolysis temperature on structural properties of sludge-based biochar and its adsorption for heavy metals. Research of Environmental Sciences. 2017;**30**(10):1637-1644

[38] Padmesh TVN, Vijayaraghavan K, Sekaran G, Velan M. Batch and column studies on biosorption of acid dyes on fresh water macro alga *Azolla filiculoides*. Journal of Hazardous Materials. 2005;**125**(1-3):121-129

[39] Vijayaraghavan K, Palanivelu K, Velan M. Biosorption of copper(II) and cobalt(II) from aqueous solutions by crab shell particles. Bioresource Technology. 2006;**97**(12):1411-1419

[40] Vijayaraghavan K, Thilakavathi M, Palanivelu K, Velan M. Continuous sorption of copper and cobalt by crab shell particles in a packed column. Environmental Technology. 2005;**26**(3):267-276

[41] Senthilkumar R, Vijayaraghavan K, Jegan J, Velan M. Batch and column removal of total chromium from aqueous solution using *Sargassum polycystum*. Environmental Progress & Sustainable Energy. 2010;**29**(3):334-341

[42] Senthilkumar R, Vijayaraghavan K, Thilakavathi M, Iyer PVR, Velan M. Application of seaweeds for the removal of lead from aqueous solution. Biochemical Engineering Journal. 2007;**33**(3):211-216

[43] Vijayaraghavan K, Jegan J, Palanivelu K, Velan M. Removal of nickel(II) ions from aqueous solution using crab shell particles in a packed bed up-flow column. Journal of Hazardous Materials. 2004;**113**(1-3):223-230

[44] Vijayaraghavan K, Yun Y-S. Chemical modification and

immobilization of *Corynebacterium glutamicum* for biosorption of reactive black 5 from aqueous solution. Industrial and Engineering Chemistry Research. 2007;**46**(2):608-617

[45] Vilvanathan S, Shanthakumar S. Column adsorption studies on nickel and cobalt removal from aqueous solution using native and biochar form of *Tectona grandis*. Environmental Progress & Sustainable Energy. 2017;**36**:1030-1038

[46] El-Hendawy ANA. Influence of HNO_3 oxidation on the structure and adsorptive properties of corncob-based activated carbon. Carbon. 2003;**41**:713-722

[47] Liu P, Liu W-J, Jiang H, Chen J-J, Li W-W, Yu H-Q. Modification of bio-char derived from fast pyrolysis of biomass and its application in removal of tetracycline from aqueous solution. Bioresource Technology. 2012;**121**:235-240

[48] Yakout SM, Daifullah AEHM, El-Reefy SA. Pore structure characterization of chemically modified biochar derived from rice straw. Environmental Engineering and Management Journal. 2015;**14**:473e480

[49] Vijayaraghavan K, Balasubramanian R. Is biosorption suitable for decontamination of metal-bearing wastewaters? A critical review on the state-of-the-art of biosorption processes and future directions. Journal of Environmental Management. 2015;**160**:283-296

[50] Xue Y, Gao B, Yao Y, Inyang M, Zhang M, Zimmerman AR, et al. Hydrogen peroxide modification enhances the ability of biochar (hydrochar) produced from hydrothermal carbonization of peanut hull to remove aqueous heavy metals: Batch and column tests. Chemical Engineering Journal. 2012;**200-202**:673-680

[51] Yao Y, Gao B, Fang J, Zhang M, Chen H, Zhou Y, et al. Characterization and environmental applications of clay-biochar composites. Chemical Engineering Journal. 2014;**242**:136-143

[52] Han Y, Cao X, Ouyang X, Sohi SP, Chen J. Adsorption kinetics of magnetic biochar derived from peanut hull on removal of Cr (VI) from aqueous solution: Effects of production conditions and particle size. Chemosphere. 2016;**145**:336-341

[53] Langmuir I. The constitution and fundamental properties of solids and liquids. Journal of the American Chemical Society. 1916;**38**:2221-2295

[54] Freundlich HMF. About the adsorption in solution. Zeitschrift für Physikalische Chemie. 1906;**57**:385-471

[55] Sips R. On the structure of a catalyst surface. The Journal of Chemical Physics. 1948;**16**:490-495

[56] Sathishkumar M, Binupriya AR, Vijayaraghavan K, Yun S-I. Two and three-parameter isothermal modeling for liquid-phase sorption of Procion blue H-B by inactive mycelial biomass of *Panus fulvus*. Journal of Chemical Technology & Biotechnology. 2007;**82**(4):389-398

[57] Toth J. State equations of the solid gas interface layer. Acta Chimica Academiae Scientiarum Hungaricae. 1971;**69**:311-317

[58] Weber WJ, Morris JC. Kinetics of adsorption on carbon solution. Journal of the Sanitary Engineering Division: American Society of Civil Engineers. 1963;**89**:31

[59] Fan J, Cai C, Chi H, Reid BJ, Coulon F, Zhang Y, et al. Remediation of cadmium and lead polluted soil

using thiol-modified biochar. Journal of Hazardous Materials. 2020;**388**:122037

[60] Xiao J, Hu R, Chen G. Micro-nano-engineered nitrogenous bone biochar developed with a ball-milling technique for high-efficiency removal of aquatic Cd(II), Cu(II) and Pb(II). Journal of Hazardous Materials. 2020;**387**:121980

Chapter 14

Biochar-Assisted Wastewater Treatment and Waste Valorization

Abhishek Pokharel, Bishnu Acharya and Aitazaz Farooque

Abstract

Biochar is the solid byproduct of pyrolysis, and its cascading use can offset the cost of the production and its use in application such as soil remediation. A wide variety of research on biochar has highlighted its ability to absorb nutrients, metal and complex compounds, filter suspended solids, enhance microorganisms' growth, retain water and nutrients as well as increasing the carbon content of the soil. Besides, sustainable biochar systems are an attractive approach for carbon sequestration and total waste management cycle. The chapter looks into such cascading use of biochar in wastewater treatment for recovering nutrients and improving the efficiency of activated sludge treatment and anaerobic digestion for producing biosolid with enhanced soil amendment properties.

Keywords: biochar, wastewater treatment, activated sludge treatment, anaerobic digestion, nutrient recovery, waste valorization

1. Introduction

Today, the global population continues to grow by 83 million annually and is predicted to be 9.8 billion in 2050 [1]. This increase in population will lead to higher demands of food, water, and energy, which have already been constrained due to the competing needs for limited resources in many parts of the world [2]. The challenges presented by climate change, pollution, and developing economy are posing significant pressure on food, water, and energy systems [3]. Efficient and integrated management of energy, food, and water resources could help address several of the biggest global challenges, such as climate change, sustainable economy, food security, environmental and social security [4, 5]. In the future, we will need increased food production, clear water sources, as well as alternative energy options with minimum resource utilization and ideally decreasing environmental impacts [6]. Work is underway to improve the food production chain as well as develop new technologies for renewable energy. So far less focus has been given to the water, especially to the management of the wastewater. There is a need for shifting the paradigm in the case of wastewater management from treatment and disposal to reuse, recycle, and resource recovery. With growing water scarcity and the fact that uncontrolled disposal of wastewater to the freshwater system is causing depletion of the system also stresses toward a change in mindset about wastewater management. This approach will prevent detrimental impacts on human health and ecosystem caused by the current handling methods. The next step toward a sustainable future will be wastewater treatment serving multiple purposes of treatment and recovery of resources like water, nutrient, and energy. The efficient wastewater management approach will see a cascaded benefit in other sectors including production of fertilizers such as nitrogen and phosphorus. Phosphorus is obtained from ore called phosphate rocks. The quality and accessibility of currently available phosphate rock reserves are declining, and the cost to mine, refine, store, and transport them is rising [7, 8]. Similarly, the production of nitrogen and other mineral fertilizers is energy intensive as well as contributes to environmental pollution [3, 9]. The nitrogen fertilizer can leach to nearby water bodies leading to the phenomenon of eutrophication. The richness of nutrients in the water results in excessive growth of macroalgae and could lead to anoxic events and loss of aquatic system. Recovery of these nutrients from wastewater helps to close the cycle and reduce the amount of chemical fertilizer, directly contributing to the sustainability of food production.

One of the first indications of intentional nutrient recycling is documented 5000 years ago in rural Asia, where human excreta was used for fertilization of fields called "night soil" [10]. In the nineteenth and twentieth century with the industrial revolution, the population density became high, which gave rise to "Sanitation Revolution," a transition from land-based to water-based disposal of human wastes. This disposal system changed the nutrient cycle from reuse to complete discard. Following the Industrial and Sanitation Revolutions, the Green Revolution that reformed agriculture largely abandoned organic fertilizers and put forth the mineral fertilizers [10, 11]. Furthermore, owing to the excessive population growth, producing enough food with only organic sources of plant nutrients has become impossible. Therefore, the need for mineral fertilizers is a true fact. Thus, many urban areas have dedicated wastewater treatment plant to remove the nuisance of human waste. But, it is becoming evident that future changes, particularly those associated with urbanization and population growth-related increase in volume of wastewater, add more stress to the wastewater system performance [12].

The greater dependency on fossil fuels in every sector is heavily contributing to global warming and climate change [13]. As an alternative, abundant biomass could play an essential role in reducing the dependency on fossil fuel as well as contribute toward sustainable development. Pyrolysis of biomass produces biochar and bio-oil. The bio-oil could be used as fuel to substitute the petroleum products with some upgrading that includes catalytic esterification and hydrogenation. The biochar could be used for energy and soil application [14]. Soil application helps in sequestration of carbon dioxide and subsequently supports food production. At present, the biochar application in soil remediation is not cost-effective. The financial feasibility could be improved by developing a cascaded use of biochar, as discussed in this chapter. The inherent properties of biochar make it suitable for (a) recovering nutrients from the wastewater, (b) improving the activated sludge treatment to reduce the energy use for aeration and to improve the settling ability of sludge, (c) increasing the energy recovery from sludge through anaerobic digestion, and (d) enhancing the quality of the biosolids for soil application. There are reports of biochar application having agronomic benefits in fertilizer management, yield, and soil biota [15–20]. Biochar, as a sound absorbent, also holds promise for low-cost wastewater treatment as an alternative to activated carbon [21–24]. The integrated use of biochar in wastewater treatment addresses the current issues with the management of wastewater. However, the benefit of using biochar varies with its type and characteristics, which depends on the biomass, and the pyrolysis conditions [25].

This chapter provides insights on the use of biochar in a wastewater treatment process to enhance the treatment as well as recover valuable byproducts. The chapter will discuss biochar production and properties, mechanisms involving removal of organic and inorganic compounds from the effluent phase, and role in activated sludge treatment and anaerobic digestion.

2. Biochar properties for wastewater treatment

Biochar is a carbon-rich solid material produced from biomass through a thermochemical process called pyrolysis. During pyrolysis, lignin, cellulose, hemicellulose, fat, and starch in the feedstock are thermally broken down forming three products: biochar (solid), bio-oil (partly condensed volatile matter), and non-condensable gases (CO_2, CO, CH_4, and H_2) [26, 27]. The bio-oil and gases can be captured to produce energy and depending on the feed valuable coproducts like wood preservatives, food flavoring, adhesive, or biochemical compounds [28]. The yield of biochar and the properties, however, depends on the pyrolysis condition. Slow pyrolysis at moderate temperature (350–500°C) and slow heating rate results in higher yield (30%) of biochar than around 10% or less yield with fast pyrolysis (600–700°C and fast heating rate) or gasification (temperature 700°C or above) [29]. The feedstock type and pyrolysis condition used during the production of biochar notably change the physiochemical properties such as surface area, polarity, atomic ratio, pH, and elemental composition [25, 30, 31]. These properties determine the effectiveness of biochar in wastewater treatment.

Biochar has wide applications in water and wastewater treatment because of its distinctive characteristics, for example, adsorption capacity, specific surface area, microporosity, and ion exchange capacity [30, 32]. The removal mechanisms of different pollutants are governed by their interactions with various attributes of biochar, which depends on pyrolysis temperature and feedstock type [33]. Pyrolysis temperature greatly affects the properties of biochar. The increase in pyrolysis temperature results in higher carbon content, hydrophobicity, aromaticity, surface area, and microporosity in biochar [34]. Similarly, the pH of the biochar increases with increasing pyrolysis temperature due to enrichment of ash content in the biochar [35, 36]. High-temperature (>500°C) biochar has low polarity and acidity due to loss of O- and H-containing functional groups [34]. Lower pyrolysis temperature (<500°C) facilitates partial carbonization, thus yielding biochar with smaller pore size, lower surface area, and high O-containing functional groups [36]. Lower temperature biochar contains a higher content of dissolved organic carbon, relatively low polarity and C/N ratio [30, 34, 37].

Biochar often compromises of both positively and negatively charged surfaces (zwitterionic) [34, 35]. The negatively charged functional groups contribute to cation exchange capacity (CEC) whereas anion exchange capacity (AEC) is also exhibited by O-containing functional groups (oxonium heterocycles) in biochar [36, 38]. Oxygen (O) containing alcohol, carbonyl, and carboxylate functional groups are generally believed to contribute to biochar cation exchange capacity because they carry a negative charge and serve as Lewis bases for the sorption of cations. Whereas, it is believed that oxonium functional groups contribute to pH-independent anion exchange and that both pyridinic functional groups and nonspecific proton adsorption by condensed aromatic rings contribute to pH-dependent anion exchange capacity in biochars [38].

Biochar derived from woody biomass and crop residues has a higher surface area compared to that of solid municipal wastes and animal manure [30]. Apart from the usual pyrolysis method, different engineering methods have been developed and used to expand biochar's applications. Engineered biochar is the derivative of biochar that is modified by physical, chemical, and biological methods to improve its physical, chemical, and biological properties (e.g., specific surface area, porosity, cation exchange capacity, surface functional group, pH etc.) and its adsorption capacity [37, 39, 40]. Some of the modification includes anaerobic digestion of feedstock before pyrolysis, steam/gas activation, pyrolysis using microwave heating, ball milling, magnetic modification, chemical modification using hydrogen peroxide, alkali or acid, and impregnation/coating with chemicals [41]. The detail about the modified biochar for wastewater treatment will be discussed in the following sections.

2.1 Biochar modification

Researchers have discussed several methods for modifying the properties of biochar [42]. These methodologies include treatments with steam, acids, bases, metal oxides, carbonaceous materials, clay minerals, organic compounds, and biofilms [43].

2.1.1 Physical activation of biochar

Physical activation methods such as steam activation involve high-temperature steam forced through the pores of the biochar. Steam activation, which is carried out after pyrolysis, is a common modification method used to increase the structural porosity of the biochar and remove impurities such as products of incomplete combustion. According to [44], higher water flow rates and longer activation times at 800°C increased the sorption of Cd, Cu, and Zn on the surface of biochar from poultry manure feedstocks pyrolyzed at 700°C. In another study, comparison of Cu^{2+} adsorption for biochar from *Miscanthus* before (500°C pyrolysis) and after (800°C) steam activation showed no significant change [45]. It was found that steam activation of the biochar increased the surface area and aromaticity alongside a decrease in the abundance of functional groups [45]. Similarly, steam-activated biochar from pine sawdust increased the surface area but had little effect on the surface functional group as a result of which adsorption capacity of biochar for phosphate was reduced due to electrostatic repulsion by the negatively charged surface of biochar [46]. The steam-activated invasive plant (*Sicyos angulatus* L.)-derived biochar produced at 700°C showed 55% increase in sorption capacity of veterinary antibiotics (sulfamethazine) compared to that of nonactivated biochar produced at the same temperature [47]. Hence, steam activation could be a process for increasing the porosity and surface area of biochar along with aromaticity to obtain better adsorption of inorganic material in the wastewater.

2.1.2 Chemical activation using acidic and alkaline solutions

The biochar activation using acidic solutions forms carboxylic groups on the biochar surface [48] and develops micropores, thus increasing the surface area [49]. The increase of oxygenated functional groups on biochar surfaces increases the potential of biochar to bind positively charged pollutants through specific adsorption chemically. The pH dependence of Cu^{2+} sorption capacity for HNO_3-activated cactus fiber biochar indicated chemical sorption on oxygen-containing functional groups on the biochar surface [48]. Higher O/C ratio in the post-activation of rice straw with H_2SO_4 and HNO_3 showed evidence of oxygen-containing functional group incorporated into the carbon structure [50]. Acid treatment of pine tree sawdust with diluted H_3PO_4 prior to pyrolysis increased the surface area, the total pore volume, and volume of micropores area along with P-O-P incorporation in the C structure [51]. This increased the Pb sorption capacity of the phosphoric-treated biochar by 20% in comparison to a nontreated sample, mainly due to phosphate precipitation and surface adsorption [51]. Similarly, almost double increase in cation exchange capacity was observed for pinewood biochar treated with 30% H_2O_2 because the oxygen-containing functional groups in the surface of biochar, which were more abundant in the activated biochar, exchanged with cations in solution [52]. Treating a hydrochar, a carbon-enriched solid produced from hydrothermal carbonization of peanut hull, with a 10% H_2O_2 solution increased Pb sorption capacity compared to the unmodified hydrochar, which can be attributed to a greater abundance of carboxyl functional groups that can form complexes with Pb [53]. However, the introduction of acid or oxidizing agents dissolves mineral components (CO_3^{2-}, SiO_4^{2-}, PO_4^{3-}) in the biochar structure and removes them from the biochar matrix. These minerals in biochar are particularly important for the removal of metal cations from water due to precipitation [54], the affinity of which could be reduced by the acid treatment.

Activation of biochar using alkali (most commonly KOH and NaOH) increases adsorption by increasing porosity, surface and oxygenated functional group at the surface. Oxygenated functional groups provide proton-donating exchange sites where cation such as Pb^{2+} adsorbs chemically [55]. The activation of ipomoea plant biochar with KOH, followed by pyrolysis (350–550°C) demonstrated an increase adsorption of Cd from aqueous solution [56]. Further evidence of kinetics of sorption fitting a pseudo-second-order model and thermodynamic studies indicating spontaneous endothermic process showed that Cu sorption on KOH-activated biochar was due to chemical adsorption [57]. The adsorption capacity of As(V) on municipal solid waste biochar was increased by 1.3 times after activation with 2 M KOH [58]. It can be concluded that activation by alkali greatly enhanced the surface area and altercation of the functional group at the surface.

2.1.3 Biochar-based composites

The biochar composites are prepared by embedding different materials into the biochar structure pre- or post-pyrolysis. Generally, biochar has a higher surface area, high pH, and a negative surface charge. This facilitates specific adsorption of metal ions via oxygenated functional groups, electrostatic attraction to aromatic groups, and precipitation on the mineral ash components of the biochar. But at the same time biochar is usually a poor adsorbent for oxy-anions contaminants like NO_3^-, PO_4^{3-}, and AsO_4^{3-} [44]. This can be improved by the homogenous spread of metal oxide on biochar surfaces. It can be done by soaking biochars or the feedstocks in a solution of metal nitrate or chloride salt solution (common examples $FeCl_3$, Fe, $Fe(NO_3)_3$, and $MgCl_2$) and heated under atmospheric condition within a temperature range of 50–300°C. This process ensures removal of nitrite and chlorine leaving behind metals in the biochar matrix. Ca-, FeO-, and Fe^{3+}-modified biochar from soaked rice husk and municipal biomass in CaO, iron powder, and $FeCl_3$ respectively, increased the capability of biochar to remove As(V), but not as high for Cr(VI), from aqueous solution [59]. Taking into consideration that one of the main mechanisms for Cr(VI) removal is the electrostatic interaction to the positively charged functional groups on the surface of adsorbents, high Cr(VI) removal is observed at low pH values [60]. It is rather possible that the high pH values of the RH-Ca2þ, RH-Fe0, and SW-Fe0 solutions are related to the deprotonation of their functional groups and the repelling of the negatively charged Cr(VI) [60]. Similarly, a 20-time increase in the sorption of As(V) was observed when corncob biochar was modified with $Fe(NO_3)_3$ [61]. Despite the lower surface area, modification of biochars from garden wood waste and wood chips as well as corncob showed the increased PO_4^{3-} sorption by a factor of 12–50% [58]. Further research has been carried out for preparing biochar-based composites by impregnation or coating the surface of the biochar with metal oxides of Al, Mn, and Mg [58]; clay minerals [62]; complex organic compounds, such as chitosan [63] or amino acids [64]; or inoculation with microorganisms [65].

Thus, the selection of biochar and modification methods for the application in wastewater treatment requires a considerable understanding of the biochar properties and mechanism by which it supports the treatment process at different stages of wastewater treatment.

3. Role of biochar use in wastewater treatment process

Biochar could be used at different stages of wastewater treatment (**Figure 1**) to improve the treatment efficiency and recovery of value-added byproducts. Biochar application in wastewater treatment could be governed by the mechanism of adsorption, buffering, and immobilization of microbial cells. If used on the treated effluents, suitably modified biochar could efficiently adsorb nutrients like nitrogen and phosphorus, which can later be used as a nutrient-enriched material for soil remediation. When used in the activated sludge treatment process, biochar could play a role for improving the treatment and settling ability of the sludge by adsorption of inhibitors and toxic compounds or provide a surface for immobilization of microbes. Addition of biochar in the biological system could eventually help to improve the soil amendment properties of the biosolid as well. As interest grows in the use of biochar in soil applications, its use in wastewater treatment could expand the value chain and create additional economic benefits [66]. The following section will discuss the role of biochar for various applications in the wastewater treatment plant.

3.1 Organic pollutant removal

In recent years, significant amount of research has been done to examine the application of biochar for removal of various organic compounds from water, which includes agrochemicals, antibiotics/drugs, polycyclic aromatic

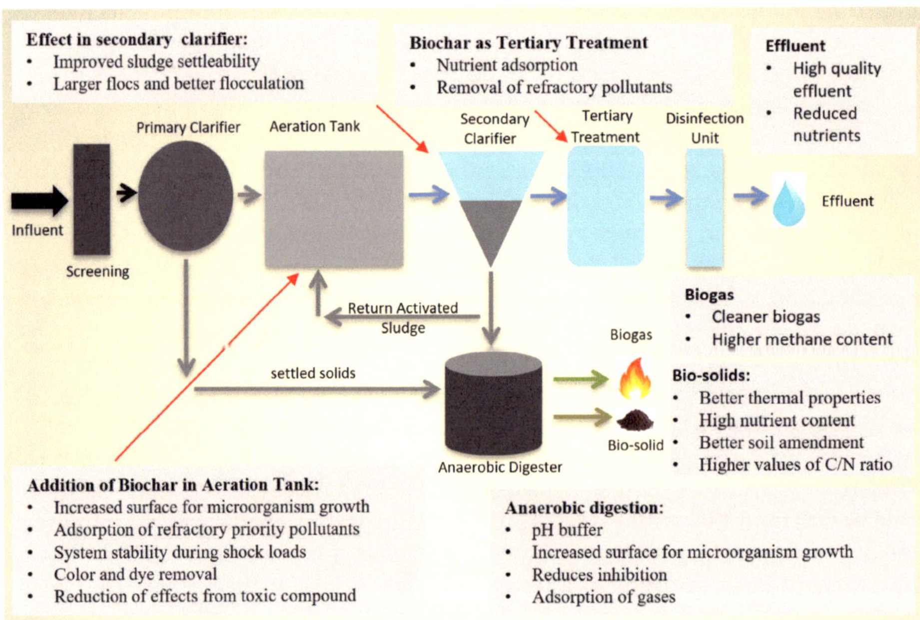

Figure 1.
Use of biochar at different stages of wastewater treatment.

hydrocarbons (PAHs), volatile organic compounds or (VOCs), cationic aromatic dyes [67–70]. Similarly, removal of organic compounds present in specific waste streams such as estrogen compounds in animal manure and sewage, inhibitory compounds of biomass degradation (furfural, hydroxymethylfurfural, phenolic compounds), and toxic organic compounds in landfill leachate has been studied using biochar [71, 72]. **Figure 2** schematically shows different interactions of the organic pollutant with biochar.

Biochar produced at higher pyrolysis temperature is found better for removal of nonpolar organic compounds due to higher surface area and microporosity [30, 73]. In contrast, biochar produced at a temperature below 500°C contains more O- and H-containing functional groups; thus, they are likely to have a high affinity to polar organic compounds [26]. For example, rice husk and soybean-derived biochar (600–700°C) facilitates removal of nonpolar carbofuran (pesticide) and trichloromethylene (VOC) from contaminated water [26]. Efficient removal of pyrimethanil and diesopropylatrazine (fungicide/pesticide) was observed with red-gum wood chips and broiler litter-derived biochar at temperature >700°C, whereas the same biochar at temperature <500°C was inefficient [74, 75]. On the other hand, removal of polar insecticide and herbicide like 1-naphthol, norflurazon, and fluridone was observed with biochar produced at <300°C, due to interaction of pollutant and the functional groups of biochar [76, 77]. Likewise, higher sorption of aromatic cationic dyes like methyl-violet and methyl-blue was observed with biochar containing more O- and H-functional groups (<400°C) but the mechanism was highly dependent on pH [70, 78]. The sorption of polar antibiotic sulfamethazine (SMZ) by hardwood/softwood-derived biochars (produced at 300–700°C) has pH-dependent

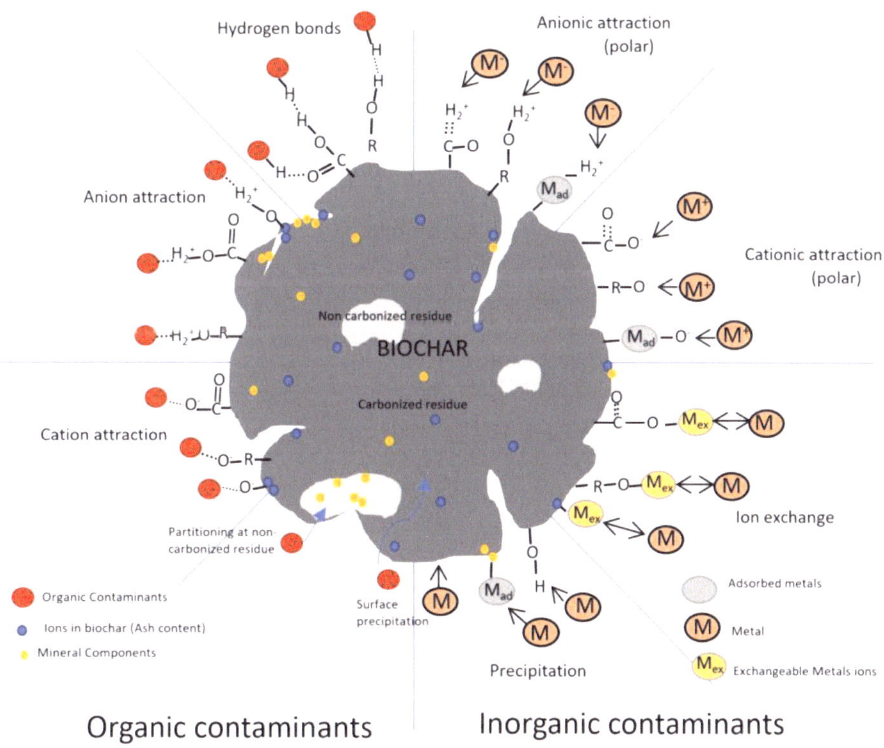

Figure 2.
Biochar interaction with organic and inorganic compounds in wastewater (adapted from Ahmad et al. [33]).

interactions [79]. It can be said that pH is the most important factor for biochar interactions and removal of polar organic pollutants.

3.2 Inorganic pollutant removal

Inorganic pollutant in wastewater includes heavy metals (Cr, Cu, Pb, Cd, Hg, Fe, Zn, and As ions) and compounds like nitrate (NO_3), nitrite (NO_2), ammonium (NH_4), phosphorus (P), and hydrogen sulfide (H_2S) that cause significant risk to public health and environment [80]. Biochar produced at lower pyrolysis temperature (<500°C) has properties that are better suited for removal of inorganic compounds. The chemical composition and the morphological structure play an important role in the sorption nature of biochar [81]. **Figure 2** summarizes the interaction methods for inorganic pollutant and biochar.

3.2.1 Heavy metals

Biochar with high organic carbon content (at non-carbonized fraction), specific porous structure, and numerous functional groups interacts with heavy metals in many ways [82]. The sorption of heavy metals by biochar is mainly by surface interaction through ion exchange and complexation between biochar functional groups (e.g., OH, COOH, R-OH) and heavy metal ions [83, 84], moreover formation of metal precipitates with inorganic constituents [83–85] and coordination of metal ions with π electrons (C=C) of biochar [74]. The physiochemical properties of biochar affect the adsorption throughout its matrix and are dependent on pyrolysis temperature, feedstock type, pH, and application rate. Cu^{2+} showed high affinity toward COOH– and OH– groups of hardwood and crop-derived biochars with dependency on pH and feedstock types [86]. Similarly, sida hermaphrodita-, guayule shrub-, soybean straw-, and wheat straw-derived biochars were effective for removal of Cd^{2+}, Ni^{2+}, and Zn^{2+} along with Cu^{2+} [87]. The higher efficiency of the above-mentioned biochar was due to high C and O contents, high O/C molar ratio, and polarity index, which were mainly regulated by pH [88, 89]. Alkaline biochars derived from various agricultural residues (e.g., soybean straw, corncob, cocoa husk, corn stover, switchgrass) and manure were efficient for Hg^{2+} removal. Animal manure-derived and cocoa husk biochar was highly effective for Hg^{2+} removal due to high sulfur (SH groups and sulfate) to precipitate 90% of Hg^{2+} as $Hg(OH)_2$ or $HgCl_2$ mainly through coprecipitation with anions (Cl, O, S) of biochar [73, 90].

For Cd^{2+}, Zn^{2+}, Pb^{2+}, and Cu^{2+} dosage of biochar also affects the removal of heavy metals. The higher removal efficiency is observed with increasing biochar loading in the aqueous system, due to increased pH and surface area with biochar addition [54, 91].

3.2.2 Nitrogen and phosphorus

The high surface charge density allows biochars to retain cations by cation exchange and the high surface area, internal porosity, and presence of both polar and nonpolar surface sites on biochar enable it to adsorb nutrients [92]. In the limited studies carried out without soil, biochar has shown the absorption NH_4^-, NO_3^-, and PO_4^{3-} despite the different charges and properties of these nutrients [93]. Some examples include digested sugar beet tailing biochar pyrolyzed at 600°C that adsorbed PO_4 ions most likely in binding sites contained in colloidal and nano sized MgO particles on the biochar surface [94]. Also, orange peel biochars pyrolyzed between 250 and 700°C removed between 8 and 83% of phosphate from solution

[95]. NH_4 was adsorbed to biochars produced from rice husk [96] and a mixture of tree trunks and branches [97], albeit weakly, as the partitioning coefficients between water and biochar were low (Freundlich coefficients of 0.251 mg g^{-1}). Similarly, NO_3^- has been adsorbed to bamboo charcoal biochar in the concentration range of 0–10 mg L^{-1} [98].

3.3 Activated sludge treatment

One of the most utilized systems for treatment of municipal wastewater is biological treatment process like activated sludge system (ASS) because of its cost-effectiveness and comparatively more straightforward operation to advance systems. Activated sludge process is a suspended growth treatment where aerobic microorganism decomposes the organic matter in wastewater, which eventually settles as solids by gravity. Currently, increasing concerns are being raised about the presence of various micro-pollutants from pharmaceuticals, personal care products (PCPs), pesticides, disinfectants, and antiseptic in domestic and municipal wastewaters. These pollutants are alien to the biota in the system, and the conventional treatment process often leads to inadequate removal of these compounds. Correspondingly, discharge requirements are currently being stringent for protection of receiving waters from possible contamination and public health hazard. There have been several modifications and changes in the activated sludge system to address the problem. One such method is AS-PACT (Activated Sludge with Powdered Activated Carbon Treatment) where powdered activated carbon is added to the aeration basin of activated sludge system. The larger surface area of carbon provides various benefits including adsorption of toxic substances such as pharmaceuticals and industrial chemicals, immobilization of bacteria, and increased sedimentation of activated sludge [99, 100]. Such system, however, requires a continuous makeup of fresh carbon [101].

Despite the benefits, the higher cost of activated carbon limits its use in municipal wastewater treatment [101]. The biochar could be a low-cost substitute to activated carbon [102], but its merits are less known. The addition of biochar to a biological treatment system, such as within the aeration tank, could result in increased process stability by (a) adsorption of inhibitors (heavy metals, polycyclic aromatic hydrocarbon), (b) increasing the buffering capacity of the system, and (c) immobilization of microbial cells [103]. Limited studies done on the use of biochar in the aeration tank showed increased settling ability of activated sludge [104]. Dissolved organic matter in the biochar could also provide additional carbon to promote denitrification [105]. The availability of organic matter, however, depends on the type of biomass and pyrolysis conditions used for producing biochar. Furthermore, the cascading benefits of using biochar in activated sludge treatment could also be seen on anaerobic digestion of the sludge and in the final quality of the biosolids.

3.4 Anaerobic digestion

In the case of anaerobic digestion, the addition of biochar has shown increases in the rate and amount of biogas production [106–108]. This is attributed to the buffering properties of biochar, promoting methanogenesis for higher biogas yield [109, 110]. Several studies have suggested increases in microbial metabolism and growth because of the support provided by the biochar [107, 111]. The biochar could also play a significant role in reducing the mobility or availability of the inhibitors like heavy metals, pesticides, antibiotics, and other organic compounds by binding them in its porous structure and maintain proper microbial activity for the digestion process [103]. Further, the adsorption of nutrients in biochar and its slow release increase the availability of nutrients to the soil while preventing leaching to surrounding water bodies, as it is prevalent in the case of biosolids [103]. Therefore, the addition of biochar in the biological system could eventually help to improve the soil amendment properties of the biosolid as well.

4. Conclusion

Biochar is a unique renewable resource, which can be used in a wide variety of applications from addressing various environmental problems like climate change, remediation of pollutants in water and soil to an alternative fuel source. The cascading use of biochar as a byproduct of pyrolysis for wastewater treatment and nutrient recycling can synergistically improve soil and water quality, carbon sequestration, greenhouse gas emissions, nutrient cycling, and fuel crisis. The approach perfectly fits the ideas of the circular economy: reuse and recycle of waste, keeping material and product in use. This approach is connected with three natural cycles: water, carbon, and nutrient and has a direct impact on energy, water, and food systems. While much work has been done in modifying the biochar for adsorption of desired organic or inorganic compounds, very less is known on its application in activated sludge treatment, anaerobic digestion, and the overall quality of the biosolids. As the benefits of the integrated use of biochar in wastewater treatment to soil application is established in this chapter, future experimental research work could verify its effectiveness.

Conflict of interest

The authors declare no conflict of interest.

Nomenclature

CO_2	carbon dioxide
CO	carbon monoxide
CH_4	methane
H_2	hydrogen
C	carbon
N	nitrogen
O	oxygen
Cu	copper
HNO_3	nitric acid
H_2SO_4	sulfuric acid
H_3PO_4	phosphoric acid
Pb	lead
H_2O_2	hydrogen peroxide
P	phosphorus
Ca	calcium
Mg	magnesium
K	potassium
Na	sodium
KOH	potassium hydroxide
NaOH	sodium hydroxide
Cd	cadmium

2M	2-molar concentration
As(V)	pentavalent arsenic
NO_3^-	nitrate
PO_4^{3-}	phosphate
AsO_4^{3-}	arsenate ion
$FeCl_3$	ferric chloride
Fe	iron
$Fe(NO_3)_3$	ferric nitrate
$MgCl_2$	magnesium chloride
Ca	calcium
FeO	ferrous oxide
Fe^{3+}	ferric ion
Cr(VI)	hexavalent chromium
Al	aluminum
Mn	manganese
Cr	chromium
Hg	mercury
Zn	zinc
As	arsenic
COOH–	carboxylic acid functional group
OH–	hydroxy functional group
R-OH	carbon chain with hydroxy functional group
$Hg(OH)_2$	mercury hydroxide (bivalent)
S	sulfur
H_2S	hydrogen sulfide
Cl	chlorine
SH	sulfanyl or thiol
$E^{+/-}$	ions of compound/element E
Abbreviations	
CEC	cation exchange capacity
AEC	anion exchange capacity
PAHs	polycyclic aromatic hydrocarbons
VOCs	volatile organic compounds
SMZ	sulfamethazine
AS-PACT	Activated Sludge-Powdered Activated Carbon Treatment

Author details

Abhishek Pokharel, Bishnu Acharya* and Aitazaz Farooque
Faculty of Sustainable Design Engineering, University of Prince Edward Island, Charlottetown, Canada

*Address all correspondence to: bacharya@upei.ca

References

[1] United Nations, Department of Economic and Social Affairs, Population Division. World Population Prospects: The 2017 Revision, Key Findings and Advance Tables. Working Paper No. ESA/P/WP/248. 2017

[2] Godfray HCJ, Beddington JR, Crute IR, Haddad L, Lawrence D, Muir JF, et al. Food security: The challenge of feeding 9 billion people. Science. 2010;**327**(5967):812-818

[3] de Amorim WS, Valduga IB, Ribeiro JMP, Williamson VG, Krauser GE, Magtoto MK, et al. The nexus between water, energy, and food in the context of the global risks: An analysis of the interactions between food, water, and energy security. Environmental Impact Assessment Review. 2018;**72**:1-11

[4] Brandoni C, Bosnjakovic B. Energy, food and water nexus in the European Union: Towards a circular economy. Proceedings of the Institution of Civil Engineers-Energy. 2018;**171**(3):140-144

[5] United Nation – General Assembly, Seventieth session. Resolution adopted by the General Assembly on 25 September 2015. Transforming Our World: The 2030 Agenda for Sustainable Development. United Nation-A/RES/70/1

[6] Foereid B. Biochar in nutrient recycling—The effect and its use in wastewater treatment. Open Journal of Soil Science. 2015;**5**(02):39

[7] Cho R. Phosphorus: Essential to life-are we running out. In: Agriculture, Earth Sciences. Columbia: Earth Institute; 2013

[8] Withers PJ, Sylvester-Bradley R, Jones DL, Healey JR, Talboys PJ. Feed the crop not the soil: Rethinking phosphorus management in the food chain. Environmental Science & Technology. 2014;48(12):6523-6530. DOI: 10.1021/es501670j

[9] Gellings CW, Parmenter KE. Efficient use and conservation of energy - energy efficiency in fertilizer production and use. UNESCO-Encyclopedia of Life Support Systems. 2016:123-136

[10] Ashley K, Cordell D, Mavinic D. A brief history of phosphorus: From the philosopher's stone to nutrient recovery and reuse. Chemosphere. 2011;**84**(6):737-746

[11] Angelakis A, Snyder S. Wastewater treatment and reuse: Past, present, and future. Water. 2015;7:4887-4895. DOI: 10.3390/w7094887

[12] Butler D, McEntee B, Onof C, Hagger A. Sewer storage tank performance under climate change. Water Science and Technology. 2007;**56**(12):29-35

[13] Bilgen S. Structure and environmental impact of global energy consumption. Renewable and Sustainable Energy Reviews. 2014;**38**:890-902

[14] Lehmann J. Bio-energy in the black. Frontiers in Ecology and the Environment. 2007;**5**(7):381-387

[15] Lehmann J, da Silva JP, Steiner C, Nehls T, Zech W, Glaser B. Nutrient availability and leaching in an archaeological Anthrosol and a Ferralsol of the Central Amazon basin: Fertilizer, manure and charcoal amendments. Plant and Soil. 2003;**249**(2):343-357

[16] Major J, Rondon M, Molina D, Riha SJ, Lehmann J. Maize yield and nutrition during 4 years after biochar application to a Colombian

savanna oxisol. Plant and Soil. 2010;**333**(1-2):117-128

[17] Knowles OA, Robinson BH, Contangelo A, Clucas L. Biochar for the mitigation of nitrate leaching from soil amended with biosolids. Science of the Total Environment. 2011;**409**(17):3206-3210

[18] Clough T, Condron L, Kammann C, Müller C. A review of biochar and soil nitrogen dynamics. Agronomy. 2013;**3**(2):275-293

[19] Zheng H, Wang Z, Deng X, Herbert S, Xing B. Impacts of adding biochar on nitrogen retention and bioavailability in agricultural soil. Geoderma. 2013;**206**:32-39

[20] Liu X, Zhang A, Ji C, Joseph S, Bian R, Li L, et al. Biochar's effect on crop productivity and the dependence on experimental conditions—A meta-analysis of literature data. Plant and soil. 2013;**373**(1-2):583-594

[21] Reddy KR, Xie T, Dastgheibi S. Evaluation of biochar as a potential filter media for the removal of mixed contaminants from urban storm water runoff. Journal of Environmental Engineering. 2014;**140**(12):04014043

[22] Perez-Mercado L, Lalander C, Berger C, Dalahmeh S. Potential of biochar filters for onsite wastewater treatment: Effects of biochar type, physical properties and operating conditions. Water. 2018;**10**(12):1835

[23] Jung C, Phal N, Oh J, Chu KH, Jang M, Yoon Y. Removal of humic and tannic acids by adsorption–coagulation combined systems with activated biochar. Journal of Hazardous Materials. 2015;**300**:808-814

[24] Ding Z, Hu X, Wan Y, Wang S, Gao B. Removal of lead, copper, cadmium, zinc, and nickel from aqueous solutions by alkali-modified biochar: Batch and column tests. Journal of Industrial and Engineering Chemistry. 2016;**33**:239-245

[25] Sun Y, Gao B, Yao Y, Fang J, Zhang M, Zhou Y, et al. Effects of feedstock type, production method, and pyrolysis temperature on biochar and hydrochar properties. Chemical Engineering Journal. 2014;**240**:574-578

[26] Suliman W, Harsh JB, Abu-Lail NI, Fortuna AM, Dallmeyer I, Garcia-Perez M. Influence of feedstock source and pyrolysis temperature on biochar bulk and surface properties. Biomass and Bioenergy. 2016;**84**:37-48

[27] Brewer CE, Hu YY, Schmidt-Rohr K, Loynachan TE, Laird DA, Brown RC. Extent of pyrolysis impacts on fast pyrolysis biochar properties. Journal of Environmental Quality. 2012;**41**(4):1115-1122

[28] Czernik S, Bridgwater AV. Overview of applications of biomass fast pyrolysis oil. Energy & Fuels. 2004;**18**(2):590-598

[29] Ahmad M, Lee SS, Dou X, Mohan D, Sung JK, Yang JE, et al. Effects of pyrolysis temperature on soybean stover-and peanut shell-derived biochar properties and TCE adsorption in water. Bioresource Technology. 2012;**118**:536-544

[30] Mohanty P, Nanda S, Pant KK, Naik S, Kozinski JA, Dalai AK. Evaluation of the physiochemical development of biochars obtained from pyrolysis of wheat straw, timothy grass and pinewood: Effects of heating rate. Journal of Analytical and Applied Pyrolysis. 2013;**104**:485-493

[31] Uchimiya M, Ohno T, He Z. Pyrolysis temperature-dependent release of dissolved organic carbon from plant, manure, and biorefinery wastes. Journal of Analytical and Applied Pyrolysis. 2013;**104**:84-94

[32] Ronsse F, Van Hecke S, Dickinson D, Prins W. Production and characterization of slow pyrolysis biochar: Influence of feedstock type and pyrolysis conditions. Gcb Bioenergy. 2013;**5**(2):104-115

[33] Ahmad M, Rajapaksha AU, Lim JE, Zhang M, Bolan N, Mohan D, et al. Biochar as a sorbent for contaminant management in soil and water: A review. Chemosphere. 2014;**99**:19-33

[34] Windeatt JH, Ross AB, Williams PT, Forster PM, Nahil MA, Singh S. Characteristics of biochars from crop residues: Potential for carbon sequestration and soil amendment. Journal of Environmental Management. 2014;**146**:189-197

[35] Keiluweit M, Nico PS, Johnson MG, Kleber M. Dynamic molecular structure of plant biomass-derived black carbon (biochar). Environmental Science & Technology. 2010;**44**(4):1247-1253

[36] Lawrinenko M, Laird DA. Anion exchange capacity of biochar. Green Chemistry. 2015;**17**(9):4628-4636

[37] Yao Y, Gao B, Chen J, Zhang M, Inyang M, Li Y, et al. Engineered carbon (biochar) prepared by direct pyrolysis of Mg-accumulated tomato tissues: Characterization and phosphate removal potential. Bioresource Technology. 2013;**138**:8-13

[38] Kong H, He J, Gao Y, Wu H, Zhu X. Cosorption of phenanthrene and mercury (II) from aqueous solution by soybean stalk-based biochar. Journal of Agricultural and Food Chemistry. 2011;**59**(22):12116-12123

[39] Rajapaksha AU, Chen SS, Tsang DC, Zhang M, Vithanage M, Mandal S, et al. Engineered/designer biochar for contaminant removal/immobilization from soil and water: Potential and implication of biochar modification. Chemosphere. 2016;**148**:276-291

[40] Wang B, Gao B, Fang J. Recent advances in engineered biochar productions and applications. Critical Reviews in Environmental Science and Technology. 2017;**47**(22):2158-2207

[41] Inyang M, Dickenson E. The potential role of biochar in the removal of organic and microbial contaminants from potable and reuse water: A review. Chemosphere. 2015;**134**:232-240

[42] Rajapaksha AU, Chen SS, Tsang DC, Zhang M, Vithanage M, Mandal S, et al. Engineered/designer biochar for contaminant 848 removal/immobilization from soil and water: Potential and implication of biochar 849 modification. Chemosphere. 2016;**148**(27):6e291

[43] Sizmur T, Fresno T, Akgül G, Frost H, Moreno-Jiménez E. Biochar modification to enhance sorption of inorganics from water. Bioresource Technology. 2017;**246**:34-47

[44] Lima IM, Marshall WE. Adsorption of selected environmentally important metals by poultry manure-based granular activated carbons. Journal of Chemical Technology & Biotechnology: International Research in Process, Environmental & Clean Technology. 2005;**80**(9):1054-1061

[45] Shim T, Yoo J, Ryu C, Park YK, Jung J. Effect of steam activation of biochar produced from a giant Miscanthus on copper sorption and toxicity. Bioresource Technology. 2015;**197**:85-90

[46] Lou K, Rajapaksha AU, Ok YS, Chang SX. Pyrolysis temperature and steam activation effects on sorption of phosphate on pine sawdust biochars in aqueous solutions. Chemical Speciation & Bioavailability. 2016;**28**(1-4):42-50

[47] Rajapaksha AU, Vithanage M, Ahmad M, Seo DC, Cho JS, Lee SE, et al. Enhanced sulfamethazine removal by

steam-activated invasive plant-derived biochar. Journal of Hazardous Materials. 2015;**290**:43-50

[48] Hadjittofi L, Prodromou M, Pashalidis I. Activated biochar derived from cactus fibres–preparation, characterization and application on Cu (II) removal from aqueous solutions. Bioresource Technology. 2014;**159**:460-464

[49] Iriarte-Velasco U, Sierra I, Zudaire L, Ayastuy JL. Preparation of a porous biochar from the acid activation of pork bones. Food and Bioproducts Processing. 2016;**98**:341-353

[50] Qian K, Kumar A, Patil K, Bellmer D, Wang D, Yuan W, et al. Effects of biomass feedstocks and gasification conditions on the physiochemical properties of char. Energies. 2013;**6**(8):3972-3986

[51] Zhao L, Zheng W, Mašek O, Chen X, Gu B, Sharma BK, et al. Roles of phosphoric acid in biochar formation: Synchronously improving carbon retention and sorption capacity. Journal of Environmental Quality. 2017;**46**(2):393-401

[52] Huff MD, Lee JW. Biochar-surface oxygenation with hydrogen peroxide. Journal of Environmental Management. 2016;**165**:17-21

[53] Xue Y, Gao B, Yao Y, Inyang M, Zhang M, Zimmerman AR, et al. Hydrogen peroxide modification enhances the ability of biochar (hydrochar) produced from hydrothermal carbonization of peanut hull to remove aqueous heavy metals: Batch and column tests. Chemical Engineering Journal. 2012;**200**:673-680

[54] Xu X, Cao X, Zhao L. Comparison of rice husk-and dairy manure-derived biochars for simultaneously removing heavy metals from aqueous solutions: Role of mineral components in biochars. Chemosphere. 2013;**92**(8):955-961

[55] Petrović JT, Stojanović MD, Milojković JV, Petrović MS, Šoštarić TD, Laušević MD, et al. Alkali modified hydrochar of grape pomace as a perspective adsorbent of Pb^{2+} from aqueous solution. Journal of Environmental Management. 2016;**182**:292-300

[56] Goswami R, Shim J, Deka S, Kumari D, Kataki R, Kumar M. Characterization of cadmium removal from aqueous solution by biochar produced from Ipomoea fistulosa at different pyrolytic temperatures. Ecological Engineering. 2016;**97**:444-451

[57] Hamid SBA, Chowdhury ZZ, Zain SM. Base catalytic approach: A promising technique for the activation of biochar for equilibrium sorption studies of copper, Cu (II) ions in single solute system. Materials. 2014;7(4):2815-2832

[58] Micháleková-Richveisová B, Frišták V, Pipíška M, Ďuriška L, Moreno-Jimenez E, Soja G. Iron-impregnated biochars as effective phosphate sorption materials. Environmental Science and Pollution Research. 2017;**24**(1):463-475

[59] Agrafioti E, Kalderis D, Diamadopoulos E. Ca and Fe modified biochars as adsorbents of arsenic and chromium in aqueous solutions. Journal of Environmental Management. 2014;**146**:444-450

[60] Saha B, Orvig C. Biosorbents for hexavalent chromium elimination from industrial and municipal effluents. Coordination Chemistry Reviews. 2010;**254**(23-24):2959-2972

[61] Frišták V, Micháleková-Richveisová B, Víglašová E, Ďuriška L, Galamboš M, Moreno-Jimenéz E, et al. Sorption separation of Eu and As from

single-component systems by Fe-modified biochar: Kinetic and equilibrium study. Journal of the Iranian Chemical Society. 2017;**14**(3):521-530

[62] Chen L, Chen XL, Zhou CH, Yang HM, Ji SF, Tong DS, et al. Environmental-friendly montmorillonite-biochar composites: Facile production and tunable adsorption-release of ammonium and phosphate. Journal of Cleaner Production. 2017;**156**:648-659

[63] Zhou Y, Gao B, Zimmerman AR, Fang J, Sun Y, Cao X. Sorption of heavy metals on chitosan-modified biochars and its biological effects. Chemical Engineering Journal. 2013;**231**:512-518

[64] Yang GX, Jiang H. Amino modification of biochar for enhanced adsorption of copper ions from synthetic wastewater. Water Research. 2014;**48**:396-405

[65] Frankel ML, Bhuiyan TI, Veksha A, Demeter MA, Layzell DB, Helleur RJ, et al. Removal and biodegradation of naphthenic acids by biochar and attached environmental biofilms in the presence of co-contaminating metals. Bioresource Technology. 2016;**216**:352-361

[66] Mumme J, Srocke F, Heeg K, Werner M. Use of biochars in anaerobic digestion. Bioresource Technology. 2014;**164**:189-197

[67] Mondal S, Bobde K, Aikat K, Halder G. Biosorptive uptake of ibuprofen by steam activated biochar derived from mung bean husk: Equilibrium, kinetics, thermodynamics, modeling and eco-toxicological studies. Journal of Environmental Management. 2016;**182**:581-594

[68] Jung C, Oh J, Yoon Y. Removal of acetaminophen and naproxen by combined coagulation and adsorption using biochar: Influence of combined sewer overflow components. Environmental Science and Pollution Research. 2015;**22**(13):10058-10069

[69] Xu RK, Xiao SC, Yuan JH, Zhao AZ. Adsorption of methyl violet from aqueous solutions by the biochars derived from crop residues. Bioresource Technology. 2011;**102**(22):10293-10298

[70] Adeel M, Song X, Wang Y, Francis D, Yang Y. Environmental impact of estrogens on human, animal and plant life: A critical review. Environment International. 2017;**99**:107-119

[71] Li Y, Shao J, Wang X, Deng Y, Yang H, Chen H. Characterization of modified biochars derived from bamboo pyrolysis and their utilization for target component (furfural) adsorption. Energy & Fuels. 2014;**28**(8):5119-5127

[72] Uchimiya M, Wartelle LH, Lima IM, Klasson KT. Sorption of deisopropylatrazine on broiler litter biochars. Journal of Agricultural and Food Chemistry. 2010;**58**(23):12350-12356

[73] Mohamed BA, Ellis N, Kim CS, Bi X, Emam AER. Engineered biochar from microwave-assisted catalytic pyrolysis of switchgrass for increasing water-holding capacity and fertility of sandy soil. Science of the Total Environment. 2016;**566**:387-397

[74] Yu X, Pan L, Ying G, Kookana RS. Enhanced and irreversible sorption of pesticide pyrimethanil by soil amended with biochars. Journal of Environmental Sciences. 2010;**22**(4):615-620

[75] Chen B, Chen Z. Sorption of naphthalene and 1-naphthol by biochars of orange peels with different pyrolytic temperatures. Chemosphere. 2009;**76**(1):127-133

[76] Sun K, Ro K, Guo M, Novak J, Mashayekhi H, Xing B. Sorption of bisphenol A, 17α-ethinyl estradiol and phenanthrene on thermally

and hydrothermally produced biochars. Bioresource Technology. 2011;**102**(10):5757-5763

[77] Li G, Zhu W, Zhang C, Zhang S, Liu L, Zhu L, et al. Effect of a magnetic field on the adsorptive removal of methylene blue onto wheat straw biochar. Bioresource Technology. 2016;**206**:16-22

[78] Teixidó M, Pignatello JJ, Beltrán JL, Granados M, Peccia J. Speciation of the ionizable antibiotic sulfamethazine on black carbon (biochar). Environmental Science & Technology. 2011;**45**(23):10020-10027

[79] Mohan D, Sarswat A, Ok YS, Pittman CU Jr. Organic and inorganic contaminants removal from water with biochar, a renewable, low cost and sustainable adsorbent—A critical review. Bioresource Technology. 2014;**160**:191-202

[80] Cao X, Ma L, Gao B, Harris W. Dairy-manure derived biochar effectively sorbs lead and atrazine. Environmental Science & Technology. 2009;**43**(9):3285-3291

[81] Abdelhafez AA, Li J. Removal of Pb (II) from aqueous solution by using biochars derived from sugar cane bagasse and orange peel. Journal of the Taiwan Institute of Chemical Engineers. 2016;**61**:367-375

[82] Oliveira FR, Patel AK, Jaisi DP, Adhikari S, Lu H, Khanal SK. Environmental application of biochar: Current status and perspectives. Bioresource Technology. 2017;**246**:110-122

[83] Lu H, Zhang W, Yang Y, Huang X, Wang S, Qiu R. Relative distribution of Pb^{2+} sorption mechanisms by sludge-derived biochar. Water Research. 2012;**46**(3):854-862

[84] Hsu NH, Wang SL, Liao YH, Huang ST, Tzou YM, Huang YM. Removal of hexavalent chromium from acidic aqueous solutions using rice straw-derived carbon. Journal of Hazardous Materials. 2009;**171**(1-3):1066-1070

[85] Ippolito JA, Strawn DG, Scheckel KG, Novak JM, Ahmedna M, Niandou MAS. Macroscopic and molecular investigations of copper sorption by a steam-activated biochar. Journal of Environmental Quality. 2012;**41**(4):1150-1156

[86] Lima IM, Boateng AA, Klasson KT. Physicochemical and adsorptive properties of fast-pyrolysis bio-chars and their steam activated counterparts. Journal of Chemical Technology & Biotechnology. 2010;**85**(11):1515-1521

[87] Lu K, Yang X, Gielen G, Bolan N, Ok YS, Niazi NK, et al. Effect of bamboo and rice straw biochars on the mobility and redistribution of heavy metals (Cd, Cu, Pb and Zn) in contaminated soil. Journal of Environmental Management. 2017;**186**:285-292

[88] Bogusz A, Oleszczuk P, Dobrowolski R. Application of laboratory prepared and commercially available biochars to adsorption of cadmium, copper and zinc ions from water. Bioresource Technology. 2015;**196**:540-549

[89] Liu P, Ptacek CJ, Blowes DW, Landis RC. Mechanisms of mercury removal by biochars produced from different feedstocks determined using X-ray absorption spectroscopy. Journal of Hazardous Materials. 2016;**308**:233-242

[90] Komkiene J, Baltrenaite E. Biochar as adsorbent for removal of heavy metal ions [Cadmium (II), Copper (II), Lead (II), Zinc (II)] from aqueous phase. International Journal of Environmental Science and Technology. 2016;**13**(2):471-482

[91] Laird D, Fleming P, Wang B, Horton R, Karlen D. Biochar impact on nutrient leaching from a Midwestern agricultural soil. Geoderma. 2010;**158**(3-4):436-442

[92] Hale SE, Alling V, Martinsen V, Mulder J, Breedveld GD, Cornelissen G. The sorption and desorption of phosphate-P, ammonium-N and nitrate-N in cacao shell and corn cob biochars. Chemosphere. 2013;**91**(11):1612-1619

[93] Yao Y, Gao B, Inyang M, Zimmerman AR, Cao X, Pullammanappallil P, et al. Removal of phosphate from aqueous solution by biochar derived from anaerobically digested sugar beet tailings. Journal of Hazardous Materials. 2011;**190**(1-3):501-507

[94] Chen B, Chen Z, Lv S. A novel magnetic biochar efficiently sorbs organic pollutants and phosphate. Bioresource Technology. 2011;**102**(2):716-723

[95] Zhu K, Fu H, Zhang J, Lv X, Tang J, Xu X. Studies on removal of NH_4^+-N from aqueous solution by using the activated carbons derived from rice husk. Biomass and Bioenergy. 2012;**43**:18-25

[96] Jones DL, Rousk J, Edwards-Jones G, DeLuca TH, Murphy DV. Biochar-mediated changes in soil quality and plant growth in a three year field trial. Soil Biology and Biochemistry. 2012;**45**:113-124

[97] Mizuta K, Matsumoto T, Hatate Y, Nishihara K, Nakanishi T. Removal of nitrate-nitrogen from drinking water using bamboo powder charcoal. Bioresource Technology. 2004;**95**(3):255-257

[98] Kalantar-Zadeh K, Gutekunst L, Mehrotra R, Kovesdy CP, Bross R, Shinaberger CS, et al. Understanding sources of dietary phosphorus in the treatment of patients with chronic kidney disease. Clinical Journal of the American Society of Nephrology. 2010;**5**(3):519-530

[99] Olmstead KP, Weber WJ. Interactions between microorganisms and activated carbon in water and waste treatment operations. Chemical Engineering Communications. 1991;**108**:113-125

[100] Bornhardt C, Drewes JE, Jekel M. Removal of organic halogens (AOX) from municipal wastewater by powdered activated carbon (PAC) activated sludge (AS) treatment. Water Science and Technology. 1997;**35**(10):147-153

[101] Jafarinejad S. Activated sludge combined with powder activated carbon (PACT process) for petroleum industry wastewater treatment: A review. Chemistry International. 2017;**3**:268-277

[102] Lehmann J, Joseph S. Chapter 1: Biochar for environmental management: An introduction. In: Biochar for Environmental Management: Science, Technology and Implementation. 2nd edn. Routledge. 2015. ISBN: 9780415704151

[103] Fagbohungbe MO, Herbert BMJ, Hurst L, Ibeto CN, Li H, Usmani SQ, et al. The challenges of anaerobic digestion and the role of biochar in optimizing anaerobic digestion. Waste Management. 2017;**61**:236-249

[104] Sima X, Li B, Jiang H. Influence of pyrolysis biochar on settleability and denitrification of activated sludge process. Chinese Journal of Chemical Physics. 2017;**30**:357-364

[105] Jamieson T, Sager E, Gueguen C. Characterization of biochar-derived dissolved organic matter using UV-visible absorption and

excitation-emission fluorescence spectroscopies. Chemosphere. 2014;**103**:197-204

[106] Luo C, Lü F, Shao L, He P. Application of eco-compatible biochar in anaerobic digestion to relieve acid stress and promote the selective colonization of functional microbes. Water Research. 2015;**68**:710-718

[107] Sunyoto NM, Zhu M, Zhang Z, Zhang D. Effect of biochar addition on hydrogen and methane production in two-phase anaerobic digestion of aqueous carbohydrates food waste. Bioresource Technology. 2016;**219**:29-36

[108] Viggi CC, Simonetti S, Palma E, et al. Enhancing methane production from food waste fermentate using biochar: The added value of electrochemical testing in pre-selecting the most effective type of biochar. Biotechnology for Biofuels. 2017;**10**:303. DOI: 10.1186/s13068-017-0994-7

[109] Cao G-L, Guo W-Q, Wang A-J, Zhao L, Xu C-J, Zhao Q-L, et al. Enhanced cellulosic hydrogen production from lime-treated cornstalk wastes using thermophilic anaerobic microflora. International Journal of Hydrogen Energy. 2012;**37**:13161-13166

[110] Zhang J, Wang Q, Zheng P, Wang Y. Anaerobic digestion of food waste stabilized by lime mud from papermaking process. Bioresource Technology. 2014;**170**:270-277

[111] Cai J, He P, Wang Y, Shao L, Lü F. Effects and optimization of the use of biochar in anaerobic digestion of food wastes. Waste Management and Research. 2016;**34**:409-416

Chapter 15

Application of Biochar for Treating the Water Contaminated with Polar Halogenated Organic Pollutants

Barbora Kamenická, Pavel Matějíček, Tomáš Weidlich and Michael Pohořelý

Abstract

Application of biochar and ionic liquid-impregnated biochar was successfully tested for removal of nonbiodegradable polar halogenated aromatic contaminants (anti-inflammatory agents diclofenac and flufenamic acid and azo dye Mordant Blue 9) from contaminated aqueous solutions. The time dependence of removal efficiencies and adsorption isotherms were evaluated, and the effect of applied ionic liquids (quaternary ammonium salts) was considered. The determined removal efficiencies of the abovementioned contaminants based on the application of biochar or biochar combined with quaternary ammonium salts were compared with the action of commercially available active carbon and/or published results obtained by the action of additional low-cost sorbents. It was demonstrated that a more laborious two-step technique, based on the initial preparation of impregnated biochar by the action of R_4NCl with subsequent application of this modified sorbent, is much less effective than simple mixing of biochar with R_4NCl directly in the treated wastewater solution.

Keywords: drug, water treatment, sorption, diclofenac, flufenamic acid, anionic azo dye, mordant blue 9

1. Introduction

An increase number of nonbiodegradable and often biologically active organic pollutants of anthropogenic origin, such as synthetic dyes and pharmaceuticals, have been detected in the natural environment and wastewaters. The occurrence of these artificial persistent or even biocidal pollutants in aqueous effluents of sewage treatment plants represents possible health hazard not only for the living aquatic organisms but also for terrestrial animals and people [1].

The representatives of these pollutants are polar and highly mobile halogenated aromatic carboxylic (e.g., diclofenac and flufenamic acid) or sulfonic acids and their salts (e.g., Mordant Blue 9 acid dye), respectively (**Table 1**). These ionizable polar compounds (their aqueous solubility strongly depends on the pH of the aqueous solution) are broadly used as remedies or colorants.

Applied contaminant (M_r [g/mol])	Chemical structure
Diclofenac sodium salt (304.1)	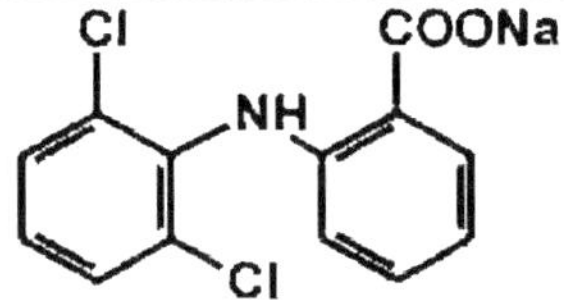
Flufenamic acid sodium salt (303.2)	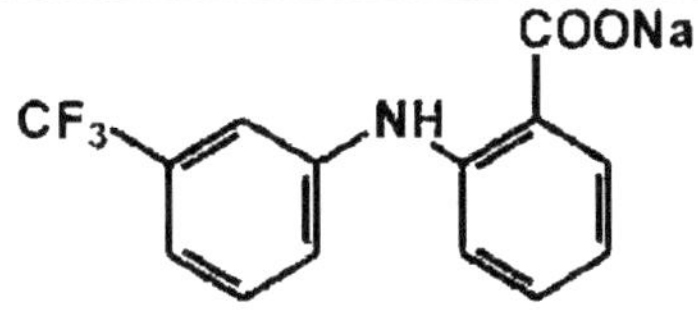
Mordant Blue 9 (551.28)	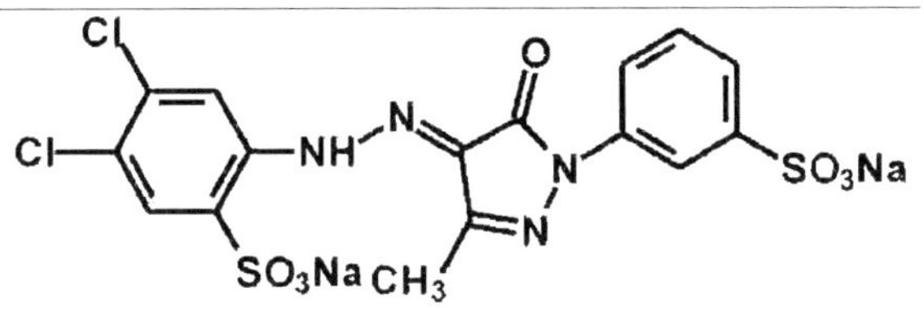

Table 1.
Structures of chlorinated aromatic acid salts.

Specifically in the case of the painkiller diclofenac (DCF), studies have shown that conventional treatment processes are relatively ineffective in removing DCF from water sources and wastewater treatment plants [2]. Kasprzyk-Hordern et al. [2] observed no diclofenac removal in wastewater treatment using trickling filters and activated sludge. Rosal et al. in [3] reported only 5% of DCF removal after biological treatment of urban wastewater.

The concentration of ionizable contaminants, especially of alkaline salts of organic acids, in aqueous mother liquors from the production of these chemicals is frequently high, especially in the case of acid dyes (including Mordant Blue 9) isolated from the aqueous reaction medium by the so-called salting-out process [4]. Adsorption serves as the common and broadly used technique applicable for the treatment of water streams contaminated with these nonbiodegradable pollutants. The adsorption technique requires, however, high quantities of sorbent for effective water treatment which is accompanied by significant material costs. Therefore, carbonaceous rest, obtained by pyrolysis of waste biomass (biochar), potentially offers possible utilization in water treatment processes based on sorption as an alternative low-cost sorbent in comparison with activated carbon. For biochar, a porous structure with a sufficient specific area (above 400 m^2/g) is typical. Biochar poses polyaromatic systems substituted with some functional groups on its surface (COOH, OH, C=O) and even inorganic impurities (metal oxides) capable of engaging in **hydrogen-bonding or π-π** interactions with polar compounds.

Another notable attribute is biochar's affinity with the cationic surfactants (quaternary ammonium salts) [5].

This ability is of great interest for us due to the feasible utilization of cationic surfactants as liquid ion exchangers for chemisorption of the abovementioned chlorinated organic acid salts [6]. Ionizable halogenated contaminants produce ion pairs with only limited aqueous solubility and a good affinity with the biochar by the action of quaternary ammonium salts [7, 8].

The price of biochar is reasonably lower (around 1000 EURO/t) in comparison with active carbon (the cost of active carbon is approximately 2000 EUROs/t [9]).

Due to the abovementioned reasons, good cationic surfactant affinity of biochar seems to be very useful for uptake of ionizable contaminants using combined ion exchange (chemisorption) and the sorption mechanism.

2. Origin of the tested biochar

Gasification is a thermochemical process that converts a (waste) woody biomass into a producer gas and ash or carbonaceous residue (char). It can be used for combined heat and power production. The carbonaceous residue of the gasification process (called char) can match requirements classified as biochar [10]. The quality of both biochar and producer gas is highly dependent on the gasification technology selected, most of all, on the design and operation parameters of the gasifier.

Gasification is a complex process that can be divided into four basic stages. The first stage is drying. The second stage, taking place in the absence of oxygen and at temperatures above 250°C, is called pyrolysis. The waste woody biomass is converted into volatile matter and a carbonaceous residue (char) during the pyrolysis. The third stage is exothermic partial oxidation of the char and the volatile matter with an oxidizing agent (basically air). This exothermic process produces heat. The last (fourth) stage, called char gasification, is the sum of the endothermic reactions of char with CO_2 and/or H_2O driven by the heat from the third stage leading to the additional formation of H_2 and CO.

The space separation of these four stages in a gasifier with proper process control can be used to optimize the outputs of the gasification process. The space separation and optimization of these four stages of the gasification process are the principle of the so-called multistage gasification. One concept used for multistage gasification is a twin-fire gasifier. The gas generator at the commercial heating plant (**Figure 1**)

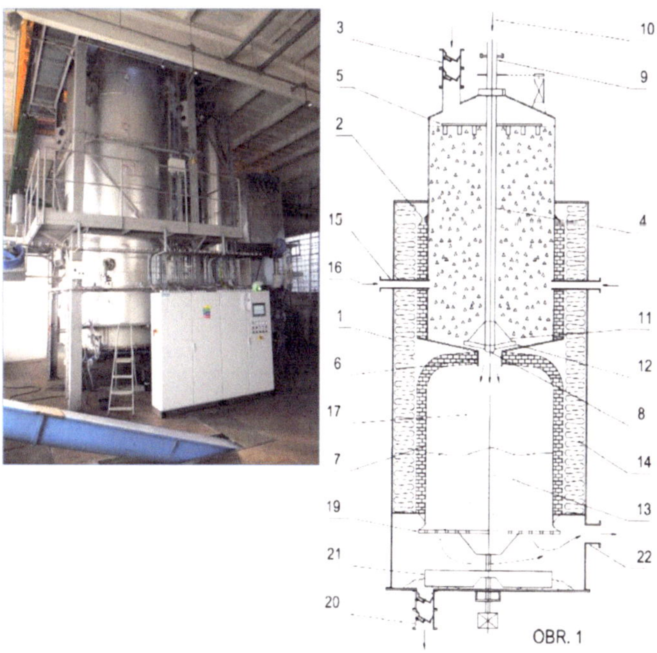

Figure 1.
Commercial twin-fire Gasifier [1].

is built on this principle [11]. This type of gasifier is composed of two autothermic stages, each with a separate air intake, creating two separate "fires," therefore "twin-fire." The first air intake used for the partial combustion of the waste biomass is directed into the upper stage of the gasifier and produces the heat necessary for "autothermal" pyrolysis. The secondary air inlet, for partial oxidation of volatile matter, is directed to the top of the lower part of the gasifier, and the temperature of its "fire" exceeds 1200°C. Partial oxidation takes place in the free space above the char bed and produces heat for char gasification at the bottom of the second stage of the gasifier, reaching a temperature of approximately 950°C. The partial oxidation stage cracks down volatile matter into gases (i.e., CO, H_2, CH_4, CO_2, and H_2O) and thereby minimalizes the tar content. In the last stage, the hot products of the partial oxidation flow through the char bed and react with it, causing a temperature drop from 950 to 750°C due to the endothermic gasifying reactions. Optimization of this step can enhance the properties of the produced solid carbonaceous residue (biochar) by increasing its specific surface and due to its "activation" by the reactions with H_2O and CO_2. Moreover, the prolonged residence time (about 1 h) of the char at a high temperature of 750°C helps to achieve the stringent requirements on the content of polycyclic aromatic hydrocarbons in the produced biochar. For safety measures (to prevent producer gas leakage), the whole gasification unit and gas

Characteristics	Value
Bulk density (g/mL)	166
Specific area, S_{BET} (m^2/g)	444
Specific total pore volume, V_{tot} (mm^3_{liq}/g)	293
Specific micropore volume, V_{micro} (mm^3_{liq}/g)	157
Specific mesopore volume, V_{meso} (m^2/g)	142
pH	11.4
Electrical conductivity, EC (μS/cm)	1450
Ash (wt. %)	9.63
Carbon content, C (wt. %)	86.8
Organic carbon content, C (wt. %)	83.8
H/C_{org} ratio	0.0890
O/C ratio	0.0205
Sulfur content, S (wt. %)	Less than 0.1
Content of combustible sulfur (mg/kg)	401
Chlorine content, Cl (mg/kg)	867
Fluorine content, F (mg/kg)	11.4
Content of P (g/kg)	0.65
Content of K (g/kg)	3.1
Content of Ca (g/kg)	15.7
Content of Mg (g/kg)	2.5
Sum PAH_{16} (mg/kg)	Less than 0.5
Water content (wt. %)	1.18

Table 2.
Characteristics of the produced biochar.

Inorganic oxide	Content of noncombustible part (wt. %)
Al_2O_3	8.89
CaO	32.6
Fe_2O_3	6.74
K_2O	6.69
MgO	5.78
MnO	1.20
Na_2O	4.49
P_2O_5	2.21
SiO_2	20.9
TiO_2	2.93
Total amount	92.5

Table 3.
Composition of noncombustible matter (ash) in the produced biochar according to XRF.

treatment line are operated at pressure slightly below atmospheric pressure (pressure difference—0.1–10 kPa) ensured by a fan (ventilator). The input material to the gasification generator is spruce wooden chips (waste biomass produced from spent packaging and pallets) with a water content max. 10 wt.%. The produced biochar (**Tables 2** and **3**) meet the European Biochar Certificate (EBC) standard of regulation (EU) 2019/1009 and the Central Institute for Supervising and Testing in Agriculture (Czech Certification Institute, the certification valid in the EU). The produced biochar can be grinded and is conveyed into a magnetic separator to eliminate any possible remaining metal particles. Lian and Xing published that increasing pyrolysis temperature above 700°C results in high aromaticity and porosity of obtained biochars structure, high carbon content, and progressive decreasing of nitrogen and oxygen content in produced biochar [12]. The described biochar produced in twin-fire gasifier is a typical high-temperature biochar with high carbon content, increased pH value, high porosity, increased ash content, and specific surface area (**Table 2**) [10]. As could be seen in **Tables 2** and **3**, elemental composition of produced highly porous biochar is mainly composed of C and very low contents of H, N, and O, with significant content of minerals. In comparison with starting wooden biomass, significant decrease in molar ratios of both H/C and O/C is observed in biochar due to the dehydration and thermolysis reactions caused by temperature above 700°C.

3. Preliminary experiments comparing sorption kinetics for removal of ionizable halogenated contaminants using biochar and quaternary ammonium chlorides in batch sorption

The above described biochar seems to be an interesting candidate for utilization in sorption processes due to the high porosity and specific surface comparable with active carbon (**Table 2**). This study is focused on the removal of ionizable organic compounds (halogenated organic acid salts) mentioned in the introduction chapter (**Table 1**). Due to the abovementioned high aqueous solubility of the studied sodium salts of chlorinated aromatic acids, the preliminary experiments were performed by the addition of high quantity (20 g/L) of the above-described biochar to the concentrated aqueous solutions of contaminants simulating effluents from industrial production sites. These preliminary experiments indicated that the maximum efficiency of contaminant removal was obtained after more or less than 90 minutes of biochar action in batch sorption under vigorous stirring (**Figure 2**). The removal efficiency for each contaminant reached more than 40% in all cases after 1 h of action. This means that these efficiencies are quite low even using this huge quantity of biochar. These results correspond with known high polarity of tested fully ionizated contaminants at pH above 8.5 and their low affinity to the low-polar surface of biochar. In contrast, however, it was published earlier that the addition of cationic surfactants to the wastewater contaminated with soluble organic acid salts can significantly improve removal efficiency due to the electrostatic attraction between negatively charged contaminant anions and positively charged cations of cationic surfactants [7, 8, 13].

Subsequently, possible enhancement of the removal efficiency of biochar caused by co-action with a cationic surfactant was tested. The sorption kinetics and removal efficiencies of the contaminant based on chlorinated carboxylic acid salts, biochar, and biochar in combined action with frequently used cationic surfactant quaternary ammonium salt (R_4NCl) benzalkonium chloride (alkylbenzyldimethylammonium chloride, $AlkBzMe_2NCl$) for removal of diclofenac sodium salt (NaDCF) and of flufenamic acid sodium salt (NaFLUFA) were compared.

The results for the removal rates of NaDCF and NaFLUFA are mentioned in **Figure 3**. As is apparent, after 30 min of action, the removal of NaDCF and NaFLUFA from aqueous solutions is completed. Whereas the application of sole biochar in quantity 20 g/L reduces the quantity of NaDCF (or NaFLUFA, respectively) with efficiency less than 45%, using a combination of cationic surfactant and biochar, the removal efficiency reaches over 65%. It should be said that the efficiency of NaDCF or NaFLUFA removal using sole cationic surfactant $AlkBzMe_2NCl$ without the addition of charcoal is much worse, below 34%. This observation could be explained by the known fact that NaDCF reacts smoothly with low-polar R_4NCl

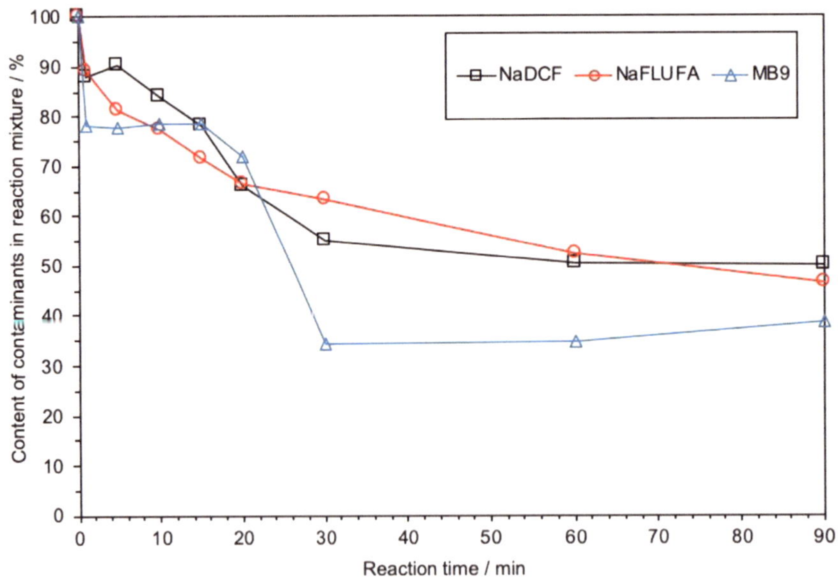

Figure 2.
Rate of removal of tested ionizable halogenated contaminants from 10 mM aqueous solutions (from 25 mM solution in case of NaDCF) using biochar in quantity 20 g/L.

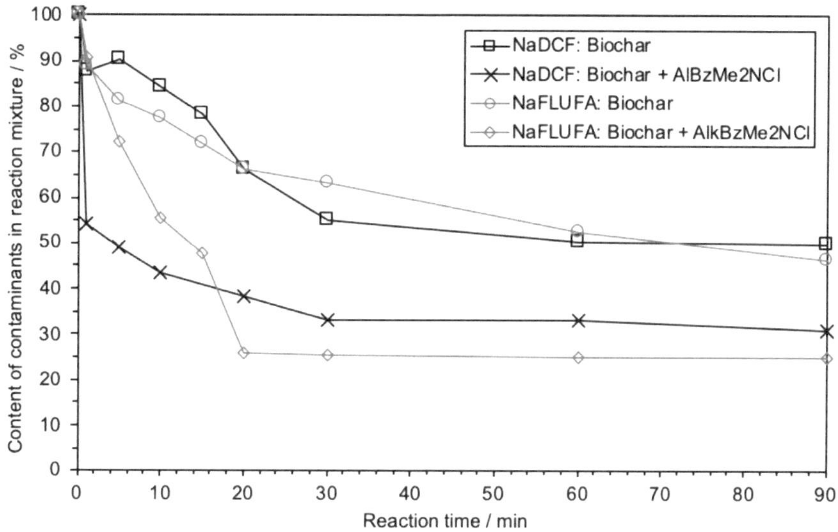

Figure 3.
Rate of NaDCF (25 mM aq. solution) and NaFLUFA (10 mM aq. solution) removal from aqueous solutions using biochar (20 g/L) or a combination of AlkBzMe2NCl (1 g/L) with biochar (20 g/L).

by ion exchange reaction producing high molecular (and less soluble) ion pairs R_4N.DCF (R_4N.FLUFA, respectively) according to the scheme:

Contaminant-COONa + R_4NCl ——> NaCl + Contaminant-COONR_4 (ion-pair).
NaDCF + R_4NCl ——> NaCl + R_4N.DCF (ion pair).
NaFLUFA + R_4NCl ——> NaCl + R_4N.FLUFA (ion pair).

Chemical structures of tested R_4NCls are depicted in **Figure 4**. The effect of cation size of different cationic surfactants R_4NCl (AlkBzMe_2NCl, hexadecyltrimethylammonium chloride (AlkMe_3NCl), and methyltrialkylammonium chloride, Aliquat 336) on removal efficiency of NaDCF and NaFLUFA is depicted in **Figures 5** and **6**. It is evident that the branched structure (quantity of long alkyl chains) of the used cationic surfactants and primarily the aqueous solubility of the produced ion pairs based on tetraalkylammonium salts R_4N^+ play an important role in the removal efficiencies of the studied contaminants. This fact could be well described by comparison of the solubility of discussed pollutants and corresponding ion pairs in water and in low-polar solvent (octan-1-ol) using distribution coefficient between these two solvents (**Figure** 7, Eq. (1)). Octan-1-ol/water distribution ratio is the most common way of expressing the lipophilicity of a compound, and it is defined as the ratio of the concentration of a solute in a water-saturated octanolic phase to its concentration in an octanol-saturated aqueous phase. $\mathbf{P}_{ow}$ is defined as in Eq. (1) [14]:

$$\log P_{ow} = \frac{c_{octanol}}{c_{aqueous\ phase}} \quad (1)$$

The observed removal efficiencies of the co-action of different R_4NCl with biochar (**Figure 5**) correspond well with the measured distribution coefficients between octan-1-ol and water of NaDCF and ion pairs R_4N.DCF produced by the ion exchange of NaDCF with R_4NCl, as could be demonstrated in **Figure** 7. The less hydrophobic quaternary cation produces less hydrophobic and more water soluble ion pair contaminant-COONR_4 (with lower value of log P_{OW},

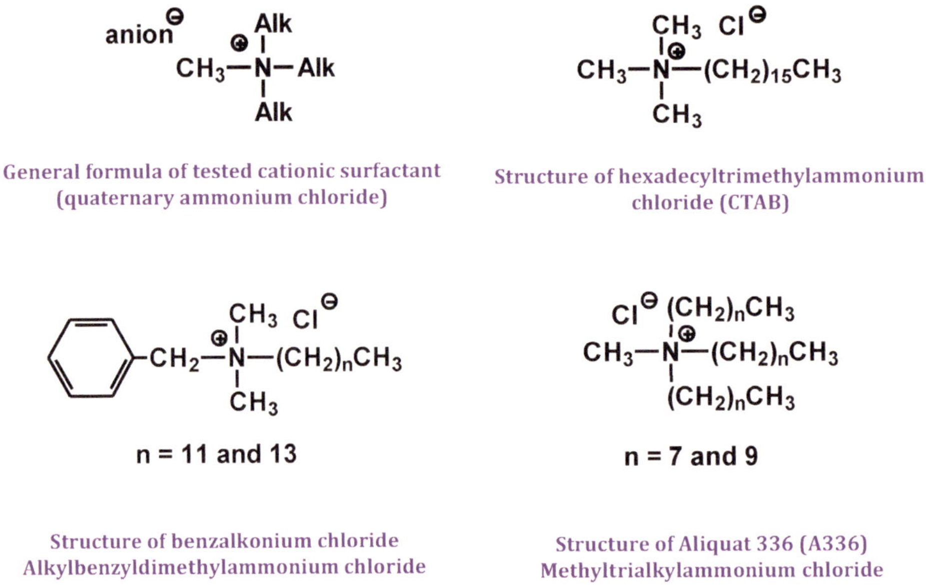

Figure 4.
Chemical structures of tested cationic surfactants (quaternary ammonium chlorides, R4NCls) with different quantity of long alkyl chains (different cation size).

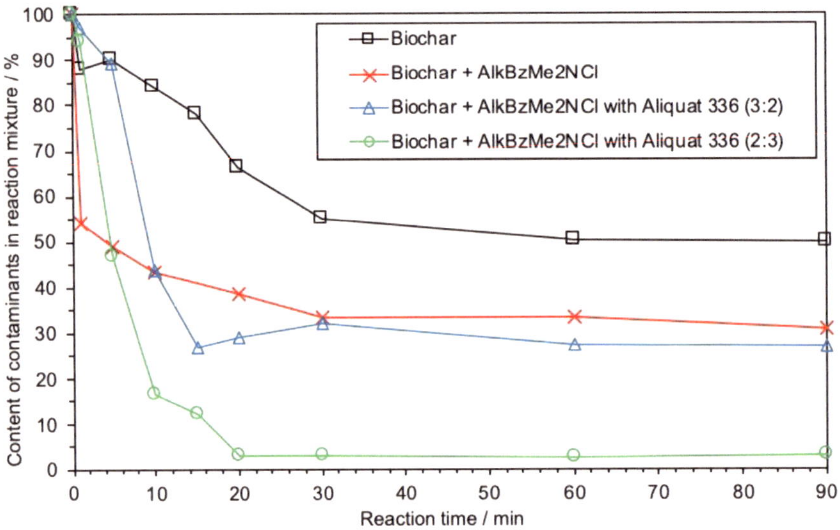

Figure 5.
Rate of NaDCF removal from 25 mM aq. Solution using biochar (20 g/L) or a combination of different R4NCls (1 g/L) and biochar (20 g/L).

Figure 5). Benzalkonium chloride ($AlkBzMe_2NCl$) which is readily soluble in water enables, for example, the worse removal of NaDCF in comparison with in-water insoluble Aliquat 336 (albeit the combination of the ion exchange with adsorption is still more effective in comparison with adsorption on sole biochar).

The highest value of log P_{ow} was determined for low-polar and water immiscible ion pairs produced by the action of Aliquat 336 (A336). In contrast, A336 is extremely viscous (honey-like) material and together with its low aqueous

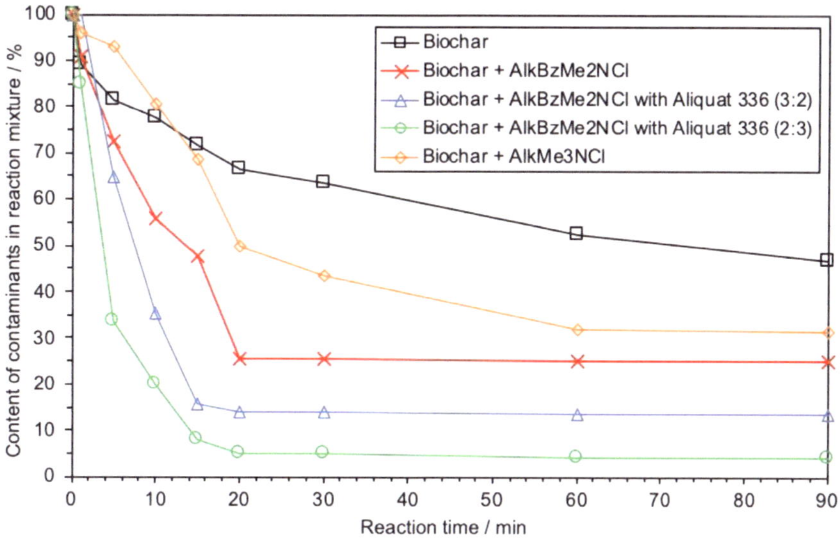

Figure 6.
Rate of NaFLUFA removal from 10 mM aqueous solution using biochar (20 g/L) or a combination of different R4NCls (1 g/L) and biochar (20 g/L).

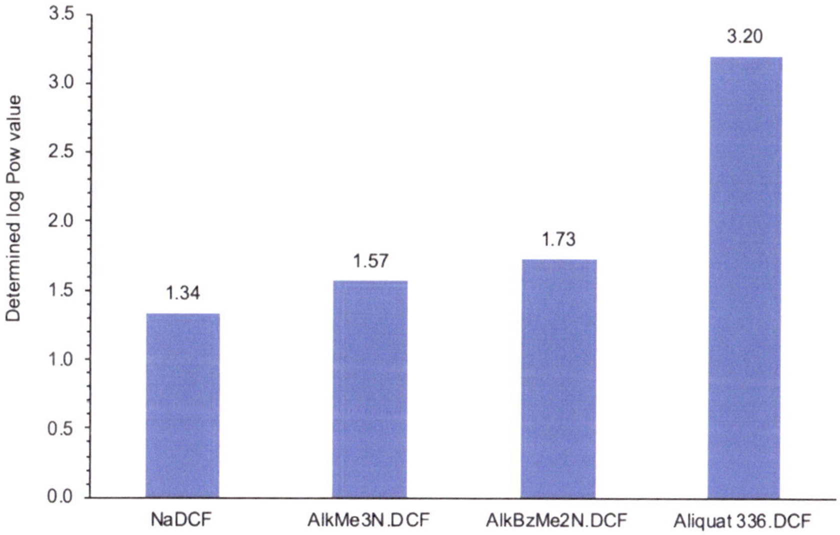

Figure 7.
Differences between the measured distribution coefficients log Pow between NaDCF and ion pairs produced by ion exchange of NaDCF with the corresponding R4NCl (R4N.DCF).

solubility is non-applicable for precise addition of appropriate quantity to treated wastewater. We tested, however, that A336 is well soluble in 50 wt.% aqueous $AlkBzMe_2NCl$ solution and the obtained mixture is less viscous and enables precise addition of R_4NCls into the stirred aqueous solution. Due to these reasons, the mixtures of A336 and 50 wt.% aqueous $AlkBzMe_2NCl$ in two different weight ratios (2/3 or 3/2) were examined (see **Figures 5** and **6**). The low solubility of ion pairs A336.DCF produced by ion exchange reaction between A336 with NaDCF enables in particular extremely effective subsequent removal from the aqueous solution by

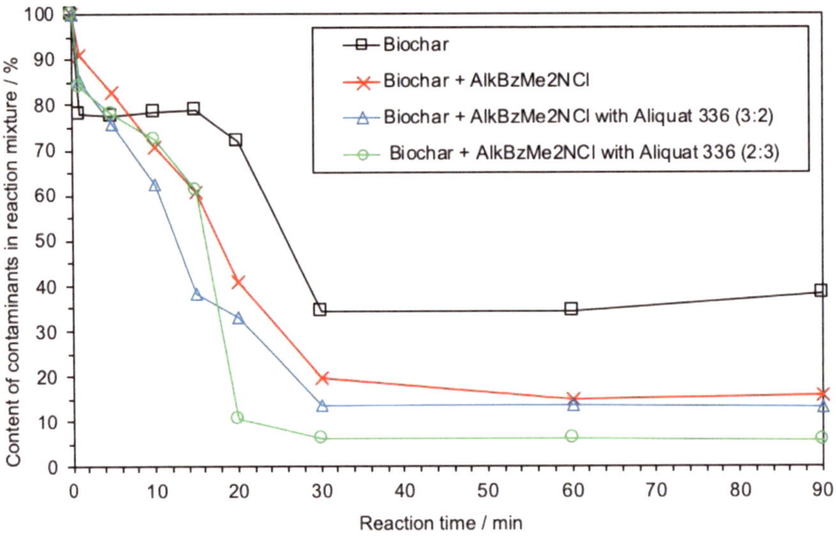

Figure 8.
Rate of mordant blue 9 removal from 10 mM aqueous solution using biochar (20 g/L) or a combination of different R4NCls (1 g/L) and biochar (20 g/L).

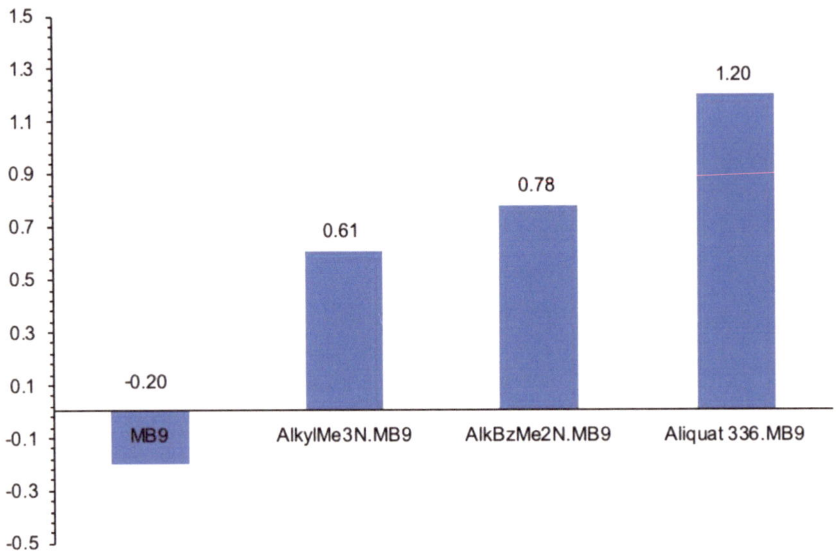

Figure 9.
Differences of measured distribution coefficients log Pow between commercial textile dye MB9 and ion pairs produced by ion exchange of MB9 with corresponding R4NCl.

the addition of biochar. The surface of biochar sorbs produced A336.DCF better than AlkBzMe2N.DCF, as could be seen in **Figure 5**.

In accordance with the above described facts, the best removal efficiency of NaFLUFA was obtained using a combination of mixture of 3/2 (w/w) of A336 and 50% aq. $AlkBzMe_2NCl$. The lowest effect of $AlkMe_3NCl$ on the removal of NaFLUFA corresponds, in contrast, with the least branched structure of $AlkMe_3N^+$ cation (**Figure 6**).

The same relationship between the structure of the used cationic surfactants and removal efficiency was observed in the case of the anionic textile dye Mordant

Blue 9 (MB9). Subsequently, the measured log P_{ow} values for MB9 and ion pairs R_4N.MB9 are in good agreement once again with the measured removal efficiencies (**Figures 8** and **9**).

4. A comparison of adsorption isotherms measured for removal efficiencies of diclofenac or flufenamic acid sodium salts using active carbon, biochar, and biochar in co-action with RNX

The abovementioned differences in removal capacity of active carbons, biochar, modified biochar, and biochar in co-action of RNCls possessing quaternary cations with different bulkiness are illustrated in **Figure 10** for removal of NaDCF. The tested sorbents (active carbons or biochar) were used in quantity 10 g/L (with appropriate co-action of 1 g/L of R_4NCl(s)).

Figure 10 illustrates that the sorption capacity (*q*) showed the following behavior: PAC > [Biochar + $AlkBzMe_2NCl$ with Aliquat 336 (2:3)] > [Biochar + $AlkBzMe_2NCl$ with Aliquat 336 (3:2)] > [Biochar + $AlkBzMe_2NCl$] > GAC > Biochar > [Modified biochar $AlkBzMe_2NCl$].

The worst removal capacity poses modified biochar$AlkBzMe_2NCl$ prepared by mixing biochar with aqueous $AlkBzMe_2NCl$ solution, subsequent washing with water, and drying [15] (for more details see Experimental section) probably due to the low concentration of $AlkBzMe_2N$-cations immobilized on surface of prepared modified biochar. **Figure 10** compared the effect of the addition of highly hydrophobic A336 added in different quantities to the hydrophilic 50% aq. $AlkBzMe_2NCl$ on the sorption capacity of biochar/ R_4NCl mixture (which means sorption capacity of in situ-prepared biochar modified with added R_4NCls). In agreement with the abovementioned effect of different R_4NCls, using a higher quantity of A336 enables an increase of sorption capacity of biochar after the addition of R_4NCls. In addition, the comparison of the effectiveness of traditional charcoal (granulated

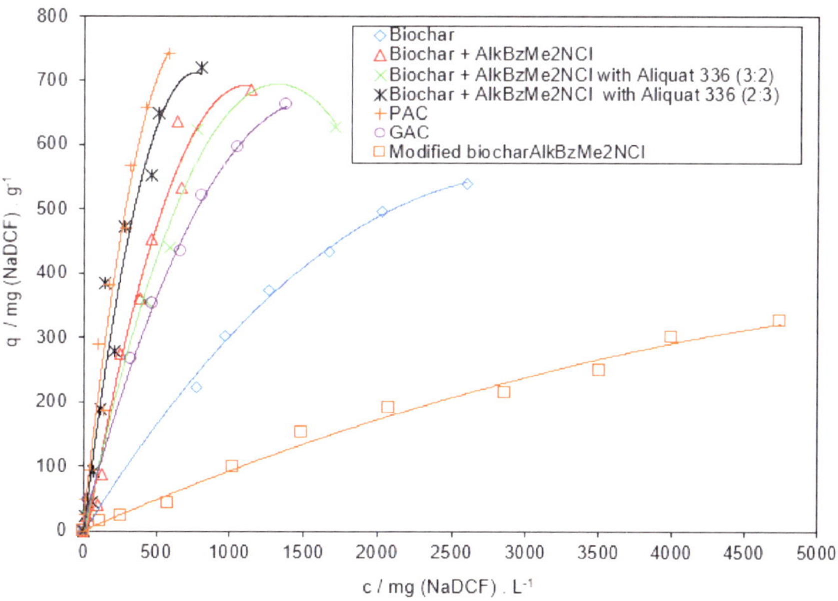

Figure 10.
The dependence of sorption capacity of powdered (PAC) and granulated (GAC) active carbons, biochar, modified biochar and biochar in co-action with R4NCls on the equilibrium concentration of the NaDCF solution.

Hydraffin CC8x30 GAC and powdered PAC Silcarbon CW20) and tested biochar and biochar with the co-action of the most effective mixture of cationic surfactants $AlkBzMe_2NCl$ with A336 is depicted in **Figure 10** using initial NaDCF concentration 0.25–8 g /L. It is evident that the combination of the aqueous surfactants mixture 50% aq. $AlkBzMe_2NCl$ and A336 in weight ratio 2/3 (used in quantity 1 g/L) with biochar (10 g/L) exhibits a similar sorption capacity as powdered active carbon Silcarbon CW20 (10 g/L) and a higher sorption capacity than granulated active carbon Hydraffin CC8x30 (10 g/L).

Similarly, studying removal efficiencies of NaFLUFA (initial concentration 0.25–7 g FLUFA/L) using active carbons (10 g/L), biochar (10 g/L), modified biochar (10 g/L), and biochar (10 g/L) with co-action of R_4NCls (1 g/L), we observed that the activity of these sorbents was similar to the abovementioned removal of NaDCF (**Figure 11**). The sorption capacity has the rank order PAC mixture Aliquat 336 in 50% aq., $AlkBzMe_2NCl$ 3/2 with biochar mixture Aliquat 336 in 50% aq., and $AlkBzMe_2NCl$ 2/3 with biochar mixture of $AlkBzMe_2NCl$ with biochar $AlkBzMe_2NCl$ biochar. This similarity with NaDCF removal is not surprising; the chemical structures of both NaDCF and NaFLUFA are very similar (see **Table 1**). On the other hand, the sorption experiments using NaDCF and NaFLUFA were performed at different pH due to the low aqueous solubility of FLUFA at pH bellow 10. High removal efficiencies of A336/$AlkBzMe_2NCl$ mixtures with biochar even at high pH values are in agreement with our observation and the observation of Kosaiyakanon that the effect of pH is not crucial using separation method based on the formation of ion pairs [7, 8, 15].

Sorption capacity q (mg/g) was calculated according to the following Eq. (2) [15]:

$$q = \frac{(c_0 - c).V}{m} \tag{2}$$

where c_0 is the initial concentration (mg/L), c is the equilibrium concentration (mg/L), m is the mass of biochar (g), and V is the volume of treated model wastewater (L).

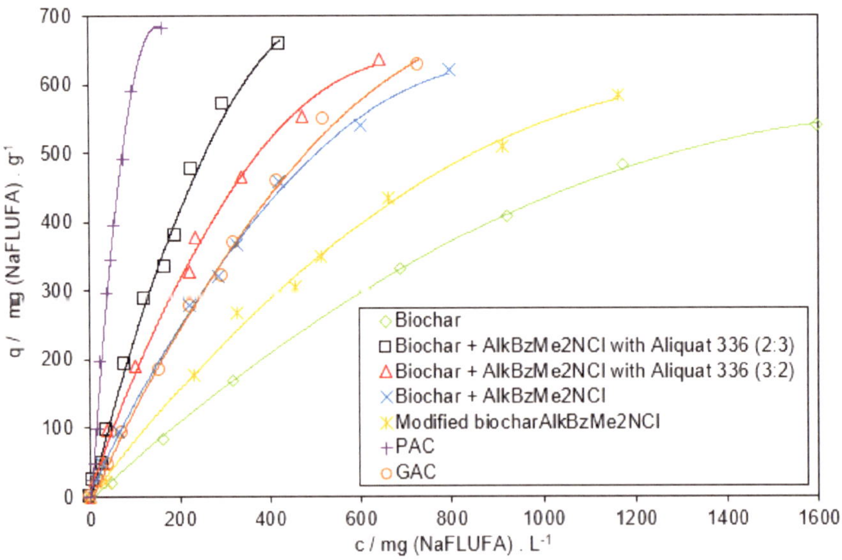

Figure 11.
The dependence of sorption capacity of biochar, in-situ modified biochar two types of active carbons (powdered (PAC) Silcarbon CW20 and granulated (GAC) Hydraffin CC8x30 on the equilibrium concentration of the NaFLUFA solution).

Langmuir and Freundlich isotherm models were fitted to the data. The Langmuir model describes monolayer adsorption on a homogenous surface. The Freundlich model describes multilayer adsorption on a heterogeneous surface. The Freundlich and Langmuir models are expressed in Eqs. (3) and (4), respectively:

$$q = k_F c^{1/n} \quad (3)$$

$$\frac{q}{q_{max}} = \frac{k_L c}{1 + k_L c} \quad (4)$$

Here, q is the amount of adsorbed contaminant on the adsorbent at equilibrium (mg/g), q_{max} is the maximum adsorption of contaminant on the adsorbent (mg/g), c is the residual contaminant concentration at equilibrium (mg/L), k_L is the Langmuir constant related to the energy of adsorption (L/mg), k_F is the Freundlich constant indicating the adsorption capacity, and n is the Freundlich exponent accounting for the adsorption intensity or the energetic heterogeneity of the adsorbing surface [15].

The correlation coefficients suggest that the Freundlich model fits the data better than the Langmuir model (**Tables 4** and **5**). This can be an indication that NaDCF and NaFLUFA and/or ion pairs R_4N.DCF (R_4N. FLUFA, respectively) were adsorbed in multilayers into the active sites of the biochar surface.

The maximum contaminant sorption capacities (q_{cont}) of the biochar, both tested active carbons, and biochar with co-action of R_4NCls at final pH of 8.7 were determined according to the Freundlich model (calculated for maximum applied concentration of contaminant using Eq. (3)). The differences in the action of sole biochar and biochar modified by different cationic surfactant(s) could be summarized by the ratio of increasing sorption capacity q_{cont}/q_{cont}^* [6] (**Table 6**).

For the tested active carbons, the q_{cont} values obtained by the same method were 661.6 mg NaDCF/g for granulated active carbon Hydraffin CC 8x30 and 742.3 mg NaDCF/g for powdered active carbon Silcarbon CW20.

Interestingly, in all the cases, modified biochar prepared independently (ex situ) by impregnation of biochar with aqueous solution of R_4NCls exhibited lower activity than the sequential addition of biochar and cationic surfactant(s) to the aqueous

	Freundlich			Langmuir		
Sorbent	**k_F (mg/g (L/mg)$^{1/n}$)**	**n**	**R^2**	**q_{max} (mg/g)**	**k_L (L/mg)**	**R^2**
PAC	2.14	1.03	0.954	833.3	0.0075	0.862
GAC	1.85	1.16	0.949	714.2	0.0034	0.811
Biochar	0.98	1.23	0.996	555.6	$6.6\ 10^{-4}$	0.886
Modified biocharAlkBzMe$_2$NCl	1.95	0.73	0.975	400.0	$5.3\ 10^{-4}$	0.714
Biochar + AlkBzMe$_2$NCl	1.25	1.01	0.978	1250.0	0.0012	0.865
Biochar +2/3 A336/ 50% aq. AlkBzMe$_2$NCl	1.22	0.93	0.986	1428.6	$5.7\ 10^{-4}$	0.418
Biochar +3/2 A336/ 50% aq. AlkBzMe$_2$NCl	1.12	0.99	0.983	1111.1	0.0022	0.577

Table 4.
Identified parameters in Langmuir and Freundlich isotherm models for NaDCF (used in initial concentration range 0.25–8 g/L) adsorbed by sorbents.

	Freundlich			Langmuir		
Sorbent	**k_F (mg/g (L/mg)$^{1/n}$)**	**n**	**R^2**	**q_{max} (mg/g)**	**k_L (L/mg)**	**R^2**
PAC	2.06	0.77	0.949	1000.0	0.010	0.702
GAC	1.13	1.01	0.997	833.3	0.0029	0.634
Biochar	0.85	1.11	0.987	666.7	0.0018	0.766
Modified biocharAlkBzMe$_2$NCl	0.95	1.07	0.996	714.3	0.0023	0.728
Biochar + AlkBzMe$_2$NCl	1.62	1.07	0.982	833.3	0.0027	0.770
Biochar +2/3 A336/ 50% aq. AlkBzMe$_2$NCl	1.88	1.04	0.968	909.1	0.0032	0.748
Biochar +3/2 A336/ 50% aq. AlkBzMe$_2$NCl	1.84	0.97	0.969	909.1	0.0044	0.635

Table 5.
Identified parameters in Langmuir and Freundlich isotherm models for NaFLUFA (used in initial concentration range 0.25–7 g/L) adsorbed by sorbents.

Contaminant	Sorbent	q_{cont} (mg/g)	Ratio of increasing capacity (q_{cont}/q_{cont}^*)
NaDCF	Biochar	539.5*	—
	Modified biocharAlkBzMe$_2$NCl	325.8	0.604
	Biochar + AlkBzMe$_2$NCl	682.2	1.264
	Biochar+2/3 A336/50% aq. AlkBzMe$_2$NCl	628.5	1.165
	Biochar +3/2 A336/50% aq. AlkBzMe$_2$NCl	719.9	1.334
NaFLUFA	Biochar	540.3*	—
	Modified biocharAlkBzMe$_2$NCl	514.6	0.952
	Biochar + AlkBzMe$_2$NCl	620.4	1.148
	biochar+2/3 A336/50% aq. AlkBzMe$_2$NCl	635.9	1.177
	biochar +3/2 A336/50% aq. AlkBzMe$_2$NCl	657.8	1.217

**Value of sorption capacity defined as q_{cont}^**

Table 6.
A comparison of surface modification technique on increasing sorption capacity of chlorinated aromatic carboxylic acid sodium salts.

solution contaminated with NaDCF (**Figure 10**). This could be explained by the possible parallel action of:

1. Rapid ion exchange reaction between the added R_4NCl and NaDCF accompanied by coagulation of the produced ion pairs $R_4N.DCF$ (this reaction was proved by isolation of mentioned $R_4N.DCF$ ion pairs by extraction and by subsequent NMR analysis)

2. The effect of polyvalent metal cations from inorganic components of biochar on additional precipitation of insoluble DCF salts (with composition $M^{+n}.(DCF)_n$) [12] (M^{+n} = Ca^{II}, Mg^{II}, Al^{III}, Fe^{III}, etc., see the content of minerals in

biochar in **Tables 2** and **3**) (effect of the well-known insolubility of DCF salts with the mentioned polyvalent metal cations)

3. The effect of negative charge-assisted H-bonds which were published as the main mechanism for sorption of ionizable organic compounds at alkaline pH area (pH above zero point of charge of biochar [12, 16, 17])

4. Subsequently the high affinity of biochar to the produced ion pair $R_4N.DCF$ [15] (caused by decreasing of polarity of produced $R_4N.DCF$ ion pairs in comparison of NaDCF with subsequent increasing of their affinity to surface of tested biochar, as we were observed)

The abovementioned results demonstrated that using a two-step procedure to enhance the biochar's adsorption capacity using $AlkBzMe_2NCl$ as a modification agent is not effective and is more laborious in comparison with the addition of a sole biochar.

5. A comparison of adsorption isotherms measured for removal efficiencies of textile dye mordant blue 9 using active carbons, biochar, and biochar in co-action with R4NX

In order to test the removal potential of active carbons (10 g/L), biochar (10 g/L), and biochar (10 g/L) with co-action of R_4NCls (1 g/L), broad concentrations (0.25–5.5 g/L) of MB9 in model aqueous solutions were chosen for the performed experiments, similar to the tested drugs. It is well-known that anionic dyes are efficiently removable from wastewater using adsorption on charcoal at low pH due to the suppression of their ionization. As the pH of the mixture increases,

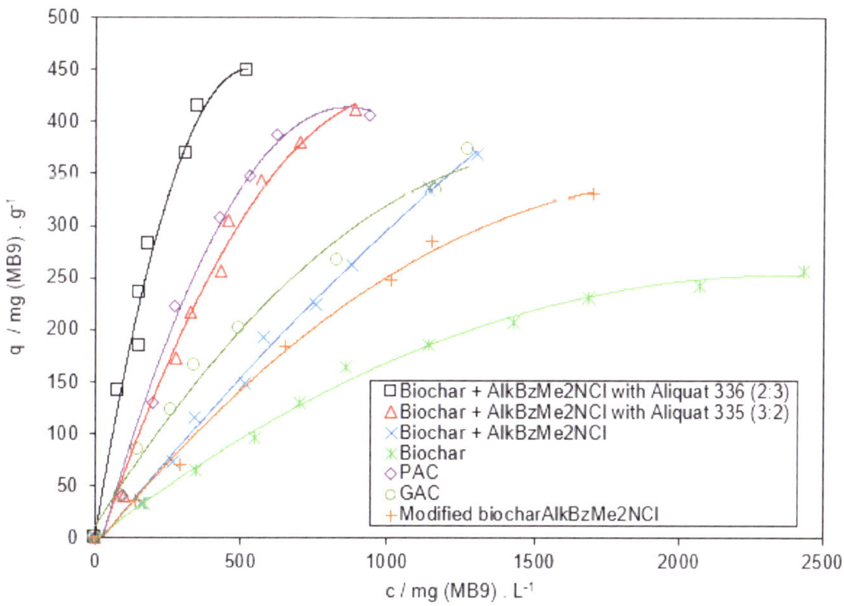

Figure 12.
The dependence of sorption capacity of powdered (PAC) and granulated (GAC) active carbons, biochar, modified biochar and biochar in co-action with R4NCls on the equilibrium concentration of the MB9 solution.

Contaminant	Sorbent	q_{cont} (mg/g)	Ratio of increasing capacity (q_{cont}/q_{cont}^*)
MB9	Biochar	257.1*	—
	Modified biocharAlkBzMe_2NCl	329.9	1.283
	Biochar + AlkBzMe_2NCl	369.6	1.437
	biochar+2/3 A336/50% aq. AlkBzMe_2NCl	411.2	1.599
	biochar +3/2 A336/50% aq. AlkBzMe_2NCl	448.2	1.743

**Value of sorption capacity defined as q_{cont}^**

Table 7.
A comparison of surface modification technique on increasing sorption capacity of MB9.

	Freundlich			Langmuir		
Sorbent	k_F (mg/g (L/mg)$^{1/n}$)	n	R^2	q_{max} (mg/g)	k_L (L/mg)	R^2
PAC	0.91	1.08	0.984	500.0	0.0042	0.772
GAC	1.03	1.17	0.995	526.3	0.0015	0.611
Biochar	0.96	1.36	0.995	384.6	$8.2\ 10^{-4}$	0.739
Modified biocharAlkBzMe_2NCl	0.90	1.24	0.995	357.1	0.0032	0.611
Biochar + AlkBzMe_2NCl	0.85	1.20	0.986	476.2	0.0014	0.550
Biochar +2/3 A336/ 50% aq. AlkBzMe_2NCl	0.88	1.09	0.985	500.0	0.0035	0.742
Biochar +3/2 A336/ 50% aq. AlkBzMe_2NCl	1.19	0.99	0.985	555.6	0.0058	0.768

Table 8.
Identified parameters in Langmuir and Freundlich isotherm models for MB9 (used in initial concentration range 0.5–5.5 g/L) adsorbed by sorbents.

the formation of negatively charged MB9 anions increases due to the ionization. The surface site of charcoal does not favor the adsorption of dye anions due to the electrostatic repulsion [18].

It has been observed, however, by Kosyiyakanon et al. that the addition of R_4NCls to the biochar enables the high efficiency of acid dyes removal even in a broad pH area from 3 to 9 [15].

This fact could be explained by the formation of low-soluble high-molecular ion pairs (dye-SO_3NR_4) by the ion exchange reaction between the added R_4NCl and -SO_3Na groups bound in the structure of dye according to the scheme:

$$\text{dye-}SO_3Na + R_4NCl \longrightarrow NaCl + \text{dye-}SO_3NR_4 \text{ (ion-pair)}.$$

As could be seen in **Figure 12**, biochar is the worst sorbent; however, using R_4NCls selected similarly to abovementioned separation of chlorinated aromatic carboxylic acids sodium salts, the sorption capacity rises efficiently.

As the sorption capacities for the removal of chlorinated aromatic sulfonic acid sodium salt MB9 were (biochar +3/2 A336/50% aq.

AlkBzMe2NCl) > PAC ~ (biochar+2/3 A336/50% aq.AlkBzMe2NCl) > GAC ~ (biochar + AlkBzMe2NCl) > (modified biocharAlkBzMe2NCl) > biochar, this indicated that the combined action of the sorted R_4NCls mixture and biochar could specifically increase the sorption capacity for the used biochar above the adsorption capacity of commercial powdered active carbon Silcarbon CW20 (**Figure 12** and **Table 7**).

Langmuir and Freundlich isotherm models were fitted to the data. Similarly, the correlation coefficients suggest that the Freundlich model fits the data better than the Langmuir model (**Table 8**). This indicates that MB9 and/or ion pairs R_4N.MB9 were adsorbed in multilayers into the active sites of the biochar surface.

In case of the tested acid dye, the sorption capacity of the ex situ-prepared modified BiocharAlkBzMe$_2$NCl is closer to the sorption capacity of in situ-mixed R_4NCls with biochar (**Figure 12**). The observed higher sorption capacity of modified biocharAlkBzMe$_2$NCl in comparison with biochar agrees with the published results by Mi et al. [13] and Kosaiyakanon [15]. This observation is in good agreement with hard and soft acids and bases (HSAB) theory [19]. According to the HSAB principle, cations of hard bases (AlkBzMe$_2N^+$) prefer to bond to anions of hard acids (dye-SO_3^-). Probably due to this reason, the observed removal efficiency of used modified biocharAlkBzMe$_2$NCl is higher in case of MB9 removal than in the application for removal of anions of soft carboxylic acids DCF and FLUFA.

6. Conclusions

Biochar obtained as a by-product in the gasification process of waste biomass was verified as a suitable sorbent for the removal of the three tested highly mobile, ionizable, and nonbiodegradable chlorinated aromatic acid sodium salts NaDCF, NaFLUFA, and MB9 from model wastewater solutions in a broad range of concentrations. For increasing biochar's removal efficiency, biochar was intentionally mixed with selected cationic surfactants to produce an in situ-modified sorbent designed for the effective removal of the abovementioned negatively charged pollutants even from alkaline aqueous solutions. The higher efficiency obtained using biochar mixed in situ with selected R_4NCls in model wastewater could be explained by the multilayer adsorption of ion pairs (contaminant-COONR$_4$ or contaminant-SO$_3$NR$_4$, respectively) on the heterogeneous biochar surface described by Freundlich isotherms.

We demonstrated that a more laborious two-step technique, based on the initial preparation of impregnated biochar by the action of R_4NCl with subsequent application of this modified sorbent, is much less effective than simple mixing of biochar with R_4NCl directly in the treated wastewater solution. According to the performed experiments, cationic surfactants based on tetraalkylammonium chloride R_4NXs carrying highly branched R_4N^+ cations were verified as very effective for the uptake of NaDCF, NaFLUFA, and MB9 from aqueous solutions by the co-action of biochar. The most effective R_4NX for the removal of these contaminants was verified Aliquat 336, which is, however, highly viscous and nonmiscible with water. Its dilution with an organic solvent is prohibited due to environmental reasons. From a practical point of view, we successfully tried and chose the application of the mixture containing three parts of A336 dissolved in two parts of 50 wt.% aqueous AlkBzMe$_2$NCl, having acceptable removal efficiency for the studied contaminants and enabling the simple and precise addition of the most effective A336 cationic surfactant together with the tested biochar for effective wastewater treatment. This technique based on joint addition of selected R_4NCls together with biochar enables attainment of removal efficiency comparable with commercial active carbons containing at least twice higher specific area as biochar. These obtained results agree with the information by Xi et al. [20] which observed that the surface area of the used sorbent by the co-action of R_4NX does not play a major role in sorption of anionic contaminants.

7. Experimental details

Additional used chemicals (benzalkonium chloride 50 wt.% aqueous solution ($AlkBzMe_2NCl$), Aliquat 336 (A336), hexadecyltrimethylammonium chloride ($AlkMe_3NCl$), diclofenac sodium salt (NaDCF, 98%+, Mr = 303.21), flufenamic acid (FLUFA, 98% + Mr = 281.23), Mordant Blue 9 (MB9, purity, 50% Mr = 551.28), and octan-1-ol, etc.) were purchased from Sigma-Aldrich, Czech Republic.

Powdered active carbon (PAC) Silcarbon CW20 (specific area 1300 m^2/g) was obtained from Brenntag Co. Granular active carbon (GAC) Hydraffin CC8x30 (specific area 1000 m^2/g) was purchased from Donau Carbon GmbH & Co.

Demineralized water was used for the preparation of the used aqueous solutions.

7.1 Preparation of used stock solutions

25 mM aqueous diclofenac stock solution was obtained by a dissolution of 8.0 g of NaDCF in 1 liter of water; the pH of the stock solution was 8.7. Aqueous 10 mM stock sol. of NaFLUFA was obtained by a dissolution of 2.81 g of flufenamic acid in 12 mM aqueous NaOH (pH = 10.3). Aqueous 25 mM stock sol. of NaFLUFA was obtained by a dissolution of 7 g of flufenamic acid in 30 mM aqueous NaOH (pH = 10.3). Aqueous 10 mM solution of MB9 was obtained by a dissolution of 11.0 g of MB9 (50% purity) in 1 liter of water (pH = 8.2).

7.2 Preparation of the mixtures of A336 with aqueous 50% $AlkBzMe_2NCl$

Solution 3/2 A336/50% aq. AlkBzMe2NCl.

30 g of Aliquat 336 was dissolved in 20 g (21 mL) of 50 wt. % aqueous $AlkBzMe_2NCl$ under stirring.

Solution 2/3 A336/50% aq. AlkBzMe2NCl.

20 g of Aliquat 336 was dissolved in 30 g (32 mL) of 50 wt. % aqueous $AlkBzMe_2NCl$ under stirring.

7.3 Preparation of ex situ modified biochar

Modified biocharAlkBzMe$_2$NCl was prepared by impregnation of biochar (20 g) using 5 wt.% aqueous $AlkBzMe_2NCl$ (100 mL) under vigorous stirring at 500 rpm overnight, subsequent filtration, filter cake washing with 400 mL of water, and drying of washed filter cake at 105°C to a constant weight.

7.4 Applied analyses

A Hach DR2800 (Austria) VIS spectrophotometer was employed for the absorbance measurements using 1 cm glass cuvettes. The concentrations of MB9 and R_4N.MB9 were determined by measuring at the maximum absorbance (A_{max}) [7, 8].

Concentration of NaDCF and NaFLUFA was determined by voltammetric determination at carbon paste electrode in situ modified by $AlkMe_3NCl$ cetyltrimethylammonium bromide (CTAB) [21]. Electrochemical measurements were carried out using an AUTOLAB analyzer (model PGSTAT-128 N; Autolab/Metrohm, the Netherlands/Switzerland), coupled with the three-electrode cell incorporating the working carbon paste electrode (CPE), containing the hand- homogenized carbon paste containing 0.5 g graphite powder (product "CR-5"; Maziva, Czech Republic) and 0.3 mL paraffin oil (Uvasol® grade; Merck, USA). This paste mixture was then manually filled into a piston-driven electrode body. The remaining electrodes were a Ag/AgCl/3.5 M KCl reference and a platinum auxiliary electrode (both from Metrohm).

It was confirmed that the anodic oxidation of NaDCF at the CPE gives rise to a well-developed signal with a peak potential of about +0.6 V vs. Ag/AgCl/3.5 M KCl (further denoted as "ref.") and anodic oxidation of NaFLUFA at the CPE gives rise to a well-developed signal with a peak potential of about +0.78 V vs. ref. Almost identical responses for measurements with differential pulse and square-wave voltammetry (DPV and SWV, respectively) have indicated that the reaction of interest is not kinetically controlled and hence suitable for (electro)analytical purposes. Therefore, measurements of all water samples were performed using DPV. It was also found that the most favorable response could be obtained in neutral media, whereas more alkaline solutions had already caused a decrease of the first peak and the total disappearance of the second one. Thus, pH 7.0 was definitely set as optimal; therefore, phosphate buffer (PBS) at pH 7.0 was used for all measurements.

Because the respective measurements had not sensitive response, electrochemical measurements were extended to the effect of a modifier on a possible enhancement of the response. Such a modification is very simple as it can be realized in situ; i.e., by adding a small amount of surfactant directly to the sample analyzed. This can be exemplarily illustrated in **Figure 13**, portraying the effect of CTAB that had

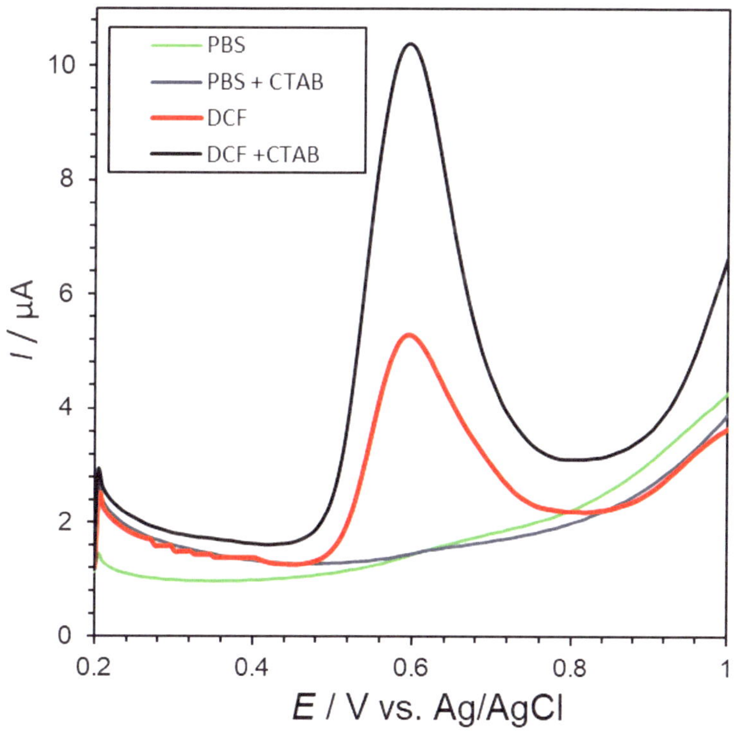

Figure 13.
The main oxidation peak of NaDCF and the effect of the presence of cetyltrimethylammonium bromide (CTAB) surfactant in the solution and phosphate buffer baseline (PBS) [19].

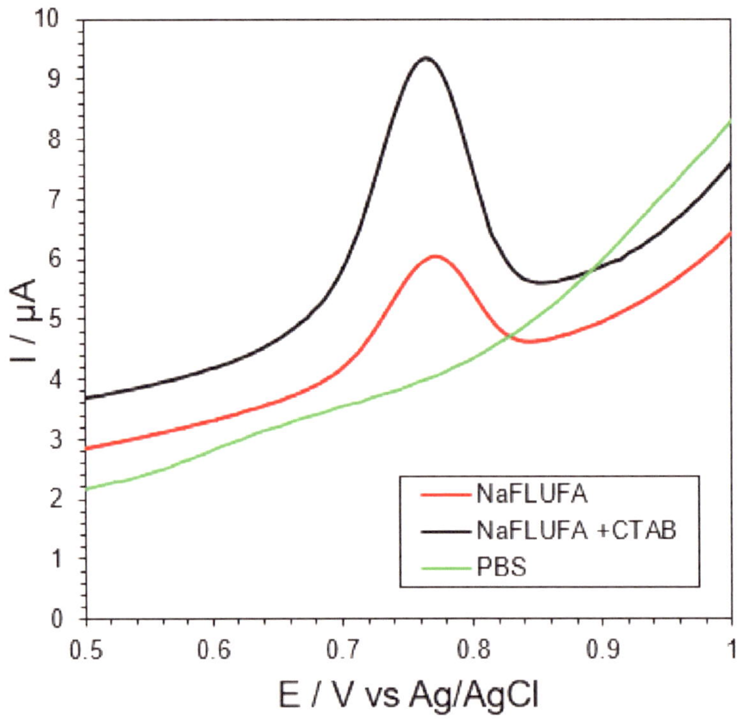

Figure 14.
The main oxidation peak of NaFLUFA and the effect of the presence of cetyltrimethylammonium bromide (CTAB) surfactant in the solution and phosphate buffer baseline (PBS).

been found the most effective for this function among all the surfactants tested. The observed benefit of CTAB was attributed to the (pre)treatment of CPE by means of "erosion effect." The same modifier effect was observed for NaFLUFA, as could be seen in **Figure 14**.

Typical experimental and instrumental conditions of NaDCF and NaFLUFA voltammetric determination (DPV) had included the following parameters: supporting electrolyte 0.1 M PBS + 0.1 mM CTAB, potential scan from +0.2 to 1.2 V vs. ref., and scan rate of 50 mV/s.

7.5 Sorption experiments

Sorption experiments were carried out in a magnetically, at 400 rpm, stirred 250 mL round-bottomed flasks at 25°C using Starfish equipment installed on an electromagnetic stirrer Heidolph-Hei-Standart with a temperature sensor Pt1000. The appropriate quantity of biochar was added to 100 mL of synthetic wastewater (possibly after the addition of $R_4NCl(s)$). In performed kinetic experiments (**Figures 2**, **3**, **5**, **6**, and **8**), the initial concentration of NaDCF in synthetic wastewater was 25 mM (8 g NaDCF/L and pH = 8.7), the concentration of FLUFA in synthetic wastewater was 10 mM NaFLUFA in 12 mM NaOH (2.81 g FLUFA/L and pH = 10.3), and the concentration of MB9 was 10 mM (5.5 g MB9/L). In a subsequent comparative study for the construction of adsorption isotherms (**Figures 10–12**), the concentrations of contaminants in starting aqueous solutions were 0.25–8 g NaDCF /L, 0.25–7 g -FLUFA/L, and 0.5–5 g MB9/L (for additional information see the text and Figures and Tables in the previous chapters). Stirred suspensions were immediately filtered and analyzed after an appropriate time period.

Experiments dealing with log P_{OW} determination were performed using the same apparatus. An aqueous solution containing 1 mmol of studied contaminant was introduced to the round-bottomed flask (in case of studying log P_{OW} of ion pairs, 1 mmol of R_4NCl per mmol of -COONa or -SO_3Na group bound in contaminant was added subsequently); the total volume of aqueous phase was adjusted to 100 mL with water saturated with octan-1-ol, and the mixture was fulfilled using 100 mL of octan-1-ol. The prepared two-phase mixture was agitated at 400 rpm overnight, and the immiscible phases were separated in a separatory funnel, and a concentration of the tested chlorinated aromatic acid sodium salt or their ion pair with R_4N^+ in the aqueous phase was analyzed using VIS spectroscopy in the case of MB9 and R_4N.MB9. In the case of NaDCF or R_4N.DCF, the concentration in the aqueous phase was analyzed using voltammetric determination.

Acknowledgements

This work was funded by the Faculty of Chemical Technology, University of Pardubice, within the financial support for the excellent team (Chemical Technology Group), Specific university research—grant No. A1_FTOP_2020_001, and by the Ministry of Agriculture of the Czech Republic—project QK1820175 and QK1910056.

Author details

Barbora Kamenická[1], Pavel Matějíček[1], Tomáš Weidlich[1*] and Michael Pohořelý[2,3]

1 Institute of Environmental and Chemical Engineering, Faculty of Chemical Technology, University of Pardubice, Czech Republic

2 Institute of Chemical Process Fundamentals of the Czech Academy of Sciences, Prague, Czech Republic

3 Department of Power Engineering, Faculty of Environmental Technology, University of Chemistry and Technology, Prague, Prague, Czech Republic

*Address all correspondence to: tomas.weidlich@upce.cz

References

[1] Schröder P, Helmreich B, Škrbić, Carballa M, Papa M, Pastore C, et al. Status of hormones and painkillers in wastewater effluents across several European states - considerations for the EU watch list concerning estradiols and diclofenac. Environmental Science and Pollution Research. 2016;**23**:12835-12866

[2] Kasprzyk-Hordern B, Dinsdale RM, Guwy AJ. The removal of pharmaceuticals, personal care products, endocrine disruptors and illicit drugs during wastewater treatment and its impact on the quality of receiving waters. Water Research. 2009;**43**:363-380. DOI: 10.1016/j.watres.2008.10.047

[3] Rosal AR, Rodríguez A, Leton P, Gomez MJ, Perdigon JA, Petre A, et al. Removal of emerging pollutants in urban wastewater through biological treatment followed by ozonation. Water Research. 2010;**44**:578-588. DOI: 10.1016/j.watres.2009.07.004

[4] Allegre C, Moulin P, Maisseu M, Charbit F. Treatment and reuse of reactive dyeing effluents. Journal of Membrane Science. 2006;**269**:15-34. DOI: 10.1016/j.memsci.2005.06.014

[5] Shi K, Xie Y, Qiu Y. Natural oxidation of a temperature series of biochars: Opposite effect on the sorption of aromatic cationic herbicides. Ecotoxicology and Environmental Safety. 2015;**114**:102-108. DOI: 10.1016/j.ecoenv.2015.01.015

[6] Mathurasa L, Damrongsiri S. Low cost and easy rice husk modification to efficiently enhance ammonium and nitrate adsorption. International Journal of Recycling of Organic Waste in Agriculture. 2018;7:143-151. DOI: 10.1007/s40093-018-0200-3

[7] Weidlich T, Martinková J. Application of tetraphenyl and ethyltriphenylphosphonium salts for separation of reactive dyes from aqueous solution. Separation Science and Technology. 2012;**47**:1310-1315. DOI: 10.1080/01496395.2012.672527

[8] Weidlich T, Fulínová T, Kovářová D, Václavíková J. Treatment of model wastewater contaminated with reactive acid dyes using ionic liquids. Waste Forum. 2018;**2**:108-116

[9] Thompson KA, Shimabuku KK, Kearns JP, Knappe DRU, Summers RS, Cook SM. Environmental comparison of biochar and activated carbon for tertiary wastewater treatment. Environmental Science & Technology. 2016;**50**:11253-11262. DOI: 10.1021/acs.est.6b03239

[10] Tomczyk A, Sokołowska Z, Boguta P. Biochar physicochemical properties: Pyrolysis temperature and feedstock kind effects. Reviews in Environmental Science and Biotechnology. 2020;**19**:191-215. DOI: 10.1007/s11157-020-09523-3

[11] Pohořelý M, Picek I, Skoblia S. Apparatus for multistage gasification of carbonaceous fules. Patent No. 306239/PV; 2015. p. 483. Applied: 15.07.09, Patented: 16.09.07.

[12] Lian F, Xing B. Black carbon (biochar) In water/soil environments: Molecular structure, sorption, stability, and potential risk. Environmental Science & Technology. 2017;**51**:13517-13532. DOI: 10.1021/acs.est.7b02528

[13] Mi X, Li G, Zhu W, Liu L. Enhanced adsorption of Orange II using cationic surfactant modified biochar pyrolyzed from cornstalk. Journal of Chemistry. 2016;**2016**:1-7. DOI: 10.1155/2016/8457030

[14] Scheytt T, Mersmann P, Lindstädt R, Heberer T. 1-Octanol/water partition

coefficients of 5 pharmaceuticals from human medical care: Carbamazepine, clofibric acid, diclofenac, ibuprofen, and propyphenazone. Water, Air, and Soil Pollution. 2005;**165**:3-11

[15] Kosaiyakanon C, Kungsanant S. Adsorption of reactive dyes from wastewater using cationic surfactant-modified coffee husk biochar. Environment and Natural Resources Journal. 2020;**18**:21-32. DOI: 10.32526/ennrj.18.1.2020.03

[16] Teixidó M, Pignatello JJ, Beltrán JL, Granados M, Peccia J. Speciation of the ionizable antibiotic sulfamethazine on black carbon (biochar). Environmental Science & Technology. 2011;**45**:10020-10027. DOI: 10.1021/es202487h

[17] Lian F, Sun B, Song Z, Zhu L, Qi X, Xing B. Physicochemical properties of herb-residue biochar and its sorption to ionizable antibiotic sulfamethoxazole. Chemical Engineering Journal. 2014;**248**:128-134. DOI: 10.1016/j.cej.2014.03.021

[18] Malik PK. Dye removal from wastewater using activated carbon developed a case study of acid yellow 36. Dyes and Pigments. 2003;**56**:239-249. DOI: 10.1016/S0143-7208(02)00159-6

[19] Chattaraj PK, Lee H, Parr RG. HSAB principle. Journal of the American Chemical Society. 1991;**113**:1855-1856

[20] Xi Y, Mallavarapu M, Naidu R. Preparation, characterization of surfactants modified clay minerals and nitrate adsorption. Applied Clay Science. 2010;**48**:92-96. DOI: 10.1016/j.clay.2009.11.047

[21] Kamenická B, Bartášková A, Švancara I, Weidlich T. Applicability of voltammetric determination of diclofenac at carbon paste electrodes to the analysis of aqueous solutions purified by adsorption and/or ionic liquid-based ion exchange. Monatshefte für Chemie. 2019;**150**:429-437. DOI: 10.1007/s00706-019-2354-8